Bitter Roots

BITTER ROOTS

The Search for Healing Plants in Africa

Abena Dove Osseo-Asare

Abena Dove Osseo-Asare (signature)

The University of Chicago Press
Chicago and London

Abena Dove Osseo-Asare is assistant professor of history at the University of California, Berkeley.

The University of Chicago Press, Chicago 60637
The University of Chicago Press, Ltd., London
© 2014 by Abena Dove Osseo-Asare
All rights reserved. Published 2014.
Printed in the United States of America

23 22 21 20 19 18 17 16 15 14   1 2 3 4 5

ISBN-13: 978-0-226-08552-4 (cloth)
ISBN-13: 978-0-226-08602-6 (paper)
ISBN-13: 978-0-226-08616-3 (e-book)
DOI: 10.7208/chicago/9780226086163.001.0001

Library of Congress Cataloging-in-Publication Data
Osseo-Asare, Abena Dove Agyepoma, author.
  Bitter roots : the search for healing plants in Africa / Abena Dove Osseo-Asare.
    pages ; cm
  Includes bibliographical references and index.
  ISBN 978-0-226-08552-4 (cloth : alkaline paper) — ISBN 978-0-226-08602-6
(paperback : alkaline paper) — ISBN 978-0-226-08616-3 (e-book)   1. Materia
medica, Vegetable—Africa. 2. Medicinal plants—Africa.   I. Title.
  RS181.O87 2014
  615.3'21–dc23

                                                                2013020550

♾ This paper meets the requirements of ANSI/NISO Z39.48-1992 (Permanence
of Paper).

For my parents, Kwadwo and Francislee Osseo-Asare

CONTENTS

Figure I.1 Exchanging plants at Alafia Bitters, Ghana. (Photo by author.)

From Plants to Pharmaceuticals

> We know the time, the persons, the circumstances, belonging to each
> step of each discovery.
> —William Whewell, *Philosophy of the Inductive Sciences*

The struggle over plants unfolds once again between scientists and healers in a humid classroom in Ghana. After tea and biscuits, the workshop
participants reconvene for the final session of the day. A middle-aged
scientist begins to plead with the healers to fill in the simple forms that
have been set before them on the long tables. "How will my colleagues
and I at the Faculty of Pharmacy complete our studies without your
help?" he implores. Near me, anxious healers debate what to do. "Look at
the banner over there," one says, nodding toward a list of research institutes and nongovernmental organizations (NGOs). "Many of the sponsors also market medicinal plants. How can we trust the organizers?"
They wisely assume that the scientists might be working with contacts
in foreign drug companies. I watch as a few healers scribble down the
names of such well-known medicinal plants as mahogany bark. No new
information here as the conflict continues between those with laboratory coats and those without.

Bitter memories shape the search for healing plants in African countries. For over a century, conflicts over rights to medicinal plants have
persisted between healers and scientists. Stolen recipes, toxic tonics,

unfulfilled promises of laboratory equipment, and personal patents usurped from the nation circulate in their recollections. This book tells the intricate stories behind six plants that competing groups of plant specialists have sought to transform into pharmaceuticals since the 1880s. The analysis centers on plants sourced in African countries, following their migration from markets to laboratories around the world. Specifically, it looks at new medications made from rosy periwinkle, Asiatic pennywort, grains of paradise, *Strophanthus*, *Cryptolepis*, and hoodia. Some of the pharmaceuticals have been profitable, most have led to patents, and all have resulted in controversies among the many people who have claimed rights to the plants and their biochemical constituents. These cases provide a new framework for understanding priority, locality, appropriation, and benefits in the quest for new medicines.

Bitter Roots is the first comparative history of the process of drug discovery from plants across different regions of Africa. Above all, it shows how chemists, healers, drugs companies, and rural communities have all contributed to the shaping of scientific knowledge on healing plants—but benefited differently. These transnational stories remind us how difficult it is to discern who was first to deduce the medicinal value of a plant. Narratives of discovery often fixate on singular inventors and original tribes, yet these accounts are often more mythic than historical. Although these different parties might seek to claim unique status in the process of discovery, the argument here is that multiple innovators participated in the shaping of drug knowledge across wide geographic regions. To aid in understanding the roots of ongoing conflicts in plant drug discovery, the book seeks to identify key moments of exchange—the time and place when people met and shared or stole information about a particular plant. These ephemeral points of connection between different stakeholders held legal implications about who was first in drug discovery. Documenting these moments of exchange allows us to better interpret how different parties have made claims to profits from successful pharmaceuticals derived from plants. This book therefore has implications for how we understand and assign credit for scientific discovery and patents, because it indicates that many people identify plant activity simultaneously.

The six plants in this book make up a metaphorical family of "bitter roots." The first four plants that I examined in my own search for healing in Africa are members of the larger botanical family *Apocynaceae*, whose bitter-tasting members with high alkaloidal content have long signaled the potential for medicinal applications and left greater historical records in their wake (see Table I.1). Periwinkle and hoodia have been absolutely central to debates about benefit sharing in drug discovery, not only in Madagascar and South Africa, but around the world. The lesser known cases of *Strophanthus* and *Cryptolepis* were critical to debates on rights to plants in the history of Ghana, where I have family ties and began my research. In "roots markets" where I sought information on healing plants, bundles of periwinkle lay side by side with stalks of pennywort, a member of the parsley or carrot botanical family. My discussions with plant experts in Madagascar led me to include this plant within my metaphorical family of bitter roots. Similarly, I added the spicy grains of paradise, a member of the ginger botanical family and a common ingredient in herbal remedies across West Africa, as a placeholder for the many healing plants for which the archival record remains silent. Although I began investigations on other useful plants—including miracle berry, mahogany bark, and a number of oil-bearing plants found in West and East Africa—I decided to limit this study to cases where African traditional remedies helped inspire new pharmaceuticals.[1]

A common impulse links the biographies of these six plants. In each, individuals pursued information on medicinal plants that might lead to new, profitable pharmaceuticals. Although the plants and actors change in each chapter, they show how different parties engaged in a unified process of drug discovery and the ways in which a common method for bringing traditional medicine into the laboratory shifted over time. By the Early Modern Period (1450s–1800s), Africans sought treatments in forests, fields, and the outskirts of farms, redistributing herbal seeds and medicinal recipes within what I term "healing plant diasporas." European colonists adapted these indigenous medicinal recipes in hospitals and laboratories while simultaneously restricting healers from practicing their trade (1800s–1950s). After independence, scientists at African research centers interacted with their own relatives and healers in rural areas in their quest to find new pharmaceuticals, taking valuable plants into their laboratories (1960s–2000s). Most recently, policy makers within the African Union have hoped to reassign profits from plant-based pharmaceuticals along ethnic lines to atone for the colonial and national appropriation of herbal remedies. Together, the cases show how class distinctions linked to distinctive historical structures in African countries allowed some parties to claim credit for their role in drug discovery at the expense of others.

The cases further emphasize the wider shift from a colonial model of open access to natural resources for scientists to a nationalist model of closed access to plants and information. Colonial occupations in Africa solidified in the late nineteenth century, providing scientists in Europe

Table I.1 Overview of Plants and Patents Profiled

Family	Genus/Species	Popular English Name	Geographic Distribution	Related Patents Discussed (for Processes of Chemical Extraction for Drug Manufacture)
Apocynaceae	*Catharanthus roseus* (formerly *Vinca rosea*)	Periwinkle	Madagascar (orig.); pan-tropical: North and South America, South and East Asia, coastal Africa	More than 100 international patents since 1960. U.S. Pat. 3,097,137 Vincaleukoblastine (filed 1960 by Beer, Cutts, and Noble of Canada) U.S. Pat. 3225030 Method of Preparing Leurosine and Vincaleukoblastine (filed 1965 by Eli Lilly and Co.)
Apocynaceae	*Strophanthus hispidus*	None. Referred to as "arrow poison plant" or *Strophanthus*.	West and Southern Africa	More than twenty since 1914; none discussed.
Apocynaceae	*Cryptolepis sanguinolenta*	None. Referred to as "bitter roots" or *Cryptolepis*.	West Africa	U.S. Pat. 5362726 Compound and method of treatment for falciparum malaria (filed 1993 by Healthsearch Inc. in the United States and the Centre for Scientific Research into Plant Medicine in Ghana) U.S. Pat. 5628999 Hypoglycemic agent from *Cryptolepis* (filed 1995 by Shaman Pharmaceuticals)

Family	Species	Common name	Geographic range	Patents
Apocynaceae (subfamily *Asclepiadoideae*)	*Hoodia gordonii*	Hoodia or ghaap	Namibia, South Africa	More than ten pending patents since 2000. U.S. Pat 6376657 Pharmaceutical compositions having appetite suppressant activity (filed 1999 by South Africa's Council for Scientific and Industrial Research) U.S. Pat 6488967 Gastric acid secretion (filed 2000 by Phytopharm PLC)
Zingiberaceae	*Aframomum melegueta*	Grains of paradise, Guinea grains, alligator peppers	West Africa, Caribbean, Brazil	U.S. Pat. 5879682 Aframomum seeds for improving penile activity (filed 1995 by Peya Biotech of Canada and Republic of Congo)
Umbelliferae	*Centella asiatica*	Indian pennywort	South Asia, West Africa, Southern Africa, Madagascar	More than 100 patents since 1960. U.S. Pat 3365442 / EC C07H15 / 256 Derivatives of Asiaticoside and Their Process of Preparation; US Pat 3366669 / EC C07J63 / 00D46 Hemisuccinates and Salts of the Hemisuccinates of Asiatic Acid (filed 1964–1968 by Ratsimamanga and Boiteau of Paris / Madagascar)

Source: Author's research.

with greater access to information on Africa's healing plants. As the instance of *Strophanthus* highlights, colonial subjects resisted efforts to relocate plants and herbal recipes from African contexts. In the mid- to late twentieth century, postindependence leadership sought to control plant information for the national good. But it turns out that investigations within African countries have depended on colonial herbarium records as well as global scientific standards that stressed the sharing of information. To manage personal gain, African scientists filed patents internationally to protect their research findings without affording benefits to herbalists or communities of plant users. Both colonial and national models of controlling access to plant medicine served to hide the many layers of plant experts whose investigations over centuries have led to recent pharmaceutical innovations.

This book's focus on the creation of scientific identity in modern African settings may be unfamiliar or uncomfortable for some readers. Analyses of bioprospecting, the search for new chemicals with industrial applications from natural sources, frequently emphasize conflicts between researchers in the Global North and small-scale communities in the Global South. My account unsettles this common narrative because it highlights the complexities of exploitation in natural products research within African countries. In particular, the cases indicate that African scientists have sought exclusive rights to drug-making processes, often without fully acknowledging healers and communities from their own countries who also helped shape information about plants. At the same time, my analysis critiques the narratives that scientists from around the world use to claim plant-based drug patents. The stories of each plant remedy destabilize hierarchies of knowledge that privilege "scientific" authority over "traditional" medical expertise.

Time and again, these case studies show that the number of potential claims among stakeholders seeking profits from plant-based therapies has depended on the relative geographic distribution of the plants. From the pantropical weed rosy periwinkle, believed to have originated in Madagascar, to the peppery cure for impotence known as "grains of paradise"—found across West Africa, the Middle East, and the Caribbean—the mobility of plants, people, and information across international borders has complicated efforts to extend benefits to specific communities. The last case study is that of hoodia, a type of milkweed found in the arid Great Karoo of South Africa that led to a miracle diet therapy and one of the first benefit-sharing agreements among scientists in Africa, drug companies, and rural communities. The initial success of the agreement depended on the relatively limited domain of the plant and the direct claim to its dietary uses made by a specific ethnic group. In most cases, however, it is difficult to find the original owners of a plant whose natural history spans countries, continents, and centuries.

These are stories that did not wish to be told. I have had to excavate these cases in contexts where the publicly available evidence was thin. Pharmaceutical companies like Eli Lilly and Bristol-Myers Squibb do not allow public access to their archives and fiercely protect what they consider their intellectual property. Neither healers nor scientists in African countries have preserved detailed written records of their investigations, and each maintains cultures of secrecy when pressed for information. I have used traces of evidence gleaned from faded pages in archives, dried plants in markets and museums, and pieces of conversations to document how various parties sought information on valuable plants. Partial data from the past alongside recent observations raise important questions about how we understand the process of plant drug discovery and rights to pharmaceuticals and profits.

A RESOURCE FOR HEALTH AND PROSPERITY

For more than a half century, scientists at African universities have devoted themselves to researching the active components of herbal cures. Plant medicine is one of the continent's most important resources. Yet, the commercialization of safe, effective plant-based therapies has remained elusive. Seldom able to realize the dream of new pharmaceuticals, these scientists competed with herbalists as they marketed medicinal plants as teas, syrups, and tonics. The stories told in this book show how conflicting claims to plant medicines have affected efforts to transform traditional therapies into viable pharmaceuticals for nearly a century. Past attempts to study herbal remedies provide a much-needed perspective on the current politics of traditional medicine promotion in African countries. Given the wide distribution of healing plants, the ongoing efforts of both scientists and healers to claim plants for themselves or their nation through patents and branded products may be unrealistic.

As I wrote this book, some politicians and scientists nonetheless held out Africa's healing plants as a means to both grow economies and combat disease. By the late twentieth century, the failure of synthetic pharmaceuticals to control drug-resistant malaria or cure new conditions like HIV/AIDS led to renewed interest in herbal therapies in urban and rural settings. Since 1997, I have participated in and observed conferences and workshops on promoting traditional medicine, held for chemists, healers, botanists, drug companies, physicians, historians, anthropologists, and other stakeholders in Ghana, South Africa, the United Kingdom, and the United States. Often, debates have centered on whether a healer has indeed discovered a cure for AIDS or malaria.[2] With high hopes, the African Union declared the opening years of the twenty-first century to be the Decade of African Traditional Medicine in order to promote research into plants and other life forms on the continent and surrounding oceans. In South Africa, the former president, Thabo Mbeki, and his health ministers notoriously claimed that roots could treat HIV, confusing the public about the proven value of pharmaceuticals over herbs. For, at the very same time that the African Union and a growing number of organizations sought to rebrand traditional medicine as the drugs of the future, there were calls for equitable distribution of pharmaceuticals readily available in wealthier countries. Complicating the issue further, many African elites, and increasingly nonelites, have rejected traditional medicine in the complex process of adopting Christian and Muslim values along with "modern" lifestyles.

But in some cases, herbal cures have genuinely offered more promise than synthetic drugs. Even as African media and the international press attacked purported herbal cures for AIDS, an ancient Chinese remedy for fever derived from wormwood emerged as the best hope for combating malaria, a major scourge on the continent. In 2005, the World Health Organization (WHO) recommended artesunate, wormwood's primary ingredient, as the first line of treatment for drug-resistant malaria. Many researchers and policy makers in Africa interpreted the WHO statement as an endorsement of their ongoing efforts to promote the commercial value, if not the spiritual claims, of traditional medicine. The World Bank began to investigate the potential of standardized herbal remedies as a way to bring African countries out of poverty by strengthening regional and global markets. In accepting foreign aid, African Ministries of Health agreed to reverse discriminatory policies against healers, conducted formal censuses of traditional medical practitioners, and held seminars to help them improve their remedies.[3]

Let us return to the humid classroom, where healers participated in a workshop on improving the quality of their herbal preparations at the University of Ghana, Legon, in 2002. These healers, most of whom specialized in herbal medicine merchandise, joined scientists from several universities as well as officials from the Ministry of Health. Because of advertisements promoting the workshop in the newspapers, healers had expected to gain useful information on ways to standardize herbal preparations.[4] But, as I observed during the proceedings, herbal manufacturers were wary of some of the other participants as well as conference sponsors. More generally, healers in Ghana were reluctant to provide information on key ingredients in their products after new regulations required them to test for toxicity at government laboratories before advertising their products. The herbal sellers wanted to keep secret formulas to themselves, whereas the laboratory researchers believed disclosure would help them test for toxic contaminants more efficiently. For scientists and health officials, the prospect of toxic Ghanaian herbal medicines on the international market led them to urge ingredient disclosure, laboratory tests, and standardization of dosages. In the years after the workshop, many healers relented and disclosed ingredients—but only after the government briefly banned all advertisements of herbal products.[5]

African herbal producers have become increasingly conversant with international preoccupations with standards, toxicity, and intellectual property rights in their efforts to expand the market for African herbal medicine products. Consider the case of a tradition Ewe tonic marketed to retailers abroad as a means of circumventing regulations for advertising in Ghana. The herbal producer R. K. Assiamah concerned himself primarily with competition from Ghanaian herbal producers, even while exploring potential markets overseas. Indeed, he applied for approval of his popular nutritive tonic by the U.S. Food and Drug Administration (FDA), but his application was denied.[6] Another version of the traditional Ewe tonic was Alafia Bitters, a preparation made by the Atiako family from medicinal roots steeped in water and alcohol (a photograph from their facilities appears as Figure I.1 at the beginning of this Introduction). Like Assiamah, the herbal manufacturer Djane Atiako explained to me that his father gleaned recipes for the Alafia brand products from family members, naming them after the Hausa phrase for "good health." He explained that his father was the "first to introduce herbs to the drugstore" before Ghana's independence. Government export figures used the names of these two leading brands for Ghanaian bitters interchangeably. In 1990, at least 1,000 bottles of Alafia Bitters (with a reported value of US$766) made their way to foreign markets. By 2000, reported exports for Alafia and Assiamah Bitters had expanded to around US$21,000; additional medicinal plant and seed exports were valued at US$467,000.[7] Although Ghanaian government workers admitted to me that there were limited incentives for accurate reporting of the real value of exports, these approximate figures hint at the wide circulation of Alafia and Assiamah Bitters in international produce shops that appeal to the African Diaspora.

Traditional medicine and its plant therapies were in transition during the time of my research. As urban areas expanded and climate change loomed, popular knowledge of plants near farms and along rural roads was dissipating, and many plants faced extinction. In Ghana, Madagascar, and South Africa, plant sellers told me their suppliers had to go deeper and deeper into the forests. At the same time, people found new sources for plants along university campus boulevards, and I saw collectors gathering overgrown flora from urban roadsides. Perhaps it will be university researchers, rather than traditional healers, who will extend the documentation and use of plant medicines into the next millennia in African countries. With the development of comprehensive courses in herbal medicine at African universities, the dream of jump-starting economic development with phytochemical industries has continued to gain supporters. Meanwhile, traditional practitioners have shown no signs of closing their shrines and clinics, increasingly hybridizing animist beliefs with Christianity, Islam, and biomedicine. The search for healing plants in Africa will continue.

PRIORITY, PATENTS, AND THE HISTORY OF SCIENCE

The problem of priority is critical to understanding conflicts over plant information. Groups of plant experts have constructed narratives of priority, omitting details on the many protagonists participating in knowledge production. Historians and sociologists of science have adopted a slightly different sense of "priority" than, say, lawyers, for whom it holds less of a philosophical gradient. In this book, my concern with priority in patent law stems from what it tells us about how societies value claims to first rights: in other words, a driving question of this book is "Who was first?" Patents hold relevance for the stories that follow, especially in the United States and Europe, where actors filed process patents for exclusive rights to derive molecules from plants with pharmaceutical applications. By the late nineteenth century, when European and North American countries coordinated national offices for inventors to register copyrights and patents through a series of international agreements, patents were awarded to those who were first, those who held priority. In patent law, generally, priority has been tied to novelty and the question of whether an individual or a company was the first to present a new technique with industrial applications. Patents provided the assignee with the exclusive right to a process or device for the claimed uses for a set number of years. In the United States, priority has been limited by what was previously written down and published; in Europe and some Asian patent laws, oral testimony or material previously published on Web sites might trump a claim for priority in a patent application. Earlier claims to a purportedly novel idea are termed "prior art" and may be used to upset a patent.[8]

The process of bioprospecting allows us to interrogate the global and social contours of invention and possibilities of prior art, especially earlier claims to purported discovery in oral cultures. The term "bioprospecting," from "biological diversity prospecting," dates to around 1992 and covers the search for new chemicals that are derived from biological matter and that have industrial applications. Bioprospecting is related to patents, as these industrial applications were increasingly the subject of exclusive legal claims in offices in North America and Europe. Around the same time that the term "bioprospecting" became popular, international structures like the World Trade Organization (WTO), the World Intellectual Property Organization, the United Nations, and the World Bank began to encourage member nations more consistently to respect patents and establish uniform patent laws.

Even with incentives for wider adoption of drug patents, by the end of the twentieth century both product patents for biologically derived compounds and process patents for their manufacture met fierce criticism. A spectrum of organizations called for drug companies to compensate communities from which plants and information were sourced. Populations living in biologically rich but economically impoverished regions made a number of explicit agreements with researchers at universities and drug companies. By the 1990s, bioprospecting and compensation had emerged as highly politicized issues. This book examines the full dimensions of possible contribution and exchange in pharmaceutical discovery. Particularly, it shows how patents for processes to create pharmaceuticals intersected with the rights of indigenous communities, scientists, and drug companies.[9]

The question of who was first can be asked of both the scientific and traditional investigations of the medicinal applications of a given plant. In all the cases considered here, bioprospecting depended on traditional plant recipes for inspiration. Histories of early modern medicine have probed the importance of folk therapies, especially as sustained through women healers in Europe, but modern pharmaceutical histories emphasize the rise of synthetic drugs and laboratory discoveries.[10] A few investigations of pharmaceuticals from the 1920s to the 1980s show their global range, with gestures toward understanding the history of drugs like penicillin or oral contraceptives in Asian, African, and Latin American contexts.[11]

Historians of science will be familiar with the question of who was first in the realm of scientific discovery. For many years, such historians focused their efforts on documenting the achievements of lone geniuses. Many of these hagiographic stories trickled down into general science education. In fact, we have been accused of overemphasizing those moments of rupture, the breakthroughs, the "eureka" instant when new information and ideas crystallized. For at least the past thirty years, however, scholars in this field have pursued a more nuanced understanding of discovery and priority claims. Science studies scholars, in particular, have assessed the extent to which scientists used a precise method to expand their understanding of natural phenomena, or how scientific advances depended on chance or broad social shifts outside the laboratory.[12] This book takes cues from studies of how assistants (often female, less educated, or nonwhite) have contributed to scientific research.[13] I contrast narratives of certitude, narratives that insist that "we know the time of each discovery," with fragmented, synchronous stories of shared creation across time and space that open up the possibility of infinite inventors. If we count initial identification of plant activity as the primary innovation, perhaps no one was first.

To understand the issue of priority in its historical contexts, I pay close attention to competing narratives of discovery among scientists. I use the broad term "scientist" to refer to people who define themselves as such, most frequently after having pursued tertiary training at the university level in Africa, Europe, or North America. I show how scientific work is socially constituted, revealing how scientists construct expert knowledge. In this book, I also use "scientist" to refer to early twentieth-century African physicians with training at colonial African secondary schools and colleges and, frequently, who completed course work in Europe. These physicians often represented the earliest form of scientific identity in their home communities.[14]

As historians have shown, practitioners of science—biologists, chemists, physicists—developed elaborate strategies for establishing authority and recognition for their innovations. Within this vein, African scientists, trained in new universities in their countries and abroad, collected signs of their contributions to science to secure careers at national institutes or to maintain consultancies with NGOs and international companies. African scientists traced their intellectual lineage to European societies, where from the 1700s a class of noblemen, primarily white, codified their ideas in an elite discourse of natural philosophy. These early "men of science," as they began to call themselves, met in salons and emerging schools to discuss new ways to organize plants and animals and to test hypotheses about minerals. As membership in the science professions grew, practitioners became less likely to share their ideas widely. Twentieth-century scientists have protected their ideas through a mix of lectures, publications, and, increasingly, patents. Even in regimes of shared knowledge, such as open-source software development, participants have developed ways to track their unique contributions.[15] Scientists in Madagascar, Cameroon, Ghana, and South Africa inherited approaches to knowledge management through school systems initiated during the colonial period, and they participated in global standards for information sharing after independence.

Simultaneously, the stories of African scientists in this book show how researchers partly rejected a close affinity to European scientific and medical cultures, with their attendant colonial implications, instead searching or researching indigenous ways of understanding the natural and metaphysical world. At the same time, the history of drug discovery from plants in African contexts indicates levels of class tension, particularly when scientists based on the continent collaborated closely with foreign drug companies and used nomenclature difficult for healers to understand.

Individuals and communities with less access to university education reported their contributions to innovation in terms similar to those of these university-trained scientists. Most of the people interviewed for this book, like the herbal manufacturer Assiamah or his competitor Atiako, told me how they or their families were the first to invent a new medicinal preparation. I use the terms "healer" and "herbalist" to refer to an assemblage of health practitioners with lay training in folk medicine and healing plants gained through family mentorship, apprenticeships at shrines, and sometimes scrutiny of books and media on plants. The names of healers and their narratives were more difficult to access than those of the scientists, given the heavy reliance in most African societies on oral transmission of recipes and plant information. Even so, the overarching story of drug prospecting between and within communities of scientists, healers, rural communities, and urban plant sellers allows us to exhume pieces of evidence from popular culture contained within herbarium records or laboratory notes. Controversies stemmed from retrospective debates about whose knowledge constituted the first innovation on a plant.

HISTORICAL GEOGRAPHIES OF TRADITIONAL MEDICINE

The question of geography complicates the process of addressing claims of priority. If traditional medicine were local and contained, then the process of identifying the "moment of discovery"—or at least the cultural source of a particular treatment—would be relatively straightforward. But traditional medicine in African contexts may be more diffuse and less tied to regional environments than has often been assumed, particularly by early colonial research that reified the static, local character of healing practices in rural African communities. Postcolonial governments in Africa similarly documented traditional medicine within the boundaries of the nation-state. International research on traditional medicine in African contexts has long emphasized culture and practice, adopting strategies of documentation implemented in former colonies during French, German, and British rule. A major contribution historians can make, I believe, is the articulation of how African traditional medicine has changed and *moved* over time. If many people in many places shared and elaborated plant medicine recipes over a span of years, then determining who can lay claim to a traditional medicament is even more complex.

A concern with historical geography emerged for me when I saw how ethnic groups sought to use claims over the unique locations of people and plants in order to gain access to drug profits. Asserting themselves as "first peoples," they fashioned narratives of primary use of an herbal remedy. Furthering their cause, proponents of indigenous knowledge and ethnobotany often emphasized the close links between people, specific environments, and natural resources. Yet the connection between plants and people has never been so close as these advocates maintain. Periwinkle has found use not only in Madagascar but also in places as far-flung as the Philippines and Jamaica. In contrast, Southern Africa, and the Kalahari Desert in particular, were arguably less networked into global regimes of trade in the past, allowing for the attenuated cultivation of knowledge of hoodia. But even in this case, recipes for hoodia were shared among several of the ethnic groups that settled in Southern Africa, including migrants from the Netherlands. In other words, plants, people, and information have moved for a long time, complicating claims to first rights to knowledge of healing plants.[16]

Internationally, the use of the term "traditional medicine" has been closely tied to early descriptions of the term in African contexts, although the phrase has since been widely adopted around the world. The WHO popularized the concept in the 1970s. Traditional medicine was a decidedly vague category, and yet those familiar with a range of healing practices, from Chinese folk medicine to Indian Ayurveda, found a common ground in it. For instance, in 1977 a group of physicians and government health experts representing a range of nations including China, Egypt,

Sri Lanka, India, Mexico, and Cameroon met to outline an international approach to integrating folk healing into biomedicine. Advising the WHO, they adopted a definition of traditional medicine that had been created by African physicians and academics the previous year—a definition that did not, it is worth noting, stress locality: "[Traditional medicine is] the sum total of all the knowledge and practices, whether explicable or not, used in diagnosis, prevention, and elimination of physical, mental, or social imbalance and relying exclusively on practical experience and observation handed down from generation to generation, whether verbally or in writing."[17] Subsequently, the WHO issued a series of international recommendations on how traditional medicine formed an important component of health provision in what were then termed "developing" countries.

One might ask several different kinds of questions when attempting to come to terms with traditional healing practices. This book does not provide extensive documentation of plant knowledge or healing beliefs. Nor does it dwell on the history of organizations of traditional medical practitioners. A question related to both issues concerns whether traditional medicine was (or is) dangerous. Missionaries and adherents to Christianity have long supposed that it is, given its attendant spiritual practices. Colonial administrators, and more recently government health officials, have stressed the tendency of traditional medicines to be poisonous or toxic in the wrong hands, the wrong concentrations, and the wrong bottles. Another central question might be whether traditional medicine has worked. In other words, did healing plants actually cure the diseases that people claimed they did? Indeed, this issue was often the impetus for scientists to investigate plants, as the stories in this book reveal.[18]

Although these are all important and worthwhile questions, this analysis leaves them to other scholars. Rather, the second central question asked of traditional medicine in this book is that of geography: was traditional medicine local? Particularly, has knowledge about healing plants long been limited to distinct people in distinct locations, or has herbal information spread widely across continents over longer periods— perhaps as much as ten centuries? The migration of plants, people, and information has led to surprising similarities in herbal recipes in widely separated locations. And it is not always possible to disaggregate knowledge by class, ethnicity, or gender in nearby locations. My investigation of the historical and geographic transmission of herbal knowledge expands previous investigations that have documented the circulation of African healing cults across towns, nations, and even continents.[19]

The idea that indigenous knowledge is *local knowledge* has recently fueled much environmental research and policy. Since the 1970s, development experts—hoping to improve the economic well-being of people living in impoverished countries—have taken up the concept of indigenous knowledge as a means to sustain agriculture and biological diversity in cultivated and wild seeds. A celebration of indigenous knowledge evolved from a rejection of modernist endeavors to shape agricultural reform, including the first so-called green revolution to increase global food production with improved seeds, fertilizers, and tractors. A burgeoning industry has grown up around querying farmers in rural areas on their opinions about climate, soil, forest cover, and related topics particular to specific environmental locations.[20]

Reports of local uses of herbal remedies are fundamental to my analysis. My use of local or indigenous knowledge, however, departs from earlier studies of ethnobotany in that I compare information across regions and time frames to show the diffusion and interconnectivity of traditional African healing. The argument here is not that there is no basis for strong diversity in African countries; the wide variety of indigenous languages attests to the high differentiation of cultures and communities over time. Rather, the book argues that traditional medicine was shared across a wide range of communities and was relatively public. A long view of plants in African history suggests their wide circulation and adaptation within overlapping systems of belief and use and within linked ecological zones. Although commentators have often associated African traditional knowledge with esoteric, local belief systems, similar recipes for the same plant across geographic areas point to more regional continuities than differences.

TRANSFORMING PLANTS INTO PHARMACEUTICALS

Regardless of where traditional knowledge comes from, the process of transforming African healing practices into pharmaceuticals is not straightforward. A third query of this book is "How did plants become pharmaceuticals?" To map this historical process, "the biographies of remedies" in this book each unravel a series of small, mundane exchanges of botanical materials that link human and plant actors across time and space.[21]

These stories of drug discovery from plants show how scientists have used a basic set of practices to transform plants into new medicines. The "taking" or appropriation of plant-based therapies depended on a complex process of creating records, experiments, explanations, products, and harvests. Using my observation of historical processes as a basis, this schematic of five steps provides a framework for comparing how scientists commercialized different plants. Moreover, it allows for the possibility of indigenous science that depended on and still depends on oral records for plant recipes, ongoing experiments on patients and animals, metaphysical explanations, traditional herbal products, and harvests from gardens and forests. The process of scientific appropriation through the laboratory, in contrast, has depended on written documentation of medicinal plants, testing of plants with isolated chemicals, esoteric language to explain experiments, expensive techniques for turning plants into pharmaceuticals, and strict control over plant harvests. Yet, not all efforts to transform plants into pharmaceuticals were complete, allowing for simultaneous histories of medicinal plants in the form of teas, bitters, syrups, capsules, and nutraceuticals.

From the late nineteenth century, European colonialism in Africa provided pharmaceutical chemists with greater access to information on herbal remedies. Written documentation replaced oral records, experiments in laboratories calculated twitches on medicated frogs, molecules stood in for spiritual explanations, dried herbal products reemerged as pills and injections, and colonial authorities relocated herbal gardens to expand harvests. Through the process of drug discovery, African traditional remedies made their way into the official pharmacopeias of European and North American countries at a time when most medicines were plant based. Later, after the demise of colonial authorities, researchers at African universities rediscovered herbal cures. Like scientists operating under colonial rule, these African scientists adopted standards for extracting active chemical ingredients through pulverizing, boiling, and reacting plant liquids with chemical reagents. They contributed to the further molecularization of African plant therapies, creating chemical explanations for pharmaceutical products backed with patents.

Of course, as those engaged in such research frequently remind policy makers and the general public, it is a huge challenge to identify which molecules in a plant allow it to act on a specific disease. Interest in this time-consuming and often fruitless process for drug discovery has waxed and waned. Moreover, the rise of twentieth-century synthetic drugs made in laboratories from chemicals not derived from remote plants has allowed the slow history of bioprospecting to often remain hidden. Cycles of drug scarcity during wartime and economic instability, epidemics of intractable diseases, and the advent of new chemical screening techniques all help account for the timing of plant-based drug discoveries.

There have been few studies of the *history* of bioprospecting in African contexts. Important work on the appropriation of pharmaceuticals in places like Uganda or Zimbabwe points to the widespread use of biomedicine, alongside continued interest in traditional therapeutics. This book proposes that biomedicine and African traditional healing have not only been complementary but were, in fact, adapted from one another. In approach, this study borrows from extensive analyses of the history of biomedicine (often referred to as "Western" and "colonial" medicine) in African countries. Of specific import here are efforts to identify how injections and pharmaceuticals gained mythic status for African consumers in the wake of epidemics of syphilis and yaws. The book also draws on studies of the history of traditional healing, which point to competition between healers and physicians.[22]

The coevolution of biochemistry and traditional medicine, as discussed here, points to the viability of both African scientific cultures and the traditional healing domains nationalist scientists have sought to rediscover. In looking at experiences within African settings, I do not see adaptations of herbal knowledge in laboratories as moments of cross-cultural exchange or situate healers outside of biomedicine and scientific practice.[23] Herbal medicine and pharmaceutical chemistry have mutually supportive, simultaneous histories up to the present. I extend Sheila Jasanoff's definition of co-production, which she defines as "the production of mutually supporting forms of knowledge and forms of life," to show how "forms of knowledge"—in my case, pharmacological knowledge and herbal medicine in Africa—have mutually supportive histories of change and interaction. These different forms of life, including plant-based medicines, transform and inform one another.[24]

Scientific practice is a normal part of daily life in African countries. Science and biomedicine are now international ways of explaining world conditions, with individuals in African countries eager to take up scientific identities. Ideas about the scientific gift economy, scientific identity, and scientific authorship or invention have operated in African settings not only within communities of healers but among scientists who took up universal standards of research.[25] Understanding African aspirations for participation in the global system of science is critical for an analysis of the social construction of plant science in Africa, with its attendant fetishes of patents and active principles. At the same time, it must be understood that these researchers have nurtured their own aspirations for pharmaceutical gain, often while operating in an impoverished setting.[26]

Not everyone has prospered equally from bioprospecting. The final question this book asks is "Who benefits when plants become pharmaceuticals, given multiple claims to priority, locality, and appropriation?" In the 1990s, latent controversies over rights to plant-based chemicals led people living in areas from which scientists and companies sourced biological resources to demand direct compensation. Environmental activists like Vandana Shiva popularized the term "biopiracy" to call attention to the longstanding theft of biological resources in the Global South.[27] The debates over benefits for drug research indicate a shift from models of open access in herbal and pharmaceutical data, as in the first case of periwinkle, to ones of closed access, as in the final case of hoodia.

Rights to pharmacological data are fundamental to dilemmas surrounding incentives for drug discovery and access to essential medicines. On one hand, in the early twentieth century medical advances—including new pharmaceuticals—assumed the contours of collective goods. Some scientists went so far as to discourage the assignment of patents to their drug-making procedures. On the other hand, the twentieth century also witnessed the growth of multibillion-dollar pharmaceutical companies whose fortunes rise and fall on intellectual property agreements. More recently, the trope of making medicines affordable and accessible to all has recurred, as, for instance, in the case of activists' demands for pharmaceutical companies to provide anti-retroviral drugs for HIV at cost. Similarly, the calls of research scientists for open access to plant therapies that might lead to cures for cancer depend on the premise that new medicines may benefit all.

By the end of the twentieth century, a major intervention in bioprospecting was the creation of benefit-sharing contracts and agreements. This approach sought to balance the efforts of researchers and companies to profit from their innovations, while allowing nations and communities to limit access to valuable natural resources and extend profits to affected groups. Benefit sharing assumed a class of individuals to whom compensation might be afforded, ideally descendants of a first-people group in a threatened ecosystem. The term gained currency in a number of arenas where stakeholders sought compensation, including law, medicine, mining, forestry, and agriculture. In the case of hoodia, the plant considered in Chapter 5, benefit sharing emerged as a catalyst for community solidarity.[28] But as this book indicates, the mobility of both biological resources and people who might use or contain them (i.e., in the case of genetic material) made it difficult to assign benefits to individuals or communities.[29]

The prospect of benefit sharing has been fundamental to international efforts to make the industry for plant-based pharmaceuticals more equitable. In 1992, the United Nations Conference on Environment and Development (or "Earth Summit") held in Rio de Janeiro launched the Convention on Biological Diversity (CBD), which proposed new ways for countries to control access to plants while extending profits to affected communities. Article 8j of the CBD recommended that nations compensate "indigenous and local communities embodying traditional lifestyles relevant for the conservation and sustainable use of biological diversity" when information they shared led to new drugs and chemicals. This was part of an effort to recognize traditional knowledge as a unique category of its own kind, or sui generis. The lifestyles and knowledge systems of indigenous and local people, activists argued, did not fit smoothly within increasingly internationalized standards for intellectual property management established through the WTO.[30]

Yet these efforts to sustain culture and biodiversity through profit sharing immediately encountered difficulties as signatory nations to the CBD grappled with ways to generate income from specific plants.[31] First, members of communities provided information on which plants might be of interest, but attempts to identify pharmaceutical uses for these plants relied on preset tests for specific bioactivities, regardless of traditional uses.[32] Communities, moreover, harbored overinflated visions of the possible benefits of laboratory research surrounding a plant. In Peru, thirteen factions of Aguaruna communities in the Amazon came into conflict as they vied for material rewards from an experimental International Cooperation Biodiversity Group (ICBG) grant scheme funded through the United States' National Institutes of Health, National Science Foundation, and Department of Agriculture.[33] In Mexico, parties to a similar pharmaceutical bioprospecting agreement sidestepped the complexities of indigenous rights by purchasing plant materials from markets and collecting samples from along highways.[34] In Indonesia, nonelites in remote areas found their efforts to secure equitable distribution of benefits thwarted by a corrupt government.[35] Communities in the Amazon rainforest anticipated the advent of clean water and new schools funded by a private company based in California, Shaman Pharmaceuticals, but when its drugs floundered, the company eventually folded, and the communities saw few benefits.[36]

Experiences specific to Africa further complicate global definitions of indigeneity and community that circulate in discussions of benefit sharing. Aside from recent instances in Southern Africa, including the hoodia case, African cases of bioprospecting have not focused on providing compensation to specific ethnic groups deemed to be holders of information. As has been the case with textile designs, herbal knowledge has until recently assumed the status of national goods. I use the alternative term "community" not to suggest specific "tribes" with shared kinship but rather multiethnic segments of nation-states that share languages or land. Many African countries have focused on downplaying differences among ethnic groups in an effort to unite citizens. Independence leaders proclaimed the importance of national identities, and narratives of multicultural inheritance have shaped later leadership regimes. For instance, Ghana's independence leader Kwame Nkrumah proposed, "By reason of our unshakable faith in African Unity, Ghana cannot support any moves designed to create disunity or fan the embers of tribalism."[37] Indeed, indigenous rights movements, such as the efforts of Native Nations in North America or Aborigines in the South Pacific, reclaimed terms like "native" or "aboriginal" that had long been rejected in African contexts as a sign of colonial racism. As Ronald Niezen expressed in his analysis of the "paradoxes of indigenous rights":

> As a liberation movement, indigenism thus stands apart from the twentieth century's most exalted freedom struggles: decolonization, antiapartheid, and civil rights. . . . The most common goals of indigenous peoples are not so much individual-oriented racial equality and liberation within a national framework as the affirmation of their collective rights, recognition of their sovereignty, and emancipation through the exercise of power.[38]

There is little precedent for indigenism in Africa. The closest approximation might be attempts of traditional kings to regain a role in civil society vis-à-vis democratic reforms. Although there have been isolated efforts to promote indigenous rights of first peoples in the context of hundreds or even thousands of different languages in each African nation, these have frequently centered on groups like the Bushmen of Southern Africa or the Pygmies of central Africa, who have long been associated with specific cultural and physical differences by speakers of Bantu and European occupiers alike. In other cases, broad religious and linguistic differences have led to secessionist movements like those in Sudan and Eritrea. In a sense, a condition of African modernity has been the simultaneous recognition of the self as a member of multiple ethnic, religious, and linguistic trajectories.

The CBD's language of promoting compensation for "indigenous and local communities embodying traditional lifestyles" therefore becomes complicated when one considers seriously the multiple identities of African citizens who participate in the constant renewal of their cultural practices, often in urban milieus detached from biologically diverse environments. A generation of Africanist scholars has shown how colonial occupations shaped current conceptions of tribal identity and homelands, often reifying differences where there were few. As recent civil struggles in Ivory Coast over "ivorité" have reminded us, everyday people in African contexts continue to seek citizenship as a proxy for indigenous rights, making the assignment of benefits for plant medicine along communal lines a complicated prospect.[39]

In my experience, as I searched for stories from Mali to Madagascar, I never spoke to local people who desperately wanted to be given money to continue living a traditional lifestyle in some economically depressed environment just to sustain wild plants. Individuals sought passage to the city or abroad, better science education for their children, maternal health care for their sisters, sturdy jeans for their bodies, or fertilizers for their fields, even as foreign researchers and tourists romanticized the sustainability of their traditional cultures. Or, they performed the channeling of outmoded spirits at theatrical religious ceremonies and pored over anthropological texts of precolonial histories to better impress visitors like myself. Given my own research, I hesitate to advocate compensation for bioprospecting on communal lines in Africa or elsewhere. As African independence leaders feared, the deepest conflicts on the continent have depended on emphasizing tribal differences; witness the struggle for oil rights during Nigeria's civil war or the tensions between Hutu and Tutsi groups in Rwanda and the subsequent genocide.

Benefit-sharing agreements with indigenous communities are especially complicated when indigenous scientists seek patents. As this book shows, African scientists have sought priority for their research through both publications and patents filed abroad. Patents for pharmaceuticals have been controversial in the Global South in recent years, including in Africa. First, the high cost of essential medicines has led to calls for the production of generics through compulsory licensing. Second, the perception that international drug companies steal indigenous knowledge to make profitable industrial and pharmaceutical chemicals has led to calls to overturn controversial patents. Nevertheless, it would be a mistake to assume that all patenting efforts in African countries are imposed from the outside. This book shows that people in African settings have participated in global patent regimes, for better or for worse.[40]

On a moral level, the creation of a new drug might benefit all of humanity, or at least that proportion of individuals afflicted with the specific ailments treatable. Indeed, drug companies frequently invoke the argument of common benefits to emphasize the necessity of their high research budgets or the need to conserve biological diversity. But in postcolonial Africa, the case for direct benefit sharing may be difficult to make, given the challenges of national regulation of widely dispersed healing therapies. This is not to say that African scientists did not exploit healers and rural communities in their bid to create new pharmaceuticals and further their careers. Nor would the question of benefit sharing between ethnic groups and scientific organizations necessarily be a bad thing. However, this book probes the many issues behind ethnicity, scientific identity, and traditional healing that have remained absent in past discussions of bioprospecting.

OF BITTER MEMORIES AND CONCEALMENT

This book is based on evidence drawn from textual and oral documentation. Each chapter details the archival and herbarium records to which I referred during my research, including letters, laboratory notes, unpublished reports, and slips of paper nestled in dried specimens of plants sourced in Africa. As more and more scanned records appeared online, my research incorporated data mining of digitized manuscripts, patents, government documents, and books and journals, as well as herbarium sheets. On occasion, the chapters detail day-to-day occurrences surrounding a plant as traced through a specific set of documents that were preserved together in the Ghana Public Records and Archives Administration Department, the National Archives of Madagascar, or the Archives of the Royal Botanic Gardens at Kew in the United Kingdom. On the whole, the book analyzes scattered evidence not for small-scale increments of social change but for large-scale patterns of how different stakeholders have sought to transform medicinal plants from African countries.

This project also benefited from hours of conversations with people engaged in the search for healing plants (whose names appear in the List of Persons Consulted in the backmatter). Oral history has been fundamental for historians of Africa confronting limited sources for the precolonial and colonial period. The technique is now widely used to reconstruct recent histories of the postcolonial period as well, where it interfaces with observational and ethnographic data collection, as well as a more abundant offering of written material. Surprisingly, oral testimonies have supplied researchers in the emerging field of pharmaceutical history with fundamental data, given that corporate archives mostly remain closed. With digital data fast replacing paper records, oral investigations will undoubtedly be critical as more historians seek user perspectives while mining Web sites, emails, and other electronic media.[41]

When I began this project, social historians of Africa typically addressed changes in the experiences of local communities within nations using colonial archives and oral history. This made their narratives quite familiar to historians who worked in areas like North American or European history, where rich archival evidence exists for the nineteenth and twentieth centuries. But as historian Dipesh Chakrabarty reminds us, dependence on colonial archives and European intellectual theory provided a comforting way for histories of non-Western settings to have the same theoretical and chronological concerns as Western accounts. Despite the best efforts at postcolonial scholarship, Chakrabarty states, " 'Europe' remains the sovereign, theoretical subject of all histories, including the ones we call 'Indian,' 'Chinese,' 'Kenyan,' and so on."[42]

This research project forced me to acknowledge that colonial archives did not always contain day-to-day accounts of bioprospecting, nor were people who recalled plant piracy under colonial occupation still alive. In addition, I wanted to examine points before and after the colonial moment to better trace networks of exchange. The project therefore left the comfort zone of European-centered narratives about colonial theft, aside from one or two caches of documents I found on pennywort or *Strophanthus* in archival collections. Instead, I depended on published records and oral accounts to probe the circulation of grains of paradise, as well as on scientific articles to investigate periwinkle, *Cryptolepis*, and hoodia. My oral interviews informed the theoretical questions that I address, including who was first, if traditional medicine was local, how people transformed plants into pharmaceuticals, and who benefited.

In conducting extended oral history interviews, I did not undertake a randomized study but employed a variety of techniques to identify possible interviewees, including scientific reputation determined through literature reviews and patent details; recommendation of government ministries, universities, and business associations; and suggestions from past interviewees. Overall, I emphasized older individuals who played leadership roles in their communities and who had a longer view of the history of science and medicine in their country. Formal interviews lasted between one and eight hours, with some leading to conversations that took place over several years. In Ghana, I formally interviewed more than fifty healers, scientists, and policy makers in Akropong, Larteh, Mampong, Sekondi-Takoradi, Kumasi, Cape Coast, and Accra during fieldwork on bioprospecting conducted intermittently between 1997 and 2005. In South Africa, I formally interviewed a total of fifteen scientists, healers, and plant sellers in Cape Town, Pretoria, and Johannesburg in 2007. In Madagascar I formally interviewed twenty plant sellers, farmers, and scientists in Tolagnaro (Fort Dauphin) and Antananarivo in 2008. Although my interview material for Ghana is more culturally nuanced and extensive, my shorter stays in South Africa and Madagascar coincided with a solidification of my research interests on specific plants and healer-scientist relations, leading to targeted, highly informative conversations. Often, interviews with the stated individuals merged into focus-group discussions when family members, customers, students, and others in the general vicinity joined into the conversation. I gave informants the option of being recorded anonymously but found that individuals were generally more than happy to go on the record with their version of events—and pleased to be remembered as leading scientists and merchants. In addition, I participated in informal activities with informants, including meals, conferences, and social events.

The two main groups of interviewees were healers and scientists. Class alliances affected my experiences with each group, as my college education placed me in closer proximity to the scientists. Generally, I found that because scientists in Africa struggled for international recognition, they carefully differentiated themselves from those in plant markets and healing shrines, whom they saw as less educated and less sophisticated in their methods. Whether I sought healers in Johannesburg, Antananarivo, or Accra, scientists there warned me that I was likely to be robbed, duped, or otherwise harassed in the plant markets. Yet, when I sat down to speak to healers about their difficulties, they invariably explained how they had limited education or money, and that scientists at the nearest universities either stole their cures or sent students to do it for them. Relief and trust shaped my discussions with healers once they realized that I did not want to know about any secret cures but rather hoped to find solutions to past grievances. Although their stories do not all directly appear in the book, these interactions shaped my understanding of how people create and share plant medicines in the present, and the ways their forebears may have exchanged valuable knowledge in the past. Nor was oral history my sole entrée into healer cultures. Healers shared with me their own written records of patients, and a surprising number of their recipes have been published in postcolonial botanical surveys, most with full attribution.

In contrast to the healers, some scientists were wary of speaking to a historian interested in observing them in their place of work and asking them questions about their personal research narrative. Most African universities abolished anthropology departments after independence, making a visit from a social researcher to their laboratories and offices especially ironic.[43] Perhaps more secretive than the famously reticent healers, the scientists would rarely reveal the names of plants they were currently researching, even if publications were out and patents filed. Thus, we danced around the topic of drug discovery and rights management through a reflection on outdated plants. I often received probing questions about exactly how much chemistry I had studied in school, as well as demands to further explain the precise aims of the history of science. In Madagascar, chemists and botanists at the University of Antananarivo and the Malagasy Institute of Applied Research were relieved to discuss a common weed like rosy periwinkle, or even pennywort. In South Africa, in contrast, my research on hoodia coincided with media frenzy over rights to the fruits of national science. Government scientists there made cautious statements and insisted that they sign forms with the clause that they could request that I redact any of their statements from my book. Similarly, I began research on *Cryptolepis* just as the two main protagonists in Ghana were close to death, in a storm of controversy over their physical and intellectual property. Tape recorders, then digital recorders, and then video recorders were reluctantly turned off during the juiciest parts of conversations. After one scientist wept profusely during a four-hour interview about his role as an intermediary exporting bundles of dried plants between foreign researchers and local healers, I realized that I needed to add a clause to my preliminary remarks and human-subject-research waiver forms that "reflecting on the past may make you sad."

My experience conducting oral history interviews speaks to the malleability of identity and the relativity of race and ethnicity. Although many historians may not spend much time considering how their gender and appearance affect the interview performance, my "positionality" had an impact on the course of my research, and I would be remiss not to mention my own status as a multiethnic historian of science. In Ghana, sometimes it was advantageous to call on the Ewe (my sister is called Asiwome, a gift from God) and Akan (I am named for my late grandmother Agyepoma) ancestry of my father. Once, I was allowed to interview a healer only when his sister recognized me from when I had read a tribute in Twi at the funeral of my grandmother's best friend (she had come from their town). I understood more clearly over the course of my research that my father's family was Guan, a first-people group found throughout Ghana that had taken up the Twi language during Akan occupations. Later, I referenced the Ga last name of my husband when I found myself among Ga-speaking farmers, who had migrated fifty years before from the coast to a forest enclave near the Ivory Coast border and were reluctant to speak to an outsider about their experiences with a valuable plant. Or, I might explain his mother was Ewe despite her Akyem last name. Or, when I needed to present further affinity to Europe and justify my interest in the French language, it might be helpful to explain that my husband grew up on the French-German border.

In other contexts, my dual position as a person of European descent affected my research. To test my character, one spiritual herbalist, a Hausa-speaker from Northern Ghana, placed several small objects under an overturned calabash, then asked me to balance on top of the rounded surface on one foot. As I stood precariously over the oracle, he said he could not tell whether I was a Ghanaian or an outsider, but he had a good research experience with a Scandinavian anthropologist. Was this the moment to explain my mother's family hailed from Norway? In South Africa, I quickly realized that I was classed as "colored" based on my appearance. Many of those interviewed assumed I spoke Afrikaans, and during a trip to Cape Town some assumed I was part of the Cape Malay community. In Madagascar, I could pass as the sister of the fiancée of my research coordinator. When we entered government offices to find statistics or people, often we were smoothly escorted through, until my American-accented French and nonexistent Malagasy showed my identity as an outsider. But then, Barack Obama had been elected president of the United States a week before my arrival in Madagascar. Obama, Abena, they sound similar, I heard on numerous occasions. Obama was featured in newspaper articles there about the rise of those of mixed ancestry. "Your father is from Ghana, your mother from America, c'est manifique!" "You are like me, my mother is Chinese, my father Malagasy." Or "My grandfather is Kenyan and her grandmother Indian. We are all mixed!"

Listening closely to the complex recollections of healers, scientists, and community members led me to the stories of the six plants in this book. Multivocal narratives suggest how many people could be first to create herbal innovations. The recipes of healers across borders show how herbal knowledge transcends locality. The sagas of scientists hoping to appropriate prior plant recipes reveal the difficulties of recording, experimenting, explaining, producing, and harvesting plants for new pharmaceuticals. And the stories of who benefits from plant-based pharmaceuticals manufactured for foreign markets indicate the difficulties of assigning credit and rewards in a global system of herbal medical exchange.

Let us begin with the fairly well-known account of Madagascar's periwinkle at the hands of the drug firm Eli Lilly. From markets and laboratories in the capital city, to ports and plantations on the southern coast, knowledge of plants on this island nation spread around the world. The wide distribution of periwinkle and the simultaneous creation of herbal remedies and laboratory investigations make it difficult to assign priority in drug discovery. Yet, Lilly charted a pathway to prosperity that many African countries have sought to emulate, at the same time that others condemned the company's profits. The issue of who should prosper when plants become pharmaceuticals remains an open question in Madagascar, with implications for the future of medicine.

Figure 1.1 Periwinkle and pennywort for sale at Analakely Market, Madagascar. (Photo by author.)

CHAPTER 1

Take Madagascar Periwinkle for Leukemia
and Pennywort for Leprosy

Manan-tsira ka mahay mahandro ny vazaha. The foreigners have salt, therefore they know how to cook.
—Malagasy proverb, in Ralibera, *Vazaha et Malgaches en Dialogue*

In 1965, the U.S. Patent Office awarded the chemist Gordon Svoboda two valuable patents for work he had conducted on behalf of the drug company Eli Lilly. The patents granted Svoboda, and therefore Lilly, exclusive rights to unique processes to extract chemicals from the plant *Vinca rosea* (now called *Catharanthus roseus*). Lilly went on to market the chemicals as drugs to treat two types of cancer, leukemia and Hodgkin's lymphoma. By the late 1980s, annual profits from the medications, marketed as Oncovin and Velban, were reportedly between US$90 and US$400 million.[1]

The plant, commonly known as "rosy periwinkle" (hereafter referred to simply as "periwinkle"), was originally from Madagascar. For two decades, periwinkle has been at the symbolic center of debates over the distribution of rewards from plant-based pharmaceuticals. Activists have argued that Lilly "owed" the citizens of Madagascar (the Malagasy) for its profits, given that the plant is widely used in herbal medicine there.[2] Not surprisingly, scientists behind the research at Lilly's laboratories in Indianapolis disagreed. Svoboda's collaborator Irving Johnson explained, "In this case I do not believe there is a compelling reason to suggest that Madagascar's role in the discovery of the pharmacological action of a few of the alkaloids from this plant represents 'easy picking' or any logical requirement for compensation. It was certainly not easy and required millions of dollars of investment."[3] Advocates for benefit-sharing agreements between drug companies and local communities retorted that regardless of whether Lilly's drugs represented a direct extension of Malagasy practice, Madagascar was in danger of losing forests that might hold other valuable cures, and compensation for periwinkle might prove an incentive for their preservation. As U.S. legal scholar James Boyle put it, Madagascar needed a way to compensate indigenous peoples who "can find no place in a legal regime constructed around a vision of individual, transformative, original genius."[4]

In fact, around the same time that Lilly researchers found new methods to extract alkaloids from periwinkle, the Malagasy physician and frequently heralded "genius" Albert Rakoto-Ratsimamanga was involved in ongoing research to develop curative acids from another plant, Indian pennywort (*Centella asiatica*). Ratsimamanga and his French collaborator Pierre Boiteau received patents in Europe and North America for a process to extract hemisuccinates from pennywort acids in 1968. They licensed the process to the French drug company La Roche, which first marketed the wound treatment as Madécassol. By 2010 the company, now La Roche-Posay, included extracts of the plant in its wonder treatment for wrinkles, Redermic. Like the Lilly scientists, Malagasy citizens sought patents for plant-based drugs to amass private wealth. Ratsimamanga's involvement with pennywort complicates the mythical story of periwinkle that centers on the exploitation of Malagasy indigenes because it shows that African scientists were involved in drug prospecting at the same time as researchers at Lilly.

The better-known periwinkle case has come to symbolize many things to different people, depending on their concerns over the environment, traditional medicine, or intellectual property rights. By the late 1980s, the world was witnessing a massive "green rush" as new laboratory techniques allowed for more rapid screening of plants to discover improved medicines, crops, and industrial chemicals.[5] This quest for new chemicals from plants became known as "biodiversity prospecting" or "bioprospecting" when a group of biologists and policy makers— including Walter Reid, Sarah A. Laird, and Calestous Juma—promoted the use of the terms in 1991 in an argument for sustainable development.[6] For environmentalists, the story of periwinkle suggested both the danger and promise of harvesting rare plants for corporate gain. Laird, an early proponent of biodiversity prospecting, worked for the Periwinkle Project, a division of the Rainforest Alliance established in New York City. Concerned that the "next periwinkle" might be destroyed by development, Laird hoped to build awareness around the drug industry's reliance on plant-based chemicals. Using WHO estimates, she claimed that 80 percent of the world's population used traditional plant-based medicines, and that the rainforests held over 70 percent of all the globe's plant biodiversity.[7] Over and over went the refrain: "Lifesaving medicinal properties are found in the plants of the Madagascar *rainforest*. Extractions from the leaves of a rosy periwinkle have been used to find a cure for childhood leukemia, and now, due to a drug developed from these extracts, we have about a 50 percent recovery rate from childhood leukemia."[8]

Ironically, periwinkle in Madagascar grew along the roads and in fields—not in threatened forests. In fact, many of the medicinal plants people have come to use in herbal remedies grew near farms; the toxicity of the valuable phytochemicals that give them medicinal power allowed them to subsist as weeds, not rare species.[9] In our conversations, the Malagasy conservationist Jean Joseph Andriamanalintsoa promptly reminded me of periwinkle's status as a field plant, as did other scientists involved environmental efforts in Tolagnaro and Ranopiso, where Lilly once sourced periwinkle harvests.[10] I timed my visit to Southern Madagascar to coincide with periwinkle's blooming cycle; I was overwhelmed to see the pink blossoms swaying in the breeze along *every* grassy highway and field. It was ubiquitous.

Other activists often retold the story of periwinkle as an argument for benefit sharing between disenfranchised communities and corporations holding valuable patents. Lawyers sought new models for acknowledging prior art in nonliterate societies. Rather than overturn patents, they hoped to extend profits or other rewards to people living in poverty who maintained ideas about the value of plants through oral communication and family apprenticeships. Yet, benefit sharing assumed a class of individuals to whom compensation might be afforded, ideally descendants of a First Peoples group in a threatened ecosystem. Moreover, the idea of indigenes with original and unique—albeit shared—communal knowledge still relied on the concept of priority so integral to the logic of patents for inventors. Controversies over the transformation of periwinkle and other plants into pharmaceuticals circulated around ideas about inventors and indigenes because each could claim to hold first rights to information.

For public health officials working in Africa more generally, the promise of turning green into gold trumped any environmental or legal controversies surrounding periwinkle and Lilly. In 2003, WHO officials in Brazzaville, the capital of the Republic of Congo, unveiled a new logo to rebrand traditional medical practices as a new engine for economic growth (see Figure 1.2). A gold ring encircled both the green continent of Africa, overlaid with a periwinkle flower, and the blue seas around it. The graphic designers behind this depiction of botanical and marine wealth did not include bones, shells, or other accoutrements of diagnosis in African healing, turning their gaze instead to periwinkle. For them, Eli Lilly's success with periwinkle suggested avenues for profitable drug discovery in African countries, not a rallying cry for retrospective benefit sharing or preservation of traditional communities.

And indeed, the little-known case of pennywort's transformation into Madécassol is an example of how scientists in African countries have created patented pharmaceuticals from plants. In this chapter, I reconsider popular understandings of the periwinkle case as one of biopiracy in light of Malagasy involvement in pennywort.[11] In particular, the

Figure 1.2 Periwinkle featured on the African Traditional Medicine Logo, unveiled October 9, 2003. (Source: World Health Organization.)

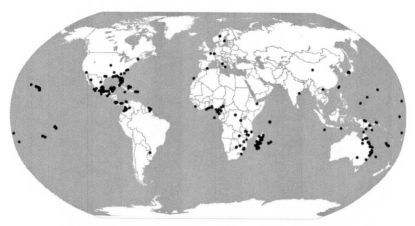

Figure 1.3 Collection points of 476 *Catharanthus roseus* samples in herbaria. Compiled from location points observed during the author's visits to the Herbarium of the Royal Botanic Garden, Kew (RBGKH) in November 2008 and from data available through the Global Biodiversity Information Facility Data Portal (GBIF), www.gbif.org, accessed March–August 2011. Plant Specimens of U.S. Department of Agriculture Plants Database; Real Jardin Botanico Vascular Plant Herbarium, Madrid; University of Kansas Biodiversity Research Center; Herbarium of Taiwan Forestry Research Institute; Fairchild Tropical Botanic Garden Virtual Herbarium; University of Alabama Herbarium; Herbario XAL del Instituto de Ecología, A.C., México; Herbario de la Universidad de Salamanca; Instituto Nacional de Biodiversidad, Costa Rica; Instituto de Ciencias Naturales; Herbier de la Guyane; National Herbarium of New South Wales; Museum National d'Histoire Naturelle et Réseau des Herbiers de France; Herbier de l'Université Louis Pasteur; Instituto de Ciencias Naturales; National Museum of Natural History; European Genetic Resources Search Catalogue; Herbaria of the University of Zürich; Leibniz Institute of Plant Genetics and Crop Plant Research; Herbario de Universidade de Santiago de Compostela; Finnish Museum of Natural History; Botanic Garden and Botanical Museum Berlin-Dahlem.

issue of priority was central to dilemmas surrounding both plants.[12] Periwinkle and pennywort have long featured in folk recipes to cure various ailments, and these recipes inspired chemists to take the plants into the laboratory to test bioactivity. Periwinkle was a pantropical weed, whereas pennywort grew in Africa and Asia (see Figures 1.3 and 1.4). The wide distribution of both plants meant that there were arguably many, many claimants to the plants, their chemicals, and related knowledge.

PERIWINKLE AND PENNYWORT IN HEALING TRADITIONS

Let us now turn to the history of priority *within* the history of traditional knowledge. Assuming we were to assign benefits to drug patents from plants along communal lines, what might have been the case for first rights to periwinkle and pennywort medications in Madagascar? And how might we prove rights to traditional knowledge claims within legal and

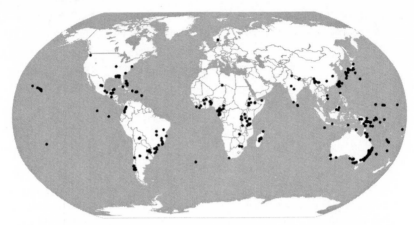

Figure 1.4 Collection points of 493 *Centella asiatica* samples in herbaria. Accessed through the Global Biodiversity Information Facility: Herbarium of Taiwan Forestry Research Institute; Herbario del Instituto de Ecología, A.C., México; Missouri Botanical Garden; National Herbarium of the Netherlands; National Museum of Nature and Science, Japan; National Institute of Genetics, Research Organization of Information and Systems (ROIS), Japan; Australian National Herbarium; National Herbarium of New South Wales; Real Jardin Botanico Vascular Plant Herbarium (Madrid); Herbarium of the New York Botanical Garden; Herbarium of the Royal Botanic Garden, Edinburgh; Herbario de Universidade de Santiago de Compostela; Harvard University Herbaria; Staatliches Museum fur Naturkunde Stuttgart, Herbarium; Fairchild Tropical Botanic Garden Virtual Herbarium.

intellectual frameworks that prioritize written knowledge? In the nineteenth century the German philosopher Georg Hegel infamously claimed that Africa had no historical consciousness because records of past events were not written down. Since then, several generations of Africanist historians have carefully documented oral traditions about past events, identified trends in historical linguistics, and examined archaeological remains to show that indeed Africans witnessed change over time. Similarly, are we to surmise that Africa had no science or innovation in medicinal plants research because little of it was recorded in textual form?[13]

Madagascar's strongest claims to periwinkle were botanical and genetic, as the plant most likely originated on the island-nation. Lilly claimed that the recipes it garnered were from the Philippines and for diabetes; thus the novelty of their invention related to both the form of the drug and its claims to efficacy—specifically, Lilly's use of extracts to treat cancer.[14] The U.S. Patent Office's grounds for novelty required that no one else had published the process to extract the alkaloids or had made the same claims about their uses. Within this legal framework, the argument for retrospective benefit sharing with Madagascar would be that Lilly's recipes from the Philippines had actually originated in Madagascar. However, I soon realized that it was literally impossible to trace how herbal recipes might have spread, given the paucity of written documentation and the pantropical distribution of this common weed.

Instead, my research—taking me to sites as far-flung as markets in the capital city of Antananarivo, farms near the southern port towns of Tolagnaro and Ranopiso, herbaria at the U.K. Royal Botanic Gardens at Kew, and the economic botany collection at Harvard—convinced me of the global history of this plant so deeply associated with Madagascar. And even when my search for periwinkle led me to pennywort, a plant that many scientists and healers in Madagascar also claimed as their own, I found a similar story. Again, traces of evidence hinted at another global history of ownership and innovation, from Ancient Sanskrit herbals to cups of *Centella* tea, to forgotten letters between scientist in Madagascar, India, France, and the United States. The process of turning plants into pharmaceuticals is never straightforward but instead requires the efforts of multiple participants.

Periwinkle in Madagascar

Madagascar's unique ecosystems led to the evolution of species of plants available only on the island. Of 13,000 plant species found there, four in five are endemic to Madagascar.[15] Although many plants and animals are unique to the island, recurring waves of migrants led to an ethnically hybridized society drawing from multiple continents. Madagascar is usually considered part of Africa, although the people who live there today trace their ancestry not only to such countries as Kenya and Mozambique in East and Southern Africa but also to the Middle East. In addition, there is a strong influence from Indonesia, Malaysia, and the Philippines, with large waves of Austronesian migrants traveling by boat to settle in Madagascar beginning in the first century before the Common Era. From 1890 to 1960, Madagascar was a colony of France. Today the official languages are French and Malagasy, a language derived from Austronesian, Swahili, and Arabic words that was reduced to written script from the 1500s. Two major ethnic divisions include the Merina, who trace their ancestry to Indonesia and reside primarily in the plateau near Antananarivo, and coastal communities of the Fianarantsoa, often of primarily continental African descent. Merina royals famously consolidated authority over the island by the mid-nineteenth century and resisted French colonization.

There are more species of periwinkle plants in Madagascar than elsewhere, indicating that periwinkle originated on the island and developed over time. Of the eight species of *Catharanthus*, seven are indigenous to different regions of Madagascar, with one unique to India. The genus also secured several different names over time, including *Vinca*, the name for European periwinkles that are similar in appearance, and *Lochnera*. Herbarium specimens stored around the world, dating to the 1770s, were riddled with the comments of botanists debating the differences among species. By the 1970s, several of the species had combined or hybridized in the national botanical gardens at Tsimbazaza Park in Antananarivo, including *C. longifolius* and *C. roseus*. Research there was critical to wider understandings of the different species of periwinkle and their varying levels of chemical activity. In particular, before 1975, confusion had surrounded the species *C. trichophyllus*, *C. ovalis*, and *C. longifolius*.[16]

Malagasy citizens nurtured continuous cultural uses of periwinkle, providing a contemporary window onto exchange of uses over time. During my visit to Analakely market stalls in 2008, a *mpivarotra-hazo* (plant seller) named Hantalalao Razaiarimanga explained, "Rich people buy from us because they eat good and fatty food, so they want to clean their insides with plants. Poor people come here, too, as well as foreigners like you." She explained that *vonenina* (periwinkle) could be thought of as wet (fresh) or dry and was extremely bitter unless boiled. Razaiarimanana's hair was smoothed into a neat ponytail, and she rested her hands in the pockets of a beige smock tied over her jeans. She explained that she had learned to sell plants from her own mother, and now her two young children sometimes joined her at the stall she had rented for a decade. Nearby, another *mpivarotra-hazo* explained that periwinkle was further divided into its parts—roots and leaves—each having different curing properties. She explained that the long, brown, carrotlike roots, for instance, were especially good to digest for a stomachache.[17]

The export of periwinkle and other plants from Madagascar each year is a major source of foreign exchange. At the Ministry for the Environment and Forestry, Ndriana Razafindratovo explained to me that in the last decade, at least 300,000 to 400,000 kilograms of periwinkle made their way overseas each year to places like France, Germany, and Mexico (see Figure 1.5). Razafindratovo showed me charts that indicated pennywort exports were fairly constant, if not increasing. However, his tables also confirmed that periwinkle exports had seen a big decrease since their heyday in the late 1970s and early 1980s, when they peaked at over 1 billion kilograms per year.[18] Ironically, it was just as Madagascar began to lose ground in the periwinkle export market, and Lilly's patents expired, that journalists began to promote the story of biopiracy of the country's herbal knowledge.

Data on the exchange of plants in herbal medicine markets was more difficult to delineate. Early on Friday mornings, wholesale medicinal

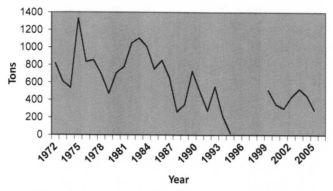

Figure 1.5 Exports of *Catharanthus roseus* from Madagascar (1972–2006). Compiled from data provided to the author at the Ministry of the Environment and Forestry in Antananarivo, and from Rasoanaivo, "Rain Forests of Madagascar," and Andria-manalintsoa, "Contribution à l'Étude de la Production." There is some discrepancy between figures, and only Andriamanalintsoa provides a breakdown between leaves and roots, using data he obtained from Promotex annual reports. Figures are unavailable for the period between 1993 and 2000.

plant sellers congregated in Antananarivo to sell newly harvested and dried flora. Their target audience was the assortment of small business-men and businesswomen who maintained plant stalls in the Analakely open air market downtown. These urban plant specialists resold their se-lections to those who passed through the narrow alleys that lay between tables mounded with dried plants. Theirs were small transactions, never written down, that passed directly from seller to client, or often via chil-dren sent to the market to run errands. Most dealers were vague in their answers if and when the tax collectors swooped in. They might report revenues of 12,000 ariary each month, when it was closer to three or four times this amount. Thus, the total profit from all sellers in the main Antananarivo plant market might be approximately 36 million ariary, or close to US$20,000 annually.

In interviews, the sellers at Analakely and, later, collectors and grow-ers near the southern port at Tolagnaro (or Fort Dauphin) revealed pop-ular understandings of periwinkle and pennywort, the two primary me-dicinal plants in the domestic and export trade. This meant that even as scientists transformed both plants into patented pharmaceuticals, plant sellers and their customers continued to experiment with popular rem-edies made from their raw botanical versions. The men and women I spoke with in Madagascar also reminded me that—as I had found in other parts of Africa—plant specialists were urban and, at the least, networked. These multiethnic merchants were not isolated members of "indigenous" groups lost to the world. These plant specialists indicated that periwinkle was not an endangered plant at the heart of rainforests but rather a valuable weed central to their medicinal herb trade. Relief filled the faces of interviewees when I explained I only wished to know about a common plant and did not hope to steal any valuable information from them about an esoteric herb. A law student assisting me in my research grimaced when I showed her some gnarled periwinkle roots, recalling the many times she had sipped bitter broths made from the pungent plant. She reminded me that anyone in Madagascar could describe uses for periwinkle.

Periwinkle Moved; Did Knowledge Travel Too?

By the beginning of the twentieth century, when we begin to see an increase in written herbal recipes for periwinkle, fishermen, merchants, and explorers traveling to and from Madagascar had transferred the plant to tropical regions around the world, transforming it into a global weed. Malagasy fishermen and mariners had historically recognized the value of periwinkle as a medicine and "chewed the leaves of the plant between their teeth to soothe the sensations of hunger and to overcome fatigue."[19] Interested in the plant's nutritive value, Malagasy fishermen transported cuttings in their boats, establishing new colonies of the plant along the sandy shores of the Indian Ocean. Some botanists argued that this was how periwinkle came to be found around the world, given the golden age of Swahili trade (c. 800–1450) during which networks of Afro-Asian merchant communities populated coasts from northeastern Madagascar to Mozambique, Kenya, Somalia, the Arabian Peninsula, and South Asia.

As part of the changes in European maritime trade, French explorers reached Madagascar in the 1600s. Their appropriation of periwinkle represented the first written documentation of both the plant's medicinal uses and its export, forever linking periwinkle to Madagascar. Later, genetic studies of different species of periwinkle would confirm this association with Madagascar as the original home of the genus. In a small museum overlooking the ocean in Tolagnaro, curator Mr. Rabenantoandro showed me photocopied pages from Étienne de Flacourt's compendium of observations on seventeenth-century Malagasy life made during the Frenchman's tenure as governor of a short-lived colony of the French East India Company. Flacourt recorded numerous plants that he observed at the southernmost tip of the island where the Antanosy people soon threw off early French occupation, including a plant they termed *tongue* or *tonga* (now *C. roseus*). Ironically, Flacourt suggested that the white periwinkle he found in Southern Madagascar had "more virtue," perhaps aesthetically but also medicinally, than the rosy version. Among the local uses he recorded for periwinkle roots were treatments for heart pangs and headaches.[20]

Before his death in 1660, Flacourt sent seeds and cuttings of the Madagascar periwinkle to France, where they were redistributed to other European botanical gardens. Flacourt's careful documentation and collection of Malagasy medicinal plants for the French Royal Gardens represented what Claude Allibert termed "one of the earliest studies in tropical ethnopharmacology," and his treatise became a valuable source for later bioprospectors.[21] European gardeners shared the seeds with friends and colleagues, assisting in the global dispersal of the *C. roseus* species from Madagascar.[22] At Kew, I peered at the shiny portrait of a periwinkle plant with white flowers in the classic work of English botanist and gardener Philip Miller that became the gold standard for scientific understandings of tropical periwinkle. Most likely, it was painted from a cutting propagated from specimens Flacourt sent to France. Miller explained, "The Seeds of this Plant were brought from Madagascar to Paris, and sown in the King's Garden at Trianon, where they succeeded; and from thence I was furnished with the Seeds, which succeeded in the Chelsea Garden. It rises with an upright branching Stalk to the Height of Three or Four Feet." Miller included the plant in his description of novel plants in Europe but did not note any medicinal uses for Madagascar periwinkle, although the Chelsea Garden specialized in herbal remedies.[23] By 1753, the Swedish botanist Carl Linnaeus associated Malagasy periwinkle with the genus *Vinca* under the species *V. major*. Not until the 1950s and 1960s was this tropical periwinkle originally from Madagascar reclassified within a separate genus as *C. roseus*.[24] Cognizant of the importance of early European collections, bioprospectors at the close of the twentieth century "beat a path" to Miller's Chelsea Garden to review periwinkle and other healing plants.[25]

Explorers circulated periwinkle specimens to and from Africa at later points as well. Some collectors referred to these specimens by common names in European languages, suggesting colonial introduction for primarily ornamental purposes. During British physician John Kirk's expedition along the Zambezi River in Southern Africa, collectors sent parts of the plants to London in 1867 with the note that periwinkle was "grown near the consulate." Collectors sent specimens to London from Martinique in November of that year. In 1912 a botanist supplied a dried pink periwinkle flower from Half Assini in the Gold Coast (now Ghana) with the note "introduced (?) coast villages," raising questions about how the plant came to grow along the West African shoreline. By 1926, botanists from Sierra Leone indicated that periwinkle was "a commonly cultivated shrubby plant with fleshy leaves. Established here in derelict gardens and along the roadside. Mauve flowers (a white-flowered variety is also cultivated commonly)." In 1931, collectors sent flowers from Sierra Leone to England with notations that periwinkle was "cultivated wildly [*sic?*] by Creoles. The mauve var is rather less common. . . . The white flowered var is called 'Waitie' by Creoles because of its white colour. The mauve-flowered var. is 'Joy-sie' in Creole on account of the pleasure it gives (flowers other than white, yellow or red are not common in cultivation)." In 1955, a collector noted that in Northern Rhodesia (now Zambia), near Lusaka, periwinkle was called "Star of Bethlehem" and was "cultivated . . . in village gardens all over the territory."[26]

Early knowledge of periwinkle's uses in treating disease spread with the roots and seeds, although it is difficult to document how or when. Handwritten notes on sheets preserving Kew herbarium specimens of *C. roseus* further indicated that knowledge of periwinkle's healing properties had spread widely. The variations in the form of the specimens—some include only flowers, while others reveal the entire plant, including the roots with notations on medicinal applications—moreover suggested that various collectors wished to emphasize the pharmaceutical, rather than ornamental properties. African names for periwinkle indicate longer experiences with the plant, perhaps through precolonial commerce. For example, a specimen from Mozambique collected in 1980 listed the local name as *guiana* and noted that the "roots [were] used in a mixture against venereal diseases."[27] Just across the Mozambique Channel from Madagascar, waves of migrants from Mozambique traveled by small boats to the large island as early as the third or fifth century before the Common Era. Thus, there could be a strong correlation between uses of the plant in both countries.

In the 1970s, after Lilly sparked interest in periwinkle as a source for cancer treatments, a U.S. medical student received sponsorship from the drug firm Smith, Kline and French (now Glaxo-Smith-Kline) to retrace Flacourt's forays through a survey of medicinal plant markets in Tolagnaro. Like Flacourt, he similarly noted that people living in the area used periwinkle for feverish pains associated with the stomach, based on the testimony of an individual "who was collecting great bales of *Catharanthus* for an American drug firm."[28] In contrast, the medicinal applications of periwinkle in such far-flung locations as Jamaica and the Philippines suggest recurring, independent discoveries of herbal recipes for the weed across time and space. But, before continuing with the periwinkle story, let us turn briefly to pennywort.

Pennywort Therapies in Madagascar and India

If periwinkle became a pantropical weed, how widely did people disperse pennywort, Madagascar's other main healing plant export? At Analakely market in Madagascar's capital city of Antananarivo, an elderly plant seller presented wads of twisted pennywort vines, or *talapetraka*. For our closer inspection, she held out the fan-shaped leaves on the fresh bundles in one hand, while the mostly shriveled stems of fully dried pale, greenish-gray ones lay in her other hand. She recommended that I steep pennywort in boiling water and sip some of the beverage each morning. When I bought a few bundles of the dried plants, she cautioned me that I would not be able to slip them safely into my suitcase, as it was illegal to export plants without proper documentation from Madagascar. I later learned that legitimate trade in pennywort, the second leading medicinal plant export, had fluctuated between 20,000 and 70,000 kilograms each year since 2000. Pennywort leaves headed from Madagascar to importers in France, Germany, Spain, Italy, and Mauritius (see Figure 1.6).

Although there are some clues as to how periwinkle may have spread from Madagascar in earlier times—including through African fishermen, Swahili merchants, and the later taking of the plant by French explorer

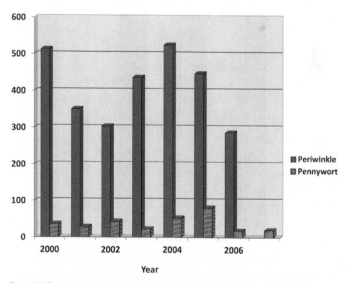

Figure 1.6 Exports of pennywort and periwinkle from Madagascar (2000–2006). Compiled from data provided to the author at the Ministry of the Environment and Forestry in Antananarivo.

Flacourt—we may never know precisely how people shared pennywort to and from Madagascar. On the island-nation, people have popularly used pennywort to treat a variety of health concerns, including infections on the skin caused by insect bites and eczema. Pennywort or *Centella asiatica* is a member of the botanical family *Umbelliferae*, also known as the carrot family, and it is unclear whether the plant originated in India or on an island within the Indian Ocean, including Madagascar. It grows wild in humid conditions and can also be cultivated. Its leaves look similar to those of the ginko tree. The appealing shape and soft texture of the leaves may have led people to independently taste pennywort and test it on their bodies. In Madagascar, people shared knowledge on the uses of pennywort through conversation and observation, in keeping with standards of oral dissemination of botanical knowledge common throughout African societies.[29]

There is also the strong possibility that given long-standing trade relations and migration in the Indian Ocean basin, remedies using pennywort created in India may have become popular in Madagascar through personal interactions. In South Asian medical traditions, people have long used pennywort to heal the body and mind. In fact, the earliest written documentation of therapeutic uses of pennywort does not locate the plant in Africa. Rather, in this case, priority for determining the medicinal value of pennywort can be afforded to Indian healers, if we go by the written evidence alone. In Ancient Sanskrit surgical manuals attributed to the Ayurveda medical practitioner Susruta and dated as early as 250 B.C.E., pennywort or *mandukaparni* was used to treat *kushtha* or leprosy and related diseases of the skin, among other health concerns.[30] Another Ancient Ayurveda herbal attributed to Caraka also extolled the virtues of pennywort, considered up to the present to be the source of an important therapeutic liquid capable of strengthening the brain.[31] By 1814, a specimen of pennywort could be found growing in Calcutta within the garden of the East India Company. Britain's primary shipping and exploration company created the experimental garden from contributions of English colonists in India to identify plants with potential commercial uses as timber, dyes, and drugs.[32]

Collection points of more than 400 pennywort specimens in a sampling of herbaria worldwide show that by the early twenty-first century, the plant thrived not only in Madagascar and India but also in places as far afield as China or Brazil (see Figure 1.4). Pennywort plants introduced to Brazil could be found growing rampant in urban areas, and plant medicine manufacturers marketed capsules made from pennywort to treat bacterial infections and to strengthen the mind. Brazilians sometimes referred to pennywort as *gotu kola*, a South Asian term traced to Sri Lanka commonly used to refer to the plant in international trade.[33] In China, pennywort was a popular ground cover that was also invasive in some ecosystems. However, a recent rise in overharvesting for use in traditional remedies there led conservationists to list it as a threatened species that might actually face extinction without concerted efforts to preserve genetic diversity through breeding programs.[34]

As we will see, Madagascar's profitable exports of pennywort relied on a longer history of herbal innovation and research on the plant across geographic regions. From the 1940s, scientists operating out of Madagascar depended on information about not only pennywort's use in Madagascar but also its simultaneous consumption in India. Similarly, scientists in North America who began to take periwinkle into the laboratories took inspiration from traditional remedies from many places.

SCIENTISTS RESEARCH PERIWINKLE

Historians and sociologists approach the issue of scientific priority with great caution. The history of science is riddled with examples of unsung heroes whose contributions were ignored because of a researcher's class, race, gender, nationality, or institutional or disciplinary affiliation. Who was the assistant to the famous surgeon? Who was the woman who helped her husband with tedious microscope work? And in a contemporary scientific culture based on multiple collaborators, joint authorship, and outsourcing, who is to say who really "came up" with a specific idea? Even Svoboda's patent—licensed immediately to his corporate employer—included the work of assistants and other technicians. A scientific approach to viewing the world and natural phenomenon has necessitated the expertise of a variety of individuals, many of whom have long been hidden within the historical record. Herbal recipes lingering in the reports of scientists provide a rare window into the further contributions of herbalists, farmers, and mothers.

Scientific interest in periwinkle initially surrounded purported cures for diabetes that circulated widely during the 1920s through 1950s. Researchers in North America appropriated periwinkle recipes in their quest for new treatments for a variety of illnesses. This practice eventually led to the creation of the cancer drugs Oncovin and Velban at Lilly. The simultaneity of herbal innovation in many parts of the world makes it difficult to determine who was first to identify the bioactivity of periwinkle. What might be the evidence for Malagasy priority to periwinkle therapies for diabetes now found around the world? And was there any evidence that the Malagasy used periwinkle to treat the kind of generalized illness and decline that contemporary Western physicians might diagnose as cancer? At the same time, patents related to anticancer alkaloids from periwinkle depended on the insight of many more than the few North American chemists who claimed priority. What might a global history of Lilly's "discovery" of periwinkle alkaloids look like?

Taking Periwinkle Recipes for Diabetes

It is unclear where periwinkle remedies for diabetes first arose. Diabetes is an old diagnosis, with the English name of the disease dating to Ancient Greek medicine. However, its definition has changed over time, according to how people interpreted and treated its symptoms. A person with diabetes lacks the ability to digest sugar, and there is a buildup of glucose evident in the liver. A sign of this sugar intolerance for centuries has been thirst, weight loss, frequent urination, and certain death. Early documentation of periwinkle remedies in Madagascar noted its importance in treating hunger pangs, which might indicate interest in using it to treat diabetic conditions.

By 1921, the development of insulin to treat diabetes, as well as its subsequent marketing by Lilly, may have led to a global surge of reports on less expensive herbal substitutes for the now treatable disease, including teas and tinctures made from periwinkle leaves.[35] Pharmaceutical history may have shifted the circulation of folk remedies in this case. In 1928, researchers in Australia reported, "It appears that this plant [*C. roseus*] was first used as a remedy for diabetes in Africa and in Queensland today a great number of persons are taking it daily in the belief that they are benefited by its use."[36] Fifteen years later, wartime shortages due to shipping blockades of drugs caused physicians and their patients who had become dependent on insulin to research herbal alternatives. Recipes available from numerous places for periwinkle teas to treat diabetes bear an uncanny resemblance to one another, such as the one from an Australian woman in South Africa who recorded the treatment as follows in 1925: "Each day boil twenty-seven leaves [of *C. roseus*] in three and a half cups of water for fifteen minutes, then strain. Take one cup after each meal; one hour afterwards as much bicarbonate of soda as can be got on a sixpence in half a glass of warm water."[37] A group of pharmacists in South Africa marketed a tea based on these remedies under the name "Covinca" and claimed that it could cure 80 percent of diabetes cases. Were these occasions of simultaneous discovery, or did a similar herbal recipe circulate somehow around the world in the early twentieth century?

These herbal recipes surfaced at a moment when physicians, chemists, and biologists increasingly used laboratory and clinical investigations to establish their authority as experts on medical therapies. Miracle drugs like the antibiotic penicillin provided physicians with new weapons against diseases by the early twentieth century. Folk remedies, often shared among women caring for their families, were open to medical suspicion; earlier plant therapies were purged from official registers of *materia medica*. In addition, small drug companies as well as traditional plant sellers came into conflict with drug companies and medical associations that increasingly sought to regulate pharmaceuticals in colonies as well as imperial powers. The South African chemist D. Epstein was suspicious of the claims that Covinca was comparable to insulin and conducted a series of tests on periwinkle using rabbits, determining that it was not effective against diabetes but that it could increase the heart rate.[38] A later brand of diabetic treatment, Vinculin, also made from periwinkle leaves, was exported from South Africa to Britain by the 1930s, suggesting that at least some patients continued to find relief from the medicine. In 1936, the American Medical Association listed Vinculin among a series of ill-advised drugs for diabetes, including "Dia-bet" and "Scheidemann's Shrub Remedy."[39]

Then, in the Philippines, an "intelligent woman" advised the physician Faustino Garcia of this treatment for diabetes after the Japanese invasion of Manila in December 1941 had resulted in drug scarcity. She explained that "the leaves of *sitsirika* [*C. roseus*] were very effective in diabetes by taking in the form of decoction nine leaves, picked by pulling the leaves downward, and taking the decoction with meals three times a day. . . . Nine leaves taken three times will be 27 leaves per day."[40] Garcia tried out the popular recipe to treat diabetes by the 1940s, noting its similarity to earlier reported recipes from South Africa. He made a tea from the periwinkle leaves but decided not to experiment with it on the diabetic patients he was treating at the North General Hospital in Manila. Garcia explained, "I smelled a very disagreeable odor, found its taste very bitter, which lingered in my mouth for a long time, and its reaction with Mayer's Reagent showed strongly positive for alkaloids." He had hoped to try it out on patients for a two-week trial but was worried about the possible side effects. In the end, Garcia abandoned the plant in favor of another, *banaba*, in his diabetic research. Garcia and his colleagues used an extract from *banaba* to treat diabetes in American soldiers in the

Philippines during the Japanese occupation, furthering interest in herbal cures for the disease in North America.[41]

Subsequently, in 1952, a Miss Farquharson, who resided in Canada, visited family in Jamaica. She sent her Canadian physician, Clark Noble, "twenty-five leaves" from periwinkle plants that friends used in Jamaica to make a tea to manage diabetes. She told Noble to get in touch with a Jamaican doctor named C. D. Johnston for further information, hoping Noble could test the plants in his laboratory. Trained in medicine at McGill University in Canada, Johnston had been experimenting with treatments for diabetes made from periwinkle. In contrast to Garcia or Epstein, Johnston was more confident in the effects of the plant. Because he had retired, Clark Noble gave the leaves to his brother Robert, who worked at the Collip Medical Research Laboratory. Robert Noble sent two researchers to Jamaica to meet with Johnston. They concluded that periwinkle might have a modest impact on blood sugar levels but were not able to replicate the effects of the plant on laboratory rats with diabetes in Canada.[42]

Who was first to use periwinkle to treat diabetes? Because the earliest folk recipe I have been able to locate dates to the 1920s and is from South Africa, the argument could be made that the recipe for the periwinkle diabetes tea first originated on the African continent. Possibly the recipe might have traveled from Madagascar, where periwinkle is widely used in healing remedies. A tenuous link for transfer of recipes could be made between early settlement of Madagascar by families from the South Pacific, including the modern-day Philippines, with perhaps diasporic knowledge circulating from Madagascar back to Asia. In the Jamaican case, it is possible that enslaved Africans familiar with the plant began using it in the New World from the eighteenth century, or that British recipes for treating bleeding with European periwinkles (*Vinca*) set a precedent for using comparable plants for self-care. The similarity in these three recipes, based on twenty-five to twenty-seven leaves, could stem from the fact that those who documented the recipes had read of earlier versions, including the South African one. Or physicians read of the periwinkle treatment in medical journals and shared it with their patients. Further evidence of diabetic treatments from periwinkle included reports from Mexico in the 1950s.[43] Similar recipes abound in recent compendiums of traditional remedies in tropical Africa. For instance, in Brazzaville, Republic of Congo, Samuel Biampamba recommended a treatment for diabetes made from boiled periwinkle in 1988.[44]

Most likely, people around the world simultaneously experimented with herbal remedies, including the ubiquitous periwinkle, attracting interest from scientists in the early twentieth century who wrote down the recipes. Scientists in North America appropriated not only folk remedies for diabetes made from periwinkle leaves, but also the prior research of scientists and physicians familiar with its use in treating diabetes. Researchers in Canada relied on the efforts of the Jamaican physician Johnston. Simultaneously, biologists and chemists at Lilly's laboratories in Indianapolis conducted research on the effects of periwinkle, hoping to find an alternative to their key product, insulin. They may have seen a report of Garcia's experiments or the Filipino botanist Eduardo Quisumbing's earlier reference to the potential of *C. roseus* for diabetes.[45]

Given this history, and Lilly's interest in treatments for diabetes, the company's research on periwinkle is perhaps not surprising. But the leap from diabetes to cancer was more serendipitous, as biologist Irving Johnson, who had been researching cancer treatments for Lilly, explained in his recollection of how Svoboda turned to periwinkle in the summer of 1956:

> One of these guys, his name was Gordon Svoboda, got a plant which you find in many parts of the world called a periwinkle. You find it in Florida. It grows wild. The plant that we got happened to be shipped in from Madagascar. Mistakenly, it has been thought that we were destroying the tropical forests.
>
> It's not a tropical forest plant; it's a ubiquitous plant that grows any place it's warm enough. So Gordon submitted extracts of the plant [to Lilly's cancer research program]. I had to actually talk him into submitting it for the cancer program.
>
> He had come across a reference, maybe it was in the Philippines: if you had diabetes you'd make a tea out of this plant for treatment. So he was hoping to find a diabetic agent. He eventually did but it turned out not to be useful. But he submitted extracts to us.[46]

In Svoboda's experiments on periwinkle, he found that periwinkle extracts could be used to affect sugar levels in experimental animals. But he determined, as had Garcia, that it was not as effective an antidiabetic, especially when compared with preexisting Lilly products.

What is critical here is how the research priorities of North American scientists shaped later exploitation of periwinkle in Madagascar. In the Philippines, Garcia had been interested in periwinkle from the point of view of a diabetes treatment only.[47] At Lilly, however, researchers quickly abandoned the possibility of periwinkle as a marketable diabetes treatment, opting to study its antitumor potential instead. Researchers harvested ethnobotanical evidence in hopes of a shortcut to identifying possible drugs.

Finding Alkaloids for Cancer

Regardless of where they gained inspiration, North American scientists used patents to establish their claim to new alkaloids derived from periwinkle, in much the same way that scientists based in Madagascar and France claimed chemicals from pennywort during the same period. Researchers in Canada and the United States secured exclusive rights to market chemical entities with clinical applications from periwinkle, as well as exclusive rights to the methods to make them. Through laboratory investigations, they transformed simple recipes for twenty-seven periwinkle leaves steeped in water to treat diabetes into complex procedures to extract alkaloids from periwinkle leaves to treat cancer. Through patents, the many contributors to the discovery of bioactivity in periwinkle disappeared.

Even though Lilly scientists recalled that they were the first to use periwinkle to treat cancer, Canadian researchers also tested the plant for applications in chemotherapy around the same time, with both teams identifying key chemicals in 1958. After Noble and his team in Canada proposed that periwinkle might suppress blood cancers, the British chemist Charles Beer used chromatography to identify alkaloids in periwinkle leaves. They singled out vincaleukoblastine (shortened to vinblastine), a compound with a molecular weight of 813.8, containing forty-six carbon, fifty-eight hydrogen, nine oxygen, and four nitrogen molecules. It could be extracted through several methods involving careful mixtures of dried leaves with water, alcohol, and acid that were then evaporated in vacuum conditions. Beginning in December 1958, Beer and Noble filed patents in the United States and Canada for the new alkaloid vincaleukoblastine with the claim that it could be used in liquid form to treat people suffering from cancer. And is so often the case, Beer and Nobel were assisted by laboratory technicians, including the female Polish émigrée Halina Czajkowski Robinson. Robinson provided assistance to Noble, who asked her to measure glucose in rats fed periwinkle solutions. Robinson decided to also measure blood counts and was the first to suggest to Noble that periwinkle might have an impact on lowering white cell counts.[48]

Meanwhile, at Lilly, chemists and biologists participated in efforts to find new cancer drugs. The North American pharmaceutical industry expanded dramatically after World War II. By the 1940s in the United States, the government supported the efforts of domestic companies to develop drugs in order to relieve a perceived overreliance on German manufacturers. U.S. pharmaceutical companies quickly expanded to sell the bulk of their products overseas. By the late 1960s, the U.S. National Institutes of Health (NIH) provided more than US$50 million each year to researchers invested in finding new drugs. Although the early emphasis was on finding techniques to synthesize chemicals for medical use, companies explored the possibilities of finding new medications from plant sources. In particular, the National Cancer Institute provided a total of US$250,000 million between 1955 to 1965 to universities and private companies, including Lilly, that were working to find cures for different forms of human cancer.[49]

In retrospective accounts, researchers who worked on the "vinca alkaloids" highlighted their innovative techniques and the priority of their findings. Within Lilly, the biologist Johnson competed with the chemist Svoboda for credit for the research on periwinkle alkaloids. At the time, it was the convention at Lilly for chemists, rather than biologists or, in Johnson's case, embryologists, to hold patents. Svoboda and Johnson collaborated on the research, bringing their different fields of expertise to run animal tests and conduct chemical analysis. Johnson explained in an oral history interview for the University of California at Berkeley how the company participated in a national cancer screening exercise, funded in part by NIH during the 1950s:

> We set up a building, and we screened five thousand things [for activity against cancer tumors] a year. . . . There was a small group of chemists in the company who were interested in what they called phytochemistry—drugs derived from plants. There's a basis for that—reserpine, and digitalis. . . .
>
> I detected activity, and Gordon [Svoboda] went through an exhaustive isolation program. This plant [periwinkle] has almost ninety very large indole-dihydroindole alkaloids. And the isolation was a terrible chore, but I detected activity, and Gordon pursued it, and we found four alkaloids were active in the P1534 [a mouse-strain-specific leukemia]. . . . Vincristine was one; the second one was velban. . . . One called leurosidine and one called leurocine. They all were there in small quantities. It took a ton of dry leaves to get an ounce of active drug (vincristine). That's an isolation problem that was pretty insurmountable.[50]

Lilly first filed U.S. patents for vinblastine in August 1958, just months ahead of the Canadian team.[51] Tests of periwinkle extracts on rats with cancer tumors led to the isolation of alkaloids that included vincristine and vinblastine, suggesting new treatments for childhood leukemia and Hodgkin's disease. The labs at Lilly used 15 tons of periwinkle to manufacture 1 ounce of vinblastine (vincaleukoblastine), in contrast to the 50 pounds of periwinkle available to Beer to isolate vinblastine. Beer used chromatography to identify vinblastine, and although he suspected another alkaloid from the chromatographic fraction, the Collip laboratories where he worked could not isolate the vincristine, which was later identified at Lilly as a dimeric indole alkaloid.[52]

Lilly assisted Noble and Beer with the patent description for vinblastine, and the Canadian group licensed it to the U.S. company after they were awarded the patent in 1963. Then, in 1965, Gordon Svoboda received a patent on behalf of Lilly for his "invention" for "a method of preparing the alkaloids, leurosine and vincaleukoblastine, in pure form from apocynaceous plant sources." He also received a complementary patent that year for methods of creating leurocristine, which he co-shared with two other Lilly researchers.[53] Lilly marketed the drugs Oncovin (vincristine) and Velban (vinblastine) for leukemia and Hodgkin's disease in humans. These U.S. patents were valid for seventeen years. By the late 1980s and early 1990s, when the original patents had expired, Lilly also released them for veterinary use and sought new methods for extraction of the alkaloids, which they patented. There were some reports of side effects, but the products continued to turn a large profit for the company.[54]

SCIENTISTS RESEARCH PENNYWORT

Even as I stressed my interest in periwinkle to plant sellers and researchers in Madagascar, our conversations invariably returned to pennywort, the country's other well-known plant medicine. Malagasy researchers proudly retold the role of their late mentor Ratsimamanga in the development of a wound treatment, Madécassol, for the French firm La Roche. Pennywort led to money from a licensing agreement linked to patents filed in France by Malagasy researchers—but periwinkle did not. As retold through conversations with Malagasy scientists, the moral is not that Eli Lilly did not give money to Madagascar for periwinkle, but that Madagascar (or Jamaica or the Philippines) only failed to beat the company to the drug. Thus, Malagasy expressed their quest for priority not only with respect to indigenous healing knowledge of periwinkle, but also in the struggle for drug patents surrounding pennywort. Yet, as with periwinkle, pennywort revealed the multiple locations of herbal knowledge that shaped drug discoveries from Madagascar to India.

For Malagasy scientists engaged in plant prospecting, Madécassol, pennywort, and Ratsimamanga were synonymous. Ratsimamanga was widely seen as the father of medicinal plant research in Madagascar and continues to be revered there among scientists and the general public. He founded the Institut Malgache de Recherches Appliquées (IMRA or Malagasy Institute for Applied Research), using royalties from his drug patents, and his portrait was mounted outside a three-story building containing a pharmacy, a herbarium, and laboratories. In the photograph, Ratsimamanga tilted his head as he looked upward to the right, smiling brightly. His dark hair made gentle waves against his head (see Figure 1.7). The former Malagasy ambassador to France, he wore a black suit adorned with no less than fourteen medals, with a broad red sash under his coat. Even though he died on September 16, 2001, his presence hovered over the grounds, and the scientists there speak of continuing his work. Isabelle Ramonta, who teaches ethnobotany at the University of Antananarivo, recalled fondly how Ratsimamanga used to stroll the

Figure 1.7 A Malagasy genius? Portrait of Ratsimamanga outside the laboratories of Malagasy Institute for Applied Research. (Photo by author.)

gardens of IMRA each morning while sipping a cup of pennywort tea. Liva Rakotomalala and Herisoa Rabarinala, curators of a sizable collection of memorabilia at IMRA documenting Ratsimamanga's career and the rise of Madécassol, pointed proudly to the vibrant green pennywort leaves stenciled along the dusty brick walls of the museum.

Taking Pennywort Recipes for Leprosy

Given this strong association between Ratsimamanga and Madécassol, I was surprised when further investigation revealed that early colonial research on pennywort was spearheaded by Pierre Boiteau, the French botanist fluent in Malagasy who had made Madagascar his adopted country. Boiteau had expanded and run Tsimbazaza Park, the botanical garden in Antananarivo, from the 1930s until he was sent into exile in France in 1947. During my visits at Tsimbazaza, images of the well-respected agricultural engineer graced the park, complete with spectacles and a conservative beard that had historically been the hallmark of foreigners or *vazaha*. His distant eyes were glassy above a curt mustache, like the busts of other white scientists sprouting from the ground at Tsimbazaza and reminiscent of an old song, "The Conquest of the Land by the Vazaha": "Their mustaches were red like peppers / Their eyes were gray like those of cats."[55] Visitors to the garden might view from afar the large home Boiteau had shared with his family, and the library, herbarium, and laboratories that still attracted Malagasy and foreign researchers. Behind the public attraction of caged lemurs, stout palm trees, and fragile orchids that the colonial government had originally established in 1925, Boiteau developed the park into a scientific laboratory. The sacred hills of the Merina kingdom cradled the rolling gardens of the park. In earlier times, King Radama I (1793–1828) had expanded a small stream to create the Tsimbazaza Lake, where he bathed and surveyed his troops. Later, royals slaughtered zebu, the Malagasy cow, for ceremonies and imperial burials at Tsimbazaza.[56] As he walked the perimeter of the lake on this historic terrain, the botanist Boiteau envisioned a program to draw valuable chemicals from Malagasy plants.

In 1937, Boiteau began a clinical trial using pennywort to treat leprosy. Boiteau studied the physical differences in the variation of pennywort that he called "*la forme asiatico-malgache*," or the Asian-Malagasy form. The plant was known in Madagascar by several names, including "Talapetraka, Viliantsahona, Viliantsahonantanety, Loviantsahona, Loviantsahonantanety, Raivolesoka."[57] As early as 1913, the Malagasy Academy, a body of colonial scientists founded in 1902 by the governor, had published a number of local recipes that used medicinal plants. They included several "remedies against leprosy," or *odi-bokal*, but none of the five plants mentioned was pennywort.[58] It is possible that Boiteau may have become enthusiastic about the plant on the basis of his conversations with healers in the 1930s and reports of the plant's use in other places, including India, East and Southern Africa, and the Americas.

Boiteau first collaborated with Charles Grimes, a physician who ran a leprosy treatment camp in the colony. They initially experimented with dried leaves supplied by Boiteau that were macerated and steeped in alcohol. They also tried fluid extracts made from fresh leaves. They found that "the treatment varied with the method of preparation, the time of year, the dosage employed; the therapeutic dose was very close to the toxic dose, which could lead to serious and fatal accidents."[59] The main struggle for the team was finding a way to extract a new glucoside, *l'asiaticoside* (or asiaticoside), from the plant without using large quantities of alcohol. They were also set back when Grimes was removed from his position as head of the colonial leprosy service in 1941. Boiteau convinced members of a scientific society affiliated with the Tsimbazaza Park to help sponsor further chemical experiments with the newly established Elementary Laboratory of Botanical and Vegetable Chemistry. Quite by accident in 1941, Boiteau dipped into an old stash of the plant product and found that he could inject a solution of asiaticoside without any side effects. By 1944, Grimes and Boiteau hoped to find ways to cultivate pennywort on a massive scale through support of the Scientific Society of the Tsimbazaza Botanical and Zoological Park.[60]

Lilly was one of the institutions that expressed an interest in early research on pennywort in Madagascar. J. H. Sandground, a parasitologist at Lilly, wrote to Boiteau in 1945, having read about the pennywort research in Madagascar in the medical journal the *Lancet*. In the letter, Sandground requested samples of asiaticoside to test in Indianapolis. He assured Boiteau that he would be more than happy to provide these "services" to the research team in the company's world-class facility. It is unclear whether or not the team took up this offer sent "very sincerely" from Lilly.[61]

In other cases, we know that Boiteau and Grimes themselves sent similarly vague, friendly letters in bids to get samples of pennywort from other countries. The *Bombay Chronicle* enthusiastically described the research out of Madagascar in March of that year. V. K. Mehendale, a leprosy specialist who ran a clinic in Sholapur, India, asked if he might receive samples of the herbal preparation and further details on its manufacture. According to his account, three percent of the people in

Sholapur were infected with leprosy. In particular, those who fell within the weaving and farming classes were disproportionately affected, and Mehendale treated more than a hundred patients daily at a "humanitarian" clinic he had set up there. In correspondence that I found in their old research files, now at the National Archives of Madagascar, Boiteau and Grimes refused to provide Mehendale with the preparation, claiming that they lacked sufficient product "for our own local needs." Instead, they requested that the director of medical and sanitary services in Madagascar relay a message that if it was possible for the Indian contact to send 60 pounds of dried pennywort, they would be willing to "study with you the possibility of having our asiaticoside prepared in India from the plant grown on the spot."[62]

Finding Sugars for Wounds

Boiteau later collaborated with Ratsimamanga, the Malagasy physician, complicating the narrative of possible colonial exploitation. Asiatic acid and asiaticoside defied wide medical use, as they did not dissolve readily in water. The chemicals cleared up skin and eye problems but were difficult to manufacture from the plant. Asiatic acid occurred naturally in pennywort bark and converted into asiaticoside in a mixture of different kinds of sugars. By 1949, asiaticoside's chemical structure had revealed itself in the laboratory experiments of Boiteau and Ratsimamanga. Together, the two scientists helped determine that the chemical compound contained forty-eight carbon atoms, seventy hydrogen atoms, and nineteen oxygen atoms.

Ratsimamanga began his undergraduate medical training in Madagascar during the colonial period. His family witnessed profound changes with the advent of French rule in 1896, a transition evident in the shift from their status as proud princes in the nineteenth century to Ratsimamanga's export as one of the exotic oddities for a World Fair in Paris in 1931. His paternal grandfather, also Ratsimamanga, was a Merina royal of the dynasty that ruled much of the island in the nineteenth century and was the uncle of Queen Ranavalona III (r. 1883–1897). The Merina leadership traced its ancestry to Indonesians who had settled in the mountains surrounding the central Antananarivo plateau between 700 and 1100. They had allied themselves with England in a bid to sustain their authority and tradition in the nineteenth century, and the scientist's grandfather wore European clothes and worshipped at Anglican services at the same time that he practiced polygamy and kept his wealth in cattle. When France went to war with the Malagasy to secure Madagascar as its own colony, Prince Ratsimamanga and Queen Ranavalona's general Rainandriamampandry resisted, and the French general Gallieni staged their execution on October 15, 1896. The younger Ratsimamanga was born just over a decade later, on December 28, 1907, to Lala, the second wife of his father Rakotomanga, and lived with the blot of this family tragedy throughout his life. Some called him "a son of the light" after the Merina belief that royals were sun kings with a spiritual connection to the heavens and gods.[63] He attended missionary schools and went on to study medicine and dentistry at the École de Médicine de Befelatanana in Antananarivo in the late 1920s.[64]

After completing his course of study in 1929, Ratsimamanga took on the unglamorous role of dentist in the rural community of Moramanga Nosy Be. He recalled that "my therapy was, in fact, simple. When there were not enough medications at the post, I treated my patients with Malagasy medicinal plants."[65] He also focused on promoting basic ideas of hygiene and sanitation to put an end to dysentery in the community. In 1931 he was selected to attend the Exposition Coloniale Internationale in Paris, an event showcasing French colonial expansion. The individuals chosen to represent Madagascar underwent thorough medical exams. The authorities in France were leery of introducing any novel diseases through participants from Madagascar. Once in Paris, they would be looked after further by a Malagasy physician and nurse. Ratsimamanga was approached because of his medical expertise, and he traveled to Europe for the first time with the group of singers, drummers, and dancers selected to perform at the fair. In a recommendation letter written on behalf of Ratsimamanga, the colonial physician Thiroux commented on his "excellent spirit" and the success he had with patients.[66] Thiroux indicated that his former student would do well to continue his medical studies the following year in France. At the exposition, the French used photographs of medical and dental classrooms, clinics, and patrols against plague and malaria to indicate how favored Malagasy subjects were being remade into doctors, dentists, and nurses, while others displayed dances and "traditional" aspects of Malagasy culture for curious Parisians.[67]

After the exposition, Ratsimamanga stayed in Paris, where he was able to hone his chemical research on plant medicine, including the uses of pennywort. Ratsimamanga took courses in tropical hygiene and bacteriology, earning both medical and scientific doctorates by 1939 from the University of Paris.[68] He was particularly interested in questions of nutrition and appropriate diet. His more than 250 published works included studies on the role of vitamin C.[69] While in France, Ratsimamanga was contracted by the drug firm La Roche in March 1943 to conduct research on the nutritive properties of different chemicals and their effects on laboratory animals for an initial fee of 2,000 francs. As legend would have it, Ratsimamanga first met Marguerite Laroche when they were working together at Necker Hospital. Her husband, Marcel, was wounded by a German soldier and was expected to die, when a hormonal drug from Ratsimamanga saved his life. Ratsimamanga also connected with colonial scientists who had ties to Madagascar, including Boiteau.[70] In 1958, Ratsimamanga established IMRA on the outskirts of Antananarivo to investigate the biological and chemical constituents of plants. He continued in his capacity as a research director at the Faculty of Medicine in Paris until 1975.[71]

In the 1960s, several U.S. and French patents protected Boiteau and Ratsimamanga's rights to specific techniques for making new pharmaceutical preparations from pennywort. First, the patents provided exclusive claim to a method for processing asiatic acid into salts that could be dissolved in water. In several elaborate recipes, the patent described how combining the plant acid with four parts of succinic anhydride, a chemical compound made from fermented sugar, resulted in more biologically useful chemicals called "hemisuccinates." After further reactions, the hemisuccinates created types of chemical salts that made ideal aqueous solutions (water mixtures) for medical use. For instance, in 1968 their first U.S. patent revealed instructions for converting asiatic acid into hemisuccinates:

> One mole of asiatic acid and 1.5 mole of succinic anhydride are dissolved in dioxane or pyridine with or without catalysts (Cl_3Zn, So_4H_2, HCl). The solution is refluxed for at least two hours and then concentrated under a vacuum until dry. The dry residue is washed with water as such, or after having been extracted with alcohol, and precipitated with water.[72]

U.S. Patent 3,366,669 included this "experiment" as the first of four procedures that amounted to very fancy recipes for skilled chemists. Boiteau and Ratsimamanga's previous patents protected the process for preparing chemicals derived from asiaticoside, a compound made from the plant acid. The additional patented procedures prevented people from merely applying one technique to a slightly different pennywort extract and calling it their own.

Ratsimamanga and Boiteau licensed their joint patents to La Roche to process pennywort into Madécassol, a pharmaceutical to treat wounds.

Ratsimamanga continued to use patents and licensing agreements as one strategy for funding his investigations, including developing a preparation from the plant *Eugenia jambolana* for which he, along with his IMRA collaborators, received a patent in 1999 from the U.S. Patent and Trademark Office.[73] As Ratsimamanga secured wealth for himself and his institute, he assumed the role of a well-connected scientific patron. His royal connections provided him with further cultural capital that healers with less education or social standing could only imagine. Over the course of his career, he developed formulations for more than fifty plant-based drugs, using herbal recipes for inspiration, and received five patents. At IMRA, I reviewed laboratory notebooks he maintained on further experiments on pennywort into the 1970s; rows of figures indicated the reactions of different laboratory animals to consumption of the plant.[74]

If Malagasy scientists sought parity in their drug discovery research, plant sellers and lay healers remained suspicious and indeed did not always see the benefits to be derived from fancy patents. Further, they fashioned their own methods for controlling secret plant therapies. At the herbal market at Analakely in Madagascar's capital city, plant medicine sellers made up a collective organization of about 100 individuals who shared plant knowledge and conferred on aspects of the plant trade at monthly meetings. The group included junior sellers (sometimes the children of herbal merchants) who had been in the business for less than fifteen years, as well as more senior sellers seasoned with forty or fifty years of experience. Those with a lifetime of marketing have sold other goods over the years, including chickens, eggs, honey, and fruits, as "each product has its period." The sellers debated and discussed the merits of various plants in everyday conversation at their stalls at Analakely and at collective meetings. But this banter was cautious, because each seller maintained several private cures as trade secrets for herself or her family. One individual, who requested anonymity, explained:

> There are many diseases that I alone can treat. All the sellers of plants
> may have their own cures. They don't show it but keep it for themselves
> in order to keep customers buying from them. For example, the
> medicine of sinusitis, I alone have it because someone brought a plant
> from Atsirabe to Tana and he demonstrated the cure at our home. No
> one has this medicine except us, so we keep it. For example, we sell it
> two hundred ariary per spoon; the medium sells for five hundred ariary
> per spoon.

Sellers found themselves in direct competition with large companies in Madagascar, including Homeopharma and IMRA, that package processed plants and botanical medicines in bottles and tubes with fancy labels and high price tags to sell in shops around the country. The seller continued, "In the last two or three years, there are many [Malagasy] companies with an interest in medicinal plants." Others explained to me that they were certain that sometimes these private organizations sent decoys to buy plants and steal information from them. For example, they recalled how IMRA had found only three cures from a particular plant, instead of the five treatments they knew. The institute was thought to have "spied and done research indirectly" to discover the two additional treatments known to the sellers. "This year, someone came to say that he does research for his thesis at Ankatso [the University of Antananarivo, Madagascar]. But if you think, he is a member of the Ratsimamanga family [founders of IMRA]."

HARVESTING PERIWINKLE AND PENNYWORT

Once scientists claimed exclusive rights to chemicals derived from periwinkle and pennywort, they sought sufficient sources of the raw plants to manufacture the new drugs. The story of plant harvesting in Madagascar is part of a larger story of peasant agriculture in African countries. People in African societies have long exploited their knowledge of plants to harvest sufficient quantities for international demand. For instance, women living along the Niger River and Atlantic Coast in West Africa harvested vast quantities of rice required to feed captives on board slave ships. Before rubber plantations were set up in Southeast Asia, men, women and children in West and Central Africa endured harrowing conditions to harvest the wild rubber used to create the first waves of bicycle and car tires at the beginning of the twentieth century. As Soto Laveaga has shown for the case of the Mexican yam barbasco, with the rise of interest in plant-based drugs after World War II, pharmaceutical companies depended on peasants to harvest plants. In the process, peasants and the local scientists with whom they communicated developed substantial techniques for harvesting suitable roots and leaves.[75] Letters from the U.K. National Archives and Malagasy exporters, alongside interviews with plant collectors and agronomists in Madagascar, point to the history of botanical harvesting on the island.

Finding Sources of Periwinkle

The transformation of periwinkle into Lily's cancer treatments Oncovin and Velban in the late 1950s and early 1960s occurred simultaneously with independence movements in Africa. This curtailed access to the periwinkle plants necessary for the creation of the lucrative drugs. Initially, the Canadian team sought plants from Jamaica. Johnston shipped dried periwinkle leaves to Noble and Beer. According to Noble, "He would send boy scouts out into the jungle to gather the leavers, which we received by mail in little packages." When drought in the West Indies made it difficult to get an adequate supply, most likely from sandy shorelines, rather than the jungles of Nobel's report, the Canadian researchers hired the Swain Brothers company to grow it in greenhouses on Lake Erie in Ontario.[76]

Subsequently, Noble sought raw materials through his network of fellow scientists working in Europe. He was keen to find a tropical source for the plant, as the tropical form appeared to be more bioactive and have higher alkaloidal content than the plants cultivated in temperate Canada.[77] As a resident of the Commonwealth, he turned to British colonial contacts. In 1951 the Department of Chemistry at the University of Manchester in the United Kingdom had requested a "sample of *Vinca rosea*" from the government chemist in Accra, the capital of the former Gold Coast colony. However, A. J. Birch, a chemist at Manchester, apparently could not find the sample when Noble requested it be sent to his Collip Medical Research Laboratory in Canada in 1958, or so he claimed:

> I am sorry to say I do not know what happened to this material. I arranged to receive it while in Cambridge but it had not arrived by the time I left there for Australia. . . . I have heard nothing further about it from Cambridge, so it may well be reposing in the cellars there.[78]

This was right around the time that Ghana gained independence from Britain in 1957, and no doubt confusion surrounded how best to secure plants from the new country. For many years, a colonial plant collection company had been commissioned by various research bodies and drug firms to obtain medicinal plants. But with the tide of independence in Africa affecting their collecting networks, scientists in Canada and England were looking for new collectors with whom they might partner.

Plant distributors were keenly aware of the competition between scientists and companies. For companies, they managed initial processing of plants with "larger scale grinding and extraction facilities." P. C. Spensley offered the services of his plant collection firm, which had long secured "from within the Commonwealth samples of the tropical plant species required by research workers." Spensley's company was privy to the research interests of competing labs:

> We have also, by virtue of both workers coming to us for the samples, managed on one or two occasions to avoid the galling situation of two people working simultaneously on the same line. With regard to *Vinca rosea* we have been connected with two separate research investigations on this species, one using material from Ghana and the other from Trinidad.[79]

Plant distributors were another node of contact, offering advice on new plants to study and letting researchers know of competing interests in a particular plant.[80]

According to different accounts, scientists at Lilly first procured samples of periwinkle from Madagascar (see Figure 1.8). Later, Lilly secured substantial periwinkle harvests from India to manufacture Velban, the first periwinkle derivative containing vinblastine released in 1961. Lilly

Figure 1.8 Drying periwinkle at Ranopiso. (Photo by author.)

reported to its shareholders that year that "supply of the raw material was posing tricky problems. It takes 5,000 pounds of crushed periwinkle leaves to produce 1 pound of the finished product." Noble noted that his team never would have been able to secure the 15 tons required to make an ounce of vinblastine at Lilly. By March of 1961, Lilly received approval of Velban from the U.S. Food and Drug Administration. Over six days of concentrated activity in Indianapolis, Lilly workers toiled around the clock to produce the first 50,000 vials of Velban so that physicians could begin to use it for cancer patients within a week of the drug's approval.[81]

The manufacture of vincristine, the second drug marketed as Oncovin, required even more periwinkle to extract. Malagasy growers took advantage of rising interest in periwinkle by the 1970s. To learn more, I traveled to find the plantations where Lilly and other drug companies had sourced periwinkle. As early as 1970, a Frenchman named Durant had hired people to cultivate periwinkle for him on a vast tract of land at Ranopiso (now over 1,000 hectares). Edmond Monja, a man who had been there from the early days, recalled that "Before, I worked here, SEAR [Société d'Exploration Agricole, Ranopiso/Society of Agricultural Exploration of Ranopiso] was in the village to the south. In 1970, before SEAR, there was Mr. Durant who had a farm here. He started SEAR to cultivate pervenche [periwinkle]. He was a European, French." Workers used their experience growing food crops in their hometowns to cultivate periwinkle for Durant. In the earliest period, about 100 men and women were contracted to plant periwinkle seeds. After that, because the plant is a perennial in Madagascar, the workers did not need to replant year after year, but only from time to time. The plantation had provided Monja's livelihood for forty years. Two brothers had joined him over the years. The sale of the Ranopiso planation to Boehringer Mann-Heim led to the creation of SEAR in 1973.[82] SEAR was not the only attempt to cultivate and export periwinkle in Madagascar. Others involved the exporter Pronotex, based in Antananarivo, as well as a 60-hectare plot in Bejofa and a project of the Department of Water and Forestry at Ambalavao in the mid-1970s.[83]

Exporters of periwinkle in southern Madagascar relied on farmers to cultivate and collect plant parts. This led to difficulties with standardization. In October 1975, major exporters of periwinkle convened a meeting at the Libanona Motel in Tolaganaro to discuss further expansion of plant cultivation. The region was ideally suited for harvesting periwinkle, with an active port at Tolagnaro providing easy access through the Indian Ocean to European and American trade routes. The exporters decided to supply people in surrounding communities with seeds of

Catharanthus roseus "to profit from the cultural experience of the peasants." The goal of the meeting was to form a new collective to coordinate periwinkle exports and to safeguard quality and quantity. Together, the exporters agreed to pay agents to collect periwinkle harvests for payments of 100 Malagasy francs, plus 20 Malagasy francs for Vatoeka (a tax to control officers) per kilogram. Of particular concern was the quality of periwinkle roots that had been collected: "For some time now, due to their incompetence and desire to gain market share, the traditional collectors gather roots of poor quality (too young, green . . . sometimes false)." Those assembled learned from representatives of laboratories about "the Quality Norms according to International Regulations" that required dry roots, at least six months old, with a weight of 4 to 5 grams. It would be necessary, participants were told, "to educate the peasants so they replant these young roots during thinning." All told, twenty-four individuals from two of the export companies attended this afternoon meeting to discuss the future of Madagascar's periwinkle industry. Those in attendance included a mix of Europeans, Asians, and Malagasy, judging from the names.[84]

Exporters came to rely on Malagasy scientists as well as peasants. The scientists took up a supporting, rather than a leading, role in the creation of Oncovin and Velban. Alongside evidence of continuous use of periwinkle and harvests for export in Madagascar, I also sought information on the contributions of Malagasy scientists to investigations that might have led to the Lilly breakthrough. Indeed, a number of Malagasy scientists with whom I consulted suggested that Phillipe Rasoanaivo, a leading chemist who had written his PhD thesis on new alkaloids from periwinkle in the 1970s, had helped shape work at Lilly. Given that Svoboda secured his patents in 1965, it seemed that these comments were merely a sign of reference to a leader in the Malagasy scientific community.[85] Rasoanaivo studied biochemistry in France at the Faculty of Science of Orsay, part of the University of Paris. By 1972 he participated in efforts to identify a new alkaloid from the *Catharanthus* genus, using methods of extraction from the work of Svoboda. In his 1974 doctoral thesis, he addressed the process of extraction of alkaloids from the plant *C. longifolius*. This flowering shrub was related to *C. roseus* and was one of the seven species of the genus original to Madagascar. Rasoanaivo continued his quest for new drugs in his capacity as professor at the University of Antananarivo and laboratory director at IMRA, which was founded by Ratsimamanga, the early proponent of pennywort. When I spoke to scientists at IMRA, they explained that Rasoanaivo had been eager to develop drugs from periwinkle but had not been able to fund all his research, necessitating the sale of his findings. During the 1980s and 1990s, Rasoanaivo filed several patents for drugs derived from additional plants and published policy suggestions on sustaining biodiversity in Madagascar.[86]

International pharmaceutical firms relied on the expertise of Malagasy agronomists to help improve periwinkle exports over the years. Rasoanaivo provided expertise to periwinkle exporters, as did Jean Joseph Andriamanalintsoa, one of the next generation of scientists, who took up the position of technical direction at SEAR, the Ranopiso plantation where first Eli Lilly, and then other international drug firms, sourced both periwinkle and pennywort. Andriamanalintsoa was picked for this post after he conducted arguably the most comprehensive study to date of the cultivation and harvesting of *C. roseus*. His thesis was littered with the chemical structures first elucidated at Lilly. Laboratories in Madagascar would have been hard pressed to fund the research for periwinkle during the mid-1960s while the Malagasy were in the process of ending French colonial rule. However, it was Andriamanalintsoa's careful monitoring of local techniques for drying racks, root and leaf harvesting, and climatic monitoring that allowed for continued export of periwinkle to companies in Europe and North America that adapted the expired patents of Lilly. As he showed me the stacks of plants ready to be sent to international researchers for their own laboratory studies, he sighed and explained that scientists in Madagascar still needed to improve their own laboratory work to allow for the better extraction of plant chemicals prior to export. A hulking machine, once used to separate essential oils from plants, was now painted white and set in the stunning gardens of Ranopiso, an eerie reminder of one of the many previous attempts to jump-start Madagascar's pharmaceutical industry.

Export companies sourced periwinkle from sustainable harvests at the same time the government restricted environmental devastation in the area. My arrival in the southern town of Fort Dauphin, now Tolagnaro, where the first documented exports of periwinkle to France began in the 1600s and where Lilly identified the best form of *C. roseus*, also coincided with the beginning of a massive destruction of forested coastland around the town. The international mining conglomerate Rio Tinto had just that week begun the excavation of illimente, a valuable mineral from beach sand used in paint. To do so, they had to obliterate entire mountains of forest. But, according to Rio Tinto employees, it was all well and good because a two-decade biological survey and impact study meant that a nursery of plants grown directly from seeds by Malagasy and collaborating researchers from England, the United States, and other countries could be used to restore the area. As the mining compactors whirred in the distance, I stood in the warehouse looking at the stunning collection of seeds, some florescent blue, others speckled with muted hues. Malagasy scientists and local farmers all wrung their hands, and everyone knew it was fruitless to try to replicate nature, but at least they had interesting research, a new school, and more jobs in the community. Not surprisingly, the small museum Rio Tinto built next to the nursery had a photograph of periwinkle with a caption describing the importance of the plant to the pharmaceutical industry. It was a sign that here, where the story of periwinkle's exportation began, the plant still held historical currency and political relevance.[87]

Finding Pennywort in Madagascar and India

SEAR, the company that exported plants from southern Madagascar, competed with the local market, especially for the second main medicinal plant export, pennywort. Alongside their private farms, SEAR relied on middlemen contracted to obtain plants for them from families in neighboring towns. During my visit, an older gentleman explained that he had played this role for SEAR for several decades. It was difficult work, as he had to take a bicycle from village to village to coordinate harvests of cultivated plants and monitor stockpiles of specimens collected from the wild. I asked if it was difficult to monitor whether people gathered the correct plants and was told that, although they had training programs, there were sometimes problems. "I go to so many villages," he explained to me. "Lately, it is difficult because my bike has broken down. It is difficult work for an old man like me. Do you really want me to name all of them? Well, then, to collect pennywort (Centella or *Mahavita*) I go to Ankaramena, Bemavoriky, and Fenoatsimo. To collect periwinkle (Pervenche or Tsahavitay), I go to [more than forty towns]."

Pennywort is generally not cultivated in Madagascar, with exports dependent on collection from the wild (see Figure 1.9). Along the road to Ranopiso, groups of young children stood with woven baskets brimming with delicate pennywort leaves. A young girl explained to me, "I collected this basket myself in the morning with my brother. It took five hours, and this basket is all of the leaves I gathered myself." Andriamanalintsoa, a technical director at SEAR, examined the basket and estimated that the leaves weighed 5 kilograms. The girl's tiny younger brother held up his basket with a mere kilogram of the leaves. Mireille Rasoanaivo, a director of research at SEAR, explained, "It is necessary to be courageous and fast. So you have a girl with a large collection, and a

Figure 1.9 Basket of freshly harvested pennywort leaves. (Photo by author.)

small boy who only has a small crop." She pointed out that the mothers of the thirty children who were gathered by the road sat nearby in the shade of the trees. Their children, wearing faded secondhand T-shirts, their bare feet and legs scratched and marked from the morning's harvest, were a contrast to a line of children in smart green uniforms walking to school.

To further complicate the story of colonial taking of folk recipes, mediated through the later collaboration of a Malagasy scientist at the expense of everyday plant sellers, Indian researchers presented competing claims to the origins and uses of pennywort.[88] Vonjy Ramarosandratana, a Malagasy biochemist who had trained in Natal in South Africa, home to a large base of South Asian researchers, explained that some scientists in India claimed that pennywort originated there and that Madagascar should not have gotten rights to a drug produced from it. Pennywort figured in Ayurvedic healing traditions and was the subject of chemical analysis in South Asia. Ramarosandratana explained that one solution people were considering was to conduct tests at the level of population genetics to determine if the species of pennywort in the northeastern part of Madagascar was more closely related to Indian species, as many people suggested. A French scientist, Pascal Danthu, coordinator of Forests and

Biodiversity at the Center for International Cooperation in Agricultural Research for Development (CIRAD), confirmed when I visited his office that genetic tests were under way on pennywort. Ratsimamanga had participated in earlier efforts to test periwinkle varieties from India and Madagascar.[89]

Ramarosandratana suggested to me that perhaps it might also be a good idea to do genetic tests on periwinkle at the level of populations in different parts of the world. This was after my conversation with the IMRA herbarium keeper, Benja Rakotonirina, which had left me unclear as to how exactly periwinkle had made its way around the world: was it the French explorers as depicted in the written record, or was it Malagasy fishermen in the Indian Ocean basin? Rakotonirina shook his head and said, "That is our problem in Africa. We do not write things down, so we must rely on the European records." And Rakotonirina wondered if we could get funding to do a population study of periwinkle. Who would provide the money? After all, Malagasy scientists had seen many researchers come and go and governments rise and fall. They remained cynical about their chances for claiming ownership of plants and knowledge.[90]

CONCLUSION: MEDICINE'S ROOTS CROSS CONTINENTS

Both periwinkle and pennywort emerged as leading medicinal plant exports for Madagascar by the late twentieth century. Pennywort relied on important investigations conducted by the French scientists Grimes and Boiteau and was later transformed into a patented method for curing wounds through the additional research of the Malagasy physician Ratsimamanga. Around the same time, periwinkle, a widely used weed, found an international market for pharmaceutical preparations after Eli Lilly laboratories identified antitumor properties in the plant. In both cases, popular knowledge of the plants—not only in Madagascar but in India, various African countries, the Caribbean, and the Philippines—led to the collection of herbal recipes by scientists.

Eli Lilly's discovery of alkaloids in Madagascar's periwinkle has become the classic example of biopiracy, but the story presents a weak case for theorizing benefit sharing, because the plant is so widely distributed and information about its uses has also coevolved with the plant's migration. Rather, the histories of drug discovery from periwinkle and pennywort show the indebtedness of scientists to one another, to early documentation of plants from colonial exploration, and to traditional plant exporters. Both plants have long featured in folk recipes to cure various ailments, and these recipes inspired physicians to conduct clinical studies and chemists to take the plants into the laboratory to test bioactivity. Periwinkle is a pantropical weed, and pennywort grows in both Africa and Asia, patterns of distribution that have led to many, many potential claimants to the plants, their chemicals, and related knowledge. Further, the shifting boundaries between scientists engaged in colonial and postcolonial research has amplified the concerns of healers that African and foreign investigators wish to usurp them of rights to medicinal plants. Their reliance on folk recipes shows the parameters of invention among physicians, biologists, and chemists in Canada, the United States, France, Madagascar, and India engaged in work on periwinkle and pennywort. Benefit sharing from the profits of patents thus becomes a symbol of the drug industry's debt to global communities rather than of retrospective justice for specific groups.

Many people—healers, scientists, elderly women, corporations, communities—help shape products of medical science, complicating claims to priority in science and traditional medicine. Because some of these groups of individuals historically did not document their research on healing plants with written records, credit for innovation has typically gone to those who could claim to be inventors of chemical extraction using complex, expensive equipment. Pharmaceuticals coexisted with viable herbal preparations. Conversations with Malagasy plant sellers and collectors indicated sustained knowledge of folk uses for the two leading herbal exports, pennywort and periwinkle, and their importance to regional and international trade networks. Importantly, in the face of difficult odds, Malagasy plant sellers used private recipes and information about plants in bids to maximize profits. Similarly, scientists at IMRA found that license agreements and patents were a useful way to maintain funding for their investigations. They also supplemented their incomes with the sale of herbal products through IMRA's commercial wing. This led to mistrust between scientists and plant sellers, who had less access to resources and education. Similar tensions surfaced, as I explore in subsequent chapters, in Cameroon, Ghana, and South Africa.

Figure 2.1 Ripe grains of paradise in Pebase, Ghana. (Photo by author.)

CHAPTER 2

Take Grains of Paradise for Love

The phenomenon of cultural exchange or borrowing makes clear that traditions have widespread roots.
—Kwame Gyekye, *Tradition and Modernity*

The oval pods swell up and turn bright red, like the glands of an animal in heat. They encase rows of pungent peppercorns wrapped in a soft, downy fiber. The peppers hide themselves close to the ground, at the base of lacy, green fronds that stretch past my waist. At first, the pods remain elusive, until we swish through the branches with our hands and sticks to find them. The wandering plants seem almost like weeds, although the farmer explains his late mother cultivated them close to the house under these trees to make a little money on the side. They intermingle now with old cans someone threw casually into the bushy embrace of their stalks. In their brilliant red cases, the grains of paradise seem for a moment to multiply into the discarded red tomato tins that surround them. The peppers remind us of the ripe promise of this hidden fruit, abandoned once more in a small settlement between two rainforest preserves far from the main road that might lead them to global markets.[1]

Grains of paradise have circulated through the history of West Africa and the world, celebrated over time as an aphrodisiac, cooking spice, and general cure-all. Historically, the peppers featured in medicinal and culinary therapies of families living along the Atlantic Coast from what is today Liberia, through Ghana, Benin, Nigeria, Cameroon, and Congo. Cultivated for markets across the Sahara Desert and Mediterranean, the spice also laced the dishes and drinks of North Africans and Europeans. Grains of paradise stimulated amorous sentiments in the works of Geoffrey Chaucer and other early European writers. Through the tragedy of the trans-Atlantic slave trade, the potent spice became important to folk remedies in the Americas. Today, readers may have sipped beverages infused with grains of paradise, including U.S. microbrewery Sam Adam's Pale Ale or Bacardi's Bombay Sapphire Gin. Most recently, the peppers were found to promote the virility of rats at university labs in Cameroon. Through such investigations, grains of paradise emerged as the subject of a process patent for a treatment to cure erectile dysfunction filed by Peya Biotech in North America and Europe. Given the wide distribution of the plant and knowledge about its uses, the pepper defied local ownership even as its chemical possibilities were secured in a patent.

Efforts to share profits from any current or future drugs derived from the plant are complicated by the fact that many different groups of people have inherited medicinal therapies for grains of paradise. To better articulate how we interpret plant ownership by ethnicity and location, this chapter provides a chronological and geospatial framework to understand the question "Was traditional medicine local?" Plants used in African traditional healing have long circulated through networks of trade and religion. Because grains of paradise have extended histories in many different places, elements of African healing are global, not local. Nor can plant-based therapies be associated with specific populations. Together, published and oral data for more than 400 medicinal therapies show similarities and differences in the use of grains of paradise over time and provide evidence of the global diffusion of herbal medicine from tropical West Africa.[2]

Grains of paradise become a metaphor through which we can view patent and benefit-sharing claims for plant-based therapies. The argument here is that West African healing was integrated through networks of trade and migration, which led to the regional spread of herbal recipes and infinite claims to ethnopharmacological uses of grains of paradise. From a legal standpoint, published evidence of herbal innovations across West Africa and the Caribbean provides evidence for prior art in the face of weak patents for processes to extract phytochemicals. This means that Peya Biotech's scientists were not the first ones to use ground grains of paradise to treat impotence, as narrated in their patent. Further, the scattered documentation shows how plant medicine incorporates multiple contributors as well as *multiple locations*. In this particular case, the Edmunds Institute in the United States has faulted Peya Biotech for not extending benefits to the Republic of Congo.[3] However, no single ethnic group can feasibly claim priority for the use of grains of paradise to treat impotence.

The geospatial component is a critical piece missing from both histories of African healing and the discussion on patents for plant-derived pharmaceuticals. This is in part due to stereotypes about the limited level of integration of rural communities in global markets. Thus, this chapter uses a broad historical survey of grains of paradise to show the creation of a wide healing plant diaspora that spans continents.[4] Grounded in studies of African mobility and migration, my approach complicates commonly held definitions suggesting that *"indigenous knowledge* (IK) refers to the unique, traditional, local knowledge existing within and developed around the specific conditions of women and men indigenous to a particular geographic area."[5] Rather, I take up geographer Doreen Massey's "view of locality which stresses its linkages with the wider world. These links exist in many ways and at many levels and they are not just products of the modern world."[6]

Grains of paradise grow not only in the Republic of Congo but also throughout West and Central Africa in the slim tropical rainforest zone that straddles the humid equatorial coast and dryer inland savannah. The rainforest region is actually quite small, at its widest only 400 km. The plant became an important commodity for families residing in areas where it could be cultivated and collected. Historically, three major areas of cultivation included the present-day Liberia/ Upper Guinea coast (former the Grain Coast, named after grains of paradise), the Western and Asante region of Ghana (former the Gold Coast), and Southern Nigeria (the Ancient Benin coast). Since at least the 1200s, the close juxtaposition of disparate ecological zones in West Africa meant that those with the pungent peppers might barter and trade them for scarce commodities prevalent in other regions: dried fish and salt from the coast or shea butter, for instance, from the arid savannahs of the Sahel bordering the Sahara desert. Families migrated within tropical rainforest areas, clearing farmland and sharing seeds for grains of paradise alongside techniques for their cultivation and use. Indeed, in West Africa, names for grains of paradise abound in virtually all languages.

A note on usage: I primarily use the term "grains of paradise" to refer to *Aframomum melegueta* (a member of the larger ginger family, *Zingiberaceae*), as the plant has been known since the nineteenth century.

However, medieval and early modern use of the term "grains of paradise" covered a range of spices with similar foliage, pods, and seeds, including other species of *Aframomum* and some cardamoms (*Amomum*). "*Aframomum*" comes from the word "Africa" and the Greek word "Αμωμων" or "Amomon," Amomon for cardamom. Other English terms for *A. melegueta* have included "guinea grains," "alligator peppers," and "malagueta" or "melegueta pepper."[7] In the very diversity of names for this group of plants, we see a hint of their distribution across time and space.

A LONG HISTORY OF EXCHANGE

The history of grains of paradise serves to complicate the stereotypes about closed, indigenous knowledge systems that have long shaped conceptualizations of African healing. When viewed at the level of the village, traditional medicine may suggest a basis for isolated, ethnic knowledge. For colonial government officials and missionaries intent on increasing their authority from the nineteenth century, oral knowledge fell flat in the face of the tomes of European science and medicine seen to have global impact. In classic anthropological texts, African healing resurfaced time and again as a deeply spiritual practice only relevant to the concerns of specific subsets of individuals. After the end of formal colonial rule, nationalist researchers perpetuated ideas of isolation in discussions of West African healing, emphasizing therapies according to tribes and national borders (see Figure 2.2).[8] Those most sympathetic to African thought have sought distinctions between science and traditional knowledge. While residing in Nigeria in the 1960s, Robin Horton, the British philosopher and proponent of African knowledge systems, nonetheless postulated that "in traditional cultures there is no developed awareness of alternatives to the established body of theoretical tenets; whereas in scientifically oriented cultures, such an awareness is highly developed. . . . Traditional cultures are 'closed' and scientifically oriented cultures 'open'."[9]

West African healing, like other facets of cultural experience in the subregion (including agriculture, textiles, cuisine, religion, and music), has long been highly mobile, mutable, and diffuse. Traveling healers transported packages of herbs on their woven cotton smocks (see Figure 2.3). Friends visiting from afar shared medicinal nuts and tonics over conversation. Women exchanged seeds to cultivate in new gardens. Merchants strung pods on strings to help them transport medicines over long distances. Fishermen transported them along the Niger and Congo

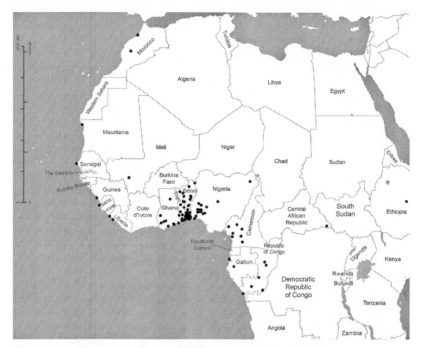

Figure 2.2 Collection points in Africa of medicinal recipes using grains of paradise.

rivers and the Atlantic coast in their wooden boats. Carrying medicinal knowledge with them, enslaved individuals traveled from *Bilad al-Sudan* ("the land of the Blacks") to work in homes and fields of North Africans. We can use grains of paradise, a critical ingredient in therapies throughout West Africa, to better understand how people created a healing plant diaspora over hundreds of years.[10]

Diffusion through West African Forest Settlement

For centuries in West Africa, two main types of pepper were cultivated and traded. One was what the Akan of the gold-rich areas in the tropical rainforest belt of modern-day Ivory Coast and Ghana called *"soro wisa"*— "sky pepper" or "heaven pepper," because it grew above the soil. The other was called *"fam wisa"* or "ground pepper" because it was harvested at the base of the plant. Grains of paradise, the pepper of the ground, are further differentiated in the Akan language into two types. *Adwoa wisa* is the wild form of the pepper (*A. hanburyi*).[11] It is named after the *adwoa* or duiker, a type of forest antelope that gnaws on its pods. (The relationship

Figure 2.3 "African man carrying pharmacy on his head." (Source: Gold Coast [now Ghana] Basel Mission Archives/Basel Mission Holdings ref. no. D-30.63.099 c. 1925–1955.)

between duiker and grains of paradise in the wild most likely informed early use and cultivation of the plant.) The second type of *fam wisa* is *wisa-pa* or "proper pepper" (*A. melegueta*). And it is this form of the pepper that is much larger in size and was cultivated on farms and within tertiary forest growth. In Ghana, to the north of the Akan, speakers of Hausa call it "*chitta*"; to the east, Ewe speakers named it "*megbedɔgbɔe*" or "cure for all sickness." In the Ga language prevalent at the Atlantic Coast, the pepper was known as "*anaivie*," or pepper of the West, possibly suggesting that people brought seeds from growing areas to the West of modern day Ghana.[12]

In the absence of early written documentation, it is unclear when people first began planting the slow-growing seeds from dried pods of grains of paradise among other food crops. In modern times, grains of paradise have been cultivated in plots in the shade of plantains and yams, in the same forested areas where agricultural settlement dates to 8000 B.C.E. Linguistic evidence points to long-standing use of *A. melegueta* by speakers of both Niger-Congo and Bantu languages, centering on regions of early yam cultivation in modern-day Nigeria, Cameroon, and the Congos, where climatic conditions would allow it to thrive. Grains of paradise grow well in humid rainforest farms in poorly drained soil and are intercropped today not only with plantains, yams, and indigenous oil palm trees but also with cassava ("manioc"), which came to West Africa from South America only after 1500 C.E.).[13]

Although wild species of grains of paradise do exist, people would have needed to carefully monitor and weed *A. melegueta* plants to nurture the larger pods with pungent seeds important to commerce.[14] Thus, rather than wander to collect surplus pods from wild sources, settled populations had to actually cultivate the plant in the rainforest belt by the time it became an important export commodity from the 1200s to 1900s, given the long period necessary to establish the pepper harvests and the requirement for constant weeding. Nevertheless, there is limited documentation on actual cultivation of grains of paradise. In my interviews with plant sellers and famers in Ghana, individuals expressed that grains of paradise plants could be started from either rhizomes, which take two weeks to develop plants, or from seeds, which take four to six months to establish seedlings. One possibility is that some farmers separated rhizomes to expand the growing area of plants, as can be done with ginger today. Similarly, people who descended from first settlers of the rainforest areas would have been familiar with rhizome propagation techniques from their experiences with yams and plantains.

A second possibility is that seeding peppers expanded through trade and slavery networks between the forest and Sahel. People familiar with seed agriculture, including rice and cotton cultivation, along the Niger River might have used this insight to grow grains of paradise. They would have needed to establish residency within tropical settlements as grains of paradise seedlings do not produce fruit for four years. In recent times farmers have weeded these established plants to maintain harvests for another three years, abandoning the plants and other crops for a seven-year fallow period before replanting. After 1600, European explorers documented the seeding of grains of paradise, comparing the practices to those used with New World crops like maize that took hold along the Atlantic coast.[15]

Diffusion from the Forest through Saharan Trade

Names for the plant in African languages indicate that use of grains of paradise in medical care extended throughout West, North, and Central Africa. The part of the plant a person incorporated into medicinal treatments depended on whether he or she lived in close proximity to where the plant was cultivated: since grains of paradise grow only in tropical conditions, traders transported the dried, lightweight fruits and seeds to more arid regions in the Sahel and the Sahara Desert and at the Atlantic and Mediterranean coasts. Thus, consumers in Morocco used seeds, whereas those in Western Nigeria might employ the leaves and ripe fruits as well as dried seeds.[16]

By 1200, signs of the wide diffusion of grains of paradise well beyond their growing zone in tropical West Africa can be found in written documentation in Arabic and early European languages. Written evidence furthermore suggests the circulation of information on the medicinal and culinary uses of the spice to Europe, including its applications as an aphrodisiac. Because grains of paradise grow best in tropical conditions, the expansive export of the plant from Egypt to England necessitated complex networks of trade between West Africans and those living to the North. Although much attention has been given to the exchange of gold and slaves with Arabia and later Europe, the circulation of medicinal plants and spices is a further avenue for exploring the connections between communities. As Paul Lovejoy's early work on the history of kola (a masticatory stimulant favored by African Muslims), and that of his student Edmund Abaka more recently, have shown, traders redistributed valued drugs from farms in the tropical forest toward the north, finding consumers among people herding animals and farmers in arid climates.[17] In contrast to kola, grains of paradise made their way even further, across the Sahara Desert and the Mediterranean Sea to Europe and the Middle East.

A main entrepôt for grains of paradise commerce was in the area of modern-day Mali. For thousands of years along the banks of the Niger River, traders have converged in the city of Jenne to sell their wares. Since the seventh century, merchants spread out dried fish, salt, cotton fabric, and forest products, including grains of paradise. Before the advent of European maritime trade on the Atlantic coast, sellers congregated along the Niger River to supply caravans moving across the Sahara Desert. For instance, traders would stop in at thriving markets like those at the oasis of Taghaza, a major salt-producing town over which leaders in Morocco and Songhay long fought for control. The written evidence situates the earliest trade in forest commodities, including perhaps grains of paradise, in the Susu kingdom, a forest empire that arose after the fall of Ancient Ghana along the Western Niger River. Tributary relationships between leadership in the Ancient Ghana, Susu, Malian, and Songhay kingdoms suggests one avenue for the redistribution of grains of paradise from the forest zones in exchange for cloth, salt, and other commodities from the Sahel and Sahara. Forest products that passed through these markets eventually turned up at Mediterranean ports.[18]

For European consumers, the cultivation of imported peppers remained mysterious for centuries, with their consumption reserved for the wealthy. The earliest references to grains of paradise can be found in old Italian descriptions of a game staged between a dozen wealthy women hiding in a mock fortress, tossing a variety of spices and perfumes at "invading" knights in Treviso in 1214. Among the flowers and perfumes thrown into the air were thought to have been grains of paradise.[19] The English name, grains of paradise, or *graine de paradis* in French, indicated associations of the pepper with the celestial realm. During the medieval period (c. 1200–1450), grains of paradise referred to a variety of spices that looked similar to European consumers unfamiliar with their habitats and cultivation in Africa and Asia. The name covered black cardamom from India (*Amomum*) as well as species of *Aframomum*, including *A. corrorima* from Ethiopia. By 1245, a list of spices sold in Lyons, France, included the rare African seeds. In Norwich, England, agents for the king collected rents in the form of herring pies made with a half ounce of grains of paradise.[20]

Early European references to grains of paradise often situated them in contexts of love, suggesting longstanding interest in the pepper as an aphrodisiac. However, there was no familiarity with the tropical environments where they grew. In 1230, the French poet Guillaume de Lorris described in *The Romance of the Rose* a heavenly garden where "fresh Grains of Paradise" or "*Graine de Paradis nouvelle*" sprouted among a variety of spices. The other plants and animals—including oak and maple trees, rabbits and squirrels, violets and periwinkle flowers—in the fantastical garden populated with singing angels and the God of Love were mainly found in temperate climates.[21] In the 1380s, the English writer Geoffrey Chaucer also described how love-struck Absolon chewed "grein and licorice" en route to his paramour in *The Canterbury Tales*. The intention seemed to be to sweeten the breath and perhaps more; "under his tonge a trewe love he bere." One argument is that Chaucer meant grains of paradise when he referenced "grein" in fourteenth-century England.[22] These could have been cardamom from Asia, or perhaps *fam wisa* from West Africa.

In addition to associations with love, medieval discussions of grains of paradise noted their medicinal and culinary uses, suggesting that African healing knowledge traveled with the plant to Europe. We know that during his reign in the early 1200s, the Byzantine Emperor John III took grains of paradise when he fell ill. The pepper was integral to "sick soups" made for those suffering from colds and fevers in France. A compendium of medieval French recipes included dozens of recipes incorporating grains of paradise to season meat and broth.[23]

In Arabic texts, references to grains of paradise usually appeared alongside discussions of other peppers traded along the Sahara Desert. In the fourteenth century, the celebrated scholar Ibn Battuta, originally from Tangiers in Morocco, made four journeys through the Muslim world. In his first trip, he journeyed to Mecca at the age of twenty-one. His further travels included a trip along the Swahili coast to what is today Somalia, excursions to South Asia, and—important for our purposes— visits to the kingdom of Mali along the Niger River from 1331 to 1335. Ibn Battuta noted that throughout the desert and particularly in Mali, aromatic spices were valued commodities. Raymond Mauny, the former economic botanist and archeologist at the French African Institute in Dakar, Senegal, argues that *al-itriyat*, the pepper observed in Mali along the Niger River by Ibn Battuta, had made its way to Cairo by the twelfth century. There it was known as *"filfil ghaina,"* or "pepper of Guinée."[24]

Perhaps the best indication of the circulation of commodities between West and North Africa and the Middle East can be seen in the travel writings of Al-Hasan al-Wazzan, sometimes known as Leo Africanus or Leon l'Africain. A wealthy merchant from Morocco, Leon documented his travels between Africa and southern Europe, where he was baptized in Italy. Like Battuta, he traveled to what was then the kingdom of Songhay in the area of present-day Mali on the banks of the Niger. He noted the value placed on peppers, and further that they might be exchanged for goods, or used as currency all the way through the eastern Sudan toward Cairo.[25]

There is even a remote possibility that grains of paradise cultivated in West Africa made their way to China during the thirteenth and fourteenth centuries. In the 1330s, Hu Sihui wrote a manual on health and diet that included references to *sharen,* which English translations have rendered as "grains of paradise."[26] Other herbal remedies in the book indicated a close familiarity with Middle Eastern therapies gleaned through Mongol incursions into Iran, Turkey, and other parts of the Muslim world. Although some have supposed that the peppers referenced by Hu might have been linked to African sources, and indeed grains of paradise covered both the pepper and related spices with similar foliage in the fourteenth century, it is more likely he recommended types of cardamom known in Chinese as *sharen* (*Amomum villosum* and *A. xanthioides*), which still grow today in China. Alternatively, related species of *Aframomum* from East Africa, such as the less spicy Ethiopian *A. corrorima*, traveled through Swahili trade on the Indian Ocean Basin. The first Chinese map of Africa made by cartographer Chu Ssu-Pen circulated by 1320, with significant commerce between China and East Africa evident by the time of the Ming dynasty in the 1400s.[27] Thus, Hu's *sharen* might have been West African *Aframomum melegueta*, Ethiopia's *A. corrorima*, or Asian *Amomum villosum or A. Xanthioides*.

Diffusion from the Forest through Atlantic Trade

When the Portuguese first traveled down the coast of West Africa in the 1440s with a cargo of grains of paradise, they were redirecting a long-standing history of Saharan commerce toward the coast. They intended to circumvent Muslim middlemen in the Middle East and North Africa so they themselves could get to the source of spices, gold, and eventually slaves. Simultaneously, communities in the tropical rainforest capitalized on this redirection of trade. Thus, once a backwater to the more vibrant trade at the ancient markets of Jenne in Mali, the Atlantic coast emerged as a hub of merchant activity with the advent of ships arriving from Europe. It is estimated that by the mid-sixteenth century 40 to 50 percent of West African trade now moved toward the south. Particularly, by 1479 farmers along the slim tropical forest belt from Liberia to Congo participated in concerted efforts to flood European markets with grains of paradise.[28]

Around 1540, in a letter to Count Raimond della Torre, a noble in Verona, the Portuguese sailor Pedro de Cintra is thought to have first coined the term "Melegueta Coast" to describe the region where the peppers were sourced. By this period, alongside the circulation of peppers from West Africa through the Middle East, we also had peppers from South Asia. Here, pepper was called "*maricha*" in Sanskrit and "*milagu*" in Tamil. From the first century, Tamil literature described how in the markets *mathuraikkanchi* or *milagu* was sold in great "sacks . . . the brokers move to and fro with steel yards and measures in their hands weighing and measuring the pepper and grains purchased by the people." *Milagu* from this time is thought to have been sold as far as Greece, and it was this spice that set the taste for pepper. Thus, when Portuguese and Italian sailors found a slightly different but similar spice, now one linked to the ginger family, they termed it *"milagu"* or *"meleguette,"* and the meleguetta coast was born.[29]

At markets near where their ships docked, European traders conversed with African merchants, gaining information on the cultivation and uses of grains of paradise (see Figure 2.4). The Dutch traveler Pieter de Marees published the first account of life on the Gold Coast, now modern-day Ghana, in 1602. He confirmed that by this point people near the Atlantic Coast were cultivating grains of paradise:

> Grain of Manigette . . . is mostly found in Africa, in an area which is named after it. It grows in Fields, like Rice, but does not become as tall. It is also sown like corn; its leaves are thin and narrow, and where the Grain grows like hazelnuts, it is as big as the cobs of Maize. It is reddish in color. Once the shells have been removed, one finds the Grain inside, covered with husks, in separate compartments, like [a] pomegranate.[30]

An engraving included with his publication portrays a collection of plants in a field, almost as if depicting a plot of intercropped vegetables, which included an accurate representation of a grains of paradise bush that de Marees may have seen himself or heard about (see Figure 2.5). De Marees also explained how women who had just given birth used grains of paradise to restore their energy. In contrast to European women who lay in bed after giving birth, women on the Gold Coast did not "use Midwives to lift and put them into the child-bed and to make them feel comfortable. They just walk away, make a mixture or spoonful of Malaguetta or Grain, which they drink."[31] While de Marees undoubtedly overestimated the constitution of African women, it does indicate the early modern use of *A. melegueta* to manage labor pains.

Through the tragedy of the trans-Atlantic slave trade, knowledge of grains of paradise extended to the Americas during the Middle Passage. The link between slavery and grains of paradise was in evidence on the slave ships themselves. In 1760, the Danish slave merchant Ludewig Ferdinand Romer explained in his *Reliable Account of the Coast of Guinea*:

> A ship's medicine chest should only contain anti-scorbutics and anti-venerics. Should the slave fall victim to the [endemic] illnesses of the land, such as worms, etc., a couple of the female slaves can be allowed to

Figure 2.4 A possible venue for exchange of grains of paradise and related information in Ghana: "Market of Cabo Corsso," from Pieter de Marees, *Description and Historical Account of the Gold Kingdom of Guinea* (1602). Key as given in de Marees's account: "*A* is the house or residence of the Captain of this place. *B* is the Hut or Barn in which the Captain stores his *Millie*. *C* is the Market for Bananas and fruits, as well as the place where they sell meat. *D* is the Lodge where Peasants come and sit in the Market with their pots of Palm wine. *E* is the Chicken Market. *F* is the Fish Market. *G* is the wood Market. *H* is the Rice Market, where *Millie* is also sold. *I* is the place where fresh water is for sale. *K* is the early Market, where they sell Sugar-cane. *L*: here Holland Linen, which the Peasants have brought ashore from the Ships lying opposite their quarter, is measured out in fathoms and remeasured. *M* is the place where women from the Castle de Mina come and sit in the market with their Kanquies. *N* is the sacrificial table of *Fetisso*, their God. *O* are the Dutchmen who come to the Market to buy something. *P* is the Captain's guardsmen, walking with their weapons. *Q* is the road to the sea-shore. *R* is the road to the Castle de Mina. *S* is the road to Foeta [Fetu] and other Inland Towns." (Note: there is no item *J* in the figure.) (Source: reproduced with permission of the British Academy.)

take over, after we have supplied them with *mallaget* and *piment*, palm oil, and citrons, from which they can prepare medicines, and the sick will feel well afterward.[32]

On the ships that carried both slaves and peppers, knowledge traveled as it was embodied in young women and men who still remembered the uses of the plant.

Seeds of the grains of paradise were made available to enslaved individuals once they landed in the Americas, allowing them to shape a healing plant diaspora on a second continent. It is unlikely that people carried the peppers on their bodies across the Atlantic, but they did know how to cultivate them. Presumably, a few coveted pepper pods in

Figure 2.5 Tropical plants illustrated in Pieter de Marees, *Description and Historical Account of the Gold Kingdom of Guinea* (1602). Key as given in de Marees's account: "*A.* is Sugarcane. *B.* is Maize or Turkish Wheat. *C.* is Rice. *D.* is *Millie*, which they use as their Corn to bake bread with. *E.*: On this shrub grow small red and black peas, which are nicely speckled with colours. *F.* is the size of Parsley. *G.* is Ginger. *H.* is a tree on which big Beans, not less than a palm in circumference, grow. *I.* is Grain or Manigette." (Source: reproduced with permission of the British Academy.)

circulation could have provided the necessary seeds to jump-start cultivation. As the Romer reference indicates, slave ships did carry the peppers on board. Geographer Judith Carney has argued that grains of paradise appeared in private plots of those of African descent in the Americas, particularly in Northeast Brazil, Martinique, and perhaps the Carolinas in the United States (where enslaved individuals cultivated rice using techniques they brought from West Africa; see Figure 2.6).[33]

In Brazil from the fifteenth century, the Portuguese forced enslaved individuals into mining and agriculture, with captives being primarily male and purchased mainly from ports in what is today Nigeria, Democratic Republic of Congo, and Angola.[34] In Brazil, the name for grains of paradise (*A. melegueta*) is "*atarê,*" derived from the Yoruba term used in Nigeria. Strong Yoruba connections with Northeast Brazil included the use of grains of paradise in animist religious ceremonies. During Candomblé events, priests and priestesses sprinkled powdered grains of paradise, *pimenta da costa,* in their intercessions with spirits. Grains of paradise were also used in Brazil to treat impotence.[35] Bahia in Brazil and the Bight of Benin in Nigeria are only 2,000 miles apart. Afro-Brazilians

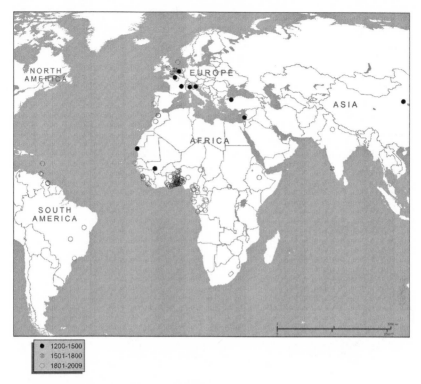

●	1200-1500
◉	1501-1800
○	1801-2009

Figure 2.6 Collection points of 423 medicinal recipes using grains of paradise. Key: Closed circle = recipe collected 1200–1500; light circle = recipe collected 1501–1800; open circle = recipe collected 1801–2009. (Source: Author's research.)

came to manage trade between these two points, including slaving, and they established permanent settlements along the West African coast. These activities could have fostered continuous knowledge of grains of paradise, commerce in African harvests of the plant, or the prevalence of the Yoruba term. (In Brazil and Portugal, the term "malagueta pepper" refers to a South American red chili pepper that grows above the ground, *Capsicum frutescens* var. *malagueta*).

In Guyana, Dutch settlers brought enslaved individuals from Africa to work on sugar plantations, as the Portuguese had in neighboring Brazil. A by-product of this activity was the cultivation of grains of paradise. The uses were medicinal as well as corporal. Slave masters in Guyana were known to cane enslaved individuals and then chew grains of paradise, which they rubbed in their victim's eye. In the nineteenth century, some of the earliest fresh samples of the pepper to make their way to England did so from Guyana, preserved in sealed bottles of alcohol. Apparently, cultivation was associated with those of African descent. One pharmacist from Guyana explained to pharmacists in England that "the plant . . . is cultivated in Demerara by many of the old African negroes, and the dried seeds sold to the Druggists in Georgetown, as grains of paradise, with which they appear to be identical. The seeds were probably carried there originally by the Africans." He further explained that an Afro-Guyanese cultivated "perhaps, a dozen plants or so in the garden attached to his hut. . . . The plants formed very pretty garden flowers."[36]

By the 1800s, the pungent peppers found markets in the United Kingdom, where brewers used them to spike beers and porters. Under King George III, grains of paradise emerged as an illegal additive in alcoholic beverage. The use of this West African spice preoccupied commissions of inquiry in the British Parliament, with brewers quizzed on their use of the imported spice through the 1820s. From Dublin to London, investigators seized quantities of drugs and other "deleterious" substances used to enhance beers, including opium, grains of paradise, and sulphate of iron. On November 24, 1813, town officials convicted one brewer of "having in his custody 44½ pounds of grains of paradise," as well as quantities of ground ginger, caraway seeds, and "orange-powder" (turmeric). They fined others for using banned ingredients thought to cover the taste of defective ales. From 1815 to 1818, brewers including Webb and Ball, Swain and Sewell, and John Mitchell paid fines of up to £200 or fled the country for using ginger, guinea-pepper, opium, honey, or Coculus India in their breweries.[37]

Despite efforts to ban the illicit spice, interest in grains of paradise in Britain did not wane in the early nineteenth century. Indications of the rise in popularity of grains of paradise in Britain can be seen from textual references. Between 1600 and 1815, the term "grains of paradise" only rarely appeared in published English-language books, but from 1815 to 1915—the period of increased colonization along the African coast—it appeared at least two to four times as frequently as "cardamon" and "tumeric" and had virtually supplanted "malegueta."[38] Botanists working in Ghana have noted a decrease in recorded exports of grains of paradise from the Gold Coast colony from the 1880s. They have argued that this decrease could be attributed to the rise in cultivation of the more lucrative crop, cocoa, which was introduced to Ghana from the late nineteenth century. The shade created by cocoa trees would have killed grains of paradise intercropped around them.[39]

Finding new markets, exports of grains of paradise have continued to this day from ports in West Africa and the Caribbean. The description for Sam Adams Summer Pale Ale makes reference to these pungent, and possibly potent peppers: "Complex yet refreshing, this tangy American wheat ale features a slight citrus note. Brewed with malted wheat, lemon zest and Grains of Paradise, a once-forgotten but flavorful African brewing spice. The perfect drink for a warm summer day."[40] Advertisements for the beer claimed that grains of paradise "according to medieval legend were thought (erroneously) to have aphrodisiac properties." Thus, the late-twentieth century marked a fourth rise in interest in grains of paradise in Europe and North America, following earlier peaks in the 1200s, 1500s, and 1800s.

These brief indications of the dispersal of grains of paradise reveal the extent to which West African medicine spread to other parts of the world through Saharan trade, Atlantic commerce, and slavery. Oral accounts and ethnographic evidence further suggest the diffusion of herbal recipes on the African continent. And, if claims in the PeyaBiotech patent discussed in the coming paragraphs are to be trusted, grains of paradise mixed with alcohol improve chances in lovemaking, providing an explanatory framework for the historical interest in beers made with the spice-drug.

TAKING HERBAL RECIPES TO MAKE DRUGS

Grains of paradise recur in treatments across continents, especially in remedies to cure impotence. Who discovered that grains of paradise reverse impotence? A deep analysis of recipes associated with grains of paradise indicates the spread of not only the pepper but also knowledge of its uses. This includes herbal inventions related to grains of paradise in areas where it grows wild and in cultivated plots, as well as herbal inventions in areas where the pepper has been acquired in markets. Grains of paradise have been used not only as cures for impotence, but also as treatments for wounds, joint pains, menstrual pain, ailments of the nose and throat, and intestinal parasites.

Within Africa, knowledge of the use of grains of paradise as an aphrodisiac followed trade routes, circulating through longstanding exchange and migration between Morocco, Mali, Ghana, Togo, Benin, Nigeria, Cameroon, Republic of Congo, and the Democratic Republic of Congo. Other important centers of use include Sierra Leone and Liberia, but modern documentation is scarce as a consequence of recent political instability from protracted wars. Is it possible to gauge where the idea that grains of paradise stimulate sexual desire began? Was it simultaneously discovered in multiple places, or did recipes travel with each market transaction and each new seedling?

To answer these questions, besides studying recipes gleaned from family, friends, informants, and early modern and colonial herbal pharmacopeia, I turned to the tremendous publications of the Scientific and Technical Research commission of the Organization of African Unity, now African Union (AU). The commission has conducted surveys of medicinal plant usage in various African countries since 1968. A first outcrop of this work was volume one of the *African Pharmacopoeia*, published in 1985, with several pharmacopoeias produced for particular countries in the 1990s and early 2000s. Edited by the Beninese botanist Edouard Adjanohoun, these surveys depended heavily on colonial botanical reports published from the late 1920s to early 1950s. However, the AU committees also introduced critical data through medicinal recipes gathered from recent field research and gave attribution to specific healers, botanists, and other plant experts, including their residential addresses. A similar set of pharmacopoeias, also edited by Adjanohoun, have been published for Francophone countries through the auspices of the Agence de Coopération Culturelle et Technique (ACCT, now Organisation International de la Francophonie, or OIF).[41]

In my analysis, I differentiated the recipes by plant part used, region, informant, place of publication, time of collection, longitude and latitude, scientific and vernacular names, medicinal claim, and actual recipe. I developed a list of fifty-one symptom codes that I used to determine which recipes could be associated with treatments for male and female reproductive disorders. The resulting maps show the wide distribution of this relatively open healing plant diaspora (Figures 2.2, 2.6, and 2.10). As we shall see, at least thirty of these therapeutic recipes specifically claim to treat impotence or sterility.

Herbal Inventions for Impotence and Infertility

The successful production of offspring is a major preoccupation in West African societies. Not surprisingly, a variety of treatments to improve sex life, with provocative names like "man power powder" and "herbal Viagra," populate the arsenals of traditional healers and herbal medicine sellers. These products are not protected through patents but rather are made from carefully guarded secret recipes. Some might be registered with national offices for managing foods and drugs, as tonics to treat general weakness, and may even have been tested at national laboratories for toxicity. They may be prepackaged in plastic, cardboard, or glass, or they may be available for purchase in powder or enema form.

Herbal medicine from different places can incorporate a universal ingredient as well as plants available at a particular location. Given the diversity of ingredients, it is a complicated thing to understand what makes a remedy "work." It is also unclear when recipes came into existence. We can control for a common additive—for instance *A. melegueta,* which herbal specialists believe indicates for erectile dysfunction. Consider, for example, the existence of twenty recipes for treating impotence collected between 1977 and 1997 from North and West Africa, all of which include grains of paradise. They point to a range of indigenous treatments that used grains of paradise, complemented with a wide variety of local botanicals.

In Salé, Morocco, the plant specialist Abdallah L'Asiri shared a purported aphrodisiac with a collaborative research team from Morocco and Japan in 1982. His remedy included seeds and fruit of seven plants: celery, onion, rape, carrot, galingale, myrobalan, and grains of paradise. He recommended that his patients mix the powder with honey and take a spoonful each night to improve their love life. Alternatively, L'Asiri recommended just the seeds of grains of paradise to treat impotence. Around the same time, Rahhal Ben El-Hajj Mohamed in Marrakech, Morocco, described his personal recipe for *ras l-hanut,* a spice mixture used frequently in North African cooking, which included slightly different ingredients, including four types of pepper, one of which was grains of paradise, as well as cardamom. He explained that it was useful in stimulating sexual desire in both men and women.[42]

In 1986, in Avévé, Togo, the traditional healer Amana Ossou Bossou revealed his treatment for encouraging sexual desire in men to fellow Togolese working with a research team of the ACCT. He also used seeds of grains of paradise but incorporated the root of *Byrsocarpus coccineus* in a mixture to be taken once a day for three days. In Ezimé, Togo, the healer Olilé Doh presented his recipe for curing male impotence: the seeds of one pod of grains of paradise, and this time, the root of *Lonchocarpus sericeus.* As opposed to a powder mixed with honey, as in Morocco, the Togolese healers recommended the aphrodisiac be served as a beverage.[43]

In 1988, in the city of Brazzaville, Republic of Congo, the plant specialist Pierre Matsimouna incorporated three or four seeds of grains of paradise in a solution of red wine, which he recommended to those who did not feel up to sexual activity. Albert Ngalouako, also from the Republic of Congo, suggested that seeds of grains of paradise be mixed together with the entire root of *Desmodium* and sipped in an infusion of red wine before a sexual encounter. In 1989, the healer Dan Benoit of Abomey, Benin, would burn kola nuts and mix them with chicken and grains of paradise to combat impotence. In 1993, in southern Nigeria, the botanist Maurice M. Iwu recommended the fresh fruit or the root of grains of paradise as an aphrodisiac and sexual stimulant. Other recipes to treat sexual incapacity collected from West Africa involved chewing seeds of grains of paradise along with several other plants, including bark of *Annona senegalensis* and *Pausinystalia johimbe*, and dried stems of *Penianthus longifolius*.[44]

This selection of recipes shows that grains of paradise featured in a variety of treatments designed for men and women who sought to improve their love life, which might have been compromised through erectile dysfunction, lack of desire, or lack of sperm (see Figure 2.10). Recipes from Morocco incorporated honey or celery, whereas those from former French colonies like the Republic of Congo infused ingredients in imported wine from red grapes. The recipes included grains of paradise as a common element, alongside other ingredients available to the healer in his location. They hint at everyday practices of real people with specialized plant knowledge. Unfortunately, we do not have written documentation of recipes as they evolved for these towns since the 1200s. Thus, such "traditional" recipes suggest how grains of paradise could be imbued with a particular medicinal meaning for many people across a wide geographic terrain at the end of the twentieth century. Although it is not possible to identify how all these individuals came to value grains of paradise for their benefits in lovemaking, the ethnomedical evidence indicates the wide diffusion of this belief over time and space. Presumably, individuals found themselves inheriting therapies from family members, which they adjusted and reevaluated. Commercial ties between ethnic communities may have fostered exchange, as well as informal conversations among friends.

Arguably, grains of paradise are integral to oral culture in West Africa, traveling in the pockets of guests and galvanizing conversation itself. Among the Igbo communities made famous in Nigerian writer Chinua Achebe's novel *Things Fall Apart*, grains of paradise were fundamental to religious and social life. They were given to women in the community at the birth of a child; men and women who came visiting were served kola nuts on the same tray with grains of paradise; and at healing shrines, it was chewed together with kola, and devotees spat the chewed cellulose onto the ground around statues of the gods and ancestors. When Okonkwo, the protagonist of Achebe's novel, partakes of both kola and alligator peppers with friends by the early twentieth century, he is implicated in larger networks of trade with a long history.[45]

A similar analysis might be made for other therapies included in the 400 recipes for grains of paradise that I collected from published and oral sources. Individual recipes provide points in a complex web of historical and geographic trends. To express the innumerable uses of the pepper in herbal medicine across the continent, the Nigerian botanist Ebenezer O. Olapade pointed to the Yoruba proverb *"Atare kii die so tire laabo,"* meaning "The fruit of *atare* always have complete seeds within its pod."[46] Next, I examine efforts to make drugs from the spicy seeds, before considering other uses for grains of paradise and the vagaries of popular and specialist knowledge.

Laboratory Rats and Male Volunteers in Paradise

At the University of Yaoundé in Cameroon during the late 1990s, a group of researchers set out to prove whether two popular aphrodisiacs really worked. Pierre Kamtchouing led the investigations at the Animal Physiology laboratory in the Faculty of Sciences. For eight days, he and his team fed a watery solution of either grains of paradise (*A. melegueta*) or West African pepper (*Piper guineense*) to a collection of male rats raised especially for the experiment. Then to see what would happen, the researchers placed the animals into a clear plastic cage with female rats injected with hormones to send them into heat. They made excruciatingly detailed observations on the behavior of the animals over the next sixty minutes. In an article published in 2002 in the *Journal of Behavioral Pharmacology*, the Cameroonian scientists showed that the male rats who digested either peppery solution rather than plain water copulated more frequently and powerfully with the female rats allocated for this most tantalizing study.[47]

The experimenters brought traditional medicine into the laboratory, a common practice for African scientists looking for inspiration. Somewhere in the capital city of Yaoundé, they acquired fresh pods of the two peppers, which were identified at the University herbarium and dried under controlled conditions. Presumably, the laboratory technicians produced dry peppers without mold, nor did the ubiquitous white powdery mildew enhance the effects of the grains of paradise. Through careful scientific language, the Cameroonian researchers further coded popular plant medicine in the guise of the laboratory. For instance, they used more than fifteen "parameters of the copulatory behaviour" of the rats to track statistical changes in "genital grooming." Despite the broad claims of the study, the researchers included only five rats in each of the groups fed distilled water, grains of paradise, or West African pepper.

During the peak sixty minutes after rat feeding, they tracked various activities, including "the mean interval separating the intromissions of a series," or, in plain language, the average amount of time between rat sex acts. Incorporating standard procedures for studying arousal in laboratory animals, the scientists developed twelve graphs to indicate differences between the rats fed water and those fed the two pepper solutions. The authors included "penile erection index" and "female ano-genital sniffing" as indicators of the enhanced virility of the doctored male rats. They devoted three graphs to aspects of ejaculation (with frequency and duration measured in seconds) and three to aspects of "mounting."

By the beginning of the twenty-first century, Cameroonian researchers had emerged as leaders in the field of African plant medicine research. In contrast to scientists in the nearby Democratic Republic of Congo, where protracted civil wars disrupted university life, Cameroonian scientists enjoyed relative stability under the autocratic rule of long-term President Paul Biya. They benefited from a legacy of European-style education and linguistic flexibility in a territory dominated by waves of German, French, and British rule between the 1880s and independence in 1960. Further, they funded their research with international support, given increasing concern about the biological diversity in the country's rainforests.[48] Closer to the University of Yaoundé campus, botanists conducted extensive surveys of medicinal plants available in urban markets. At a cost of about 135 French francs for 42 kg of plants, such investigations were relatively inexpensive, yet they produced a wealth of information on patterns of plant use. In one study conducted in 2002, Jean Betti of the Ministry of Environment and Forests worked with Mr. Koufani at the University herbarium to identify the Latin names for the thirty main herbs traded in Yaoundé, the political capital nestled in the rainforest region. They found that the majority of plant medicine sellers at the Elig-Edzoa, Central Post, Mvog-mbi, Nfoundi, Longkak, and Mokolo markets traced their lineage to the Ewonda ethnic community. The researchers commonly found grains of paradise for sale, at approximately 0.80 French African francs for each gram. The botanists learned that this plant was integral to treatments for male infertility and impotence. Indeed, their surveys revealed that plant sellers treated primarily malaria, back pains, and male sexual dysfunction with their merchandise.[49]

In Cameroon, then, people used grains of paradise in popular therapies for male impotence. During the AU survey of traditional medicine in the country, scientists, healers, and everyday people supplied more than 1,000 herbal recipes for several hundred ailments.[50] The healer Augustine Engole, for instance, submitted a recipe incorporating the spicy peppers to improve the love life of men: "Chew 2 leaves of Dipcadi sp. and 9 grains of *Aframomum melegueta* and swallow. Treat for three days and repeat if necessary until erection becomes normal."[51] Other healers and informants provided additional recipes incorporating the barks, leaves, and roots of twenty-five other plants to treat male sexual impotence. Whether wittingly or not, these plant specialists participated in the transparent project "to give information to African scientists which would lead towards the development of new drugs from our abundant natural resources."[52]

From July 11 to August 18, 1995, the healers participated in a bioprospecting initiative whose final aims were unclear. As Assistant Secretary General Chief E. Inoni reiterated on behalf of the office of the President of Cameroon, "It is also my wish to see African Governments build on this survey by encouraging scientific collaboration between our traditional medical practitioners and conventional medical scientists so that jointly they can develop and promote the production of properly standardized and quality-controlled drugs from our medicinal plants."[53] The healers codified as the originators of what may have been common recipes might even find their names immortalized beyond paper in a proposed "continental electronic consultative data bank."[54] Yet, beyond name recognition, the healers failed to be assigned clear benefits. They received the usual glib statements, such as "There can be no doubt, that the economic benefits to be derived from such collaborative efforts would result in a significant reduction in the billions of dollars spent annually by African governments to import drugs that would otherwise have been produced on the continent."[55]

As in Cameroon, biologists and chemists in the Republic of Congo also documented widespread use of grains of paradise. In 1993, the biochemist A. Diafouka observed the peppers for sale in markets of the capital city Brazzaville, noting that several plants, including grains of paradise, also circulated in herbal markets in the capital cities of Benin and Zaire (now the Democratic Republic of Congo).[56] In his 1997 PhD thesis completed at the Free University of Brussels, he addressed the use of grains of paradise in Republic of Congo for the treatment of impotence.[57]

Although some scientists like Diafouka were content to report the medicinal uses of plants in their countries, others worked to establish patents overseas and financial gain for private companies. Around the time that research on grains of paradise was done in Cameroon, Congolese scientists with ties to North America established a patent for the treatment of impotence from various species of *Aframomum*. Victor Ngoka and Simon Ossawa, based in Poto-Poto, Republic of Congo, collaborated with

Michel Ibea and Neil Hartman in Quebec, Canada, for the Canadian dietary supplement company Peya Biotech. With the assistance of a French law firm, they received U.S. Patent number 5,879,682 in 1999, and the World Intellectual Property Organization published WO/2000/035466 the following year.[58]

It is important that the inventors explicitly indicated they had discovered not an aphrodisiac but a medical treatment to improve "penile rigidity" and prevent premature ejaculation in male mammals. They specified the actions of the pepper's seeds on humans and animals, noting that "the pharmaceutical composition of the present invention is not an aphrodisiac." In their patent, they provided brief instructions on how grains of paradise would be reconfigured into a pharmaceutical capsule. Basically, the inventors suggested that seeds imported from Brazzaville should be cleaned, ground, and transferred to a premade capsule purchased from the Baker Company in Maine. Or alternatively, they claimed that a similar action might be produced with seeds compressed in a tablet or pill form. To manufacture pills, they suggested combining milled seeds or seed extracts with a substance—such as tragacanth mucilage and colloidal silicon dioxide—to thicken the powder or liquid benzenoids. The Peya Biotech researchers furthermore included the possibility of combining processed *Aframomum* seeds with alcoholic beverages or testosterone to improve libido, or with ointments to make a body lotion (see Figure 2.7).

Although the Yaoundé-based research team used rats as an indicator of virility, the impotence-treatment inventors based their new medication on studies in humans. They recruited male volunteers (apparently with willing female partners) to test their concoction. The Peya Biotech patents on "*Aframomum* Seeds for improving penile rigidity" cited one heteronormative study conducted with fifteen men between the ages of twenty-three and forty, nine of whom were coded as Black and six as White. During the research, ten of the tested volunteers allowed their blood pressure to be taken before, during, and after consuming capsules of *A. stipulatum*, a species of the plant genus. Overall, these men experienced a surge in their blood pressure during the six hours while digesting the seeds. Further, male and female "testimony" reported above-average increases in penile rigidity and, in most cases, delay in ejaculation.

The existence of these patents forces the obvious question: did Peya Biotech really invent a new drug? The company claimed that "[a]ntimicrobial activity have also been reported for seed constituents of *Aframomum danielli*, but no species of the genus has ever been reported . . . to reestablish erectile function and/or to improve penile rigidity in men."

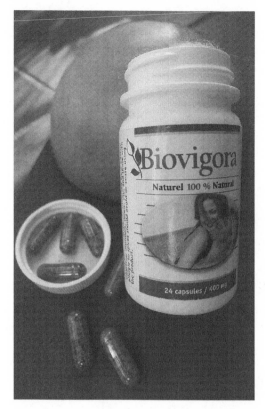

Figure 2.7 Biovigora. (Photo by author, with permission of Option Biotech.)

Lawyers for Peya Biotech concocted a statement that was patently false, as the Cameroonian healer Engol's recipe to use *A. melegueta* "until erection becomes normal" bears witness. The Peya Biotech inventors filed their first patent in the United States by November 24, 1995, and Engol presumably provided his recipe to the AU team by August 18, 1995; at least one earlier ethnobotanical survey reported the common use of the plant for male impotence in Cameroon by 1962.[59] But aside from these specific documents, the account just provided here should make clear that herbal recipes for treating impotence with *A. melegueta* (in association with other plants) proliferated in Morocco, Togo, Benin, Nigeria, and Brazil and were documented before the 1995 patent through publications.

In terms of innovation, a concentrated, ground form of grains of paradise seeds is not a major jump from someone quickly chewing through a whole pepper pod. Perhaps swallowing, rather than chewing, would allow greater consumption of the fiery hot seeds. But overall, this simple drug recapitulates what is a common occurrence in folk-medicine-based inventions: a process patent that is just a fancy recipe dependent on preexisting indigenous techniques.[60] The clear capsules filled with dried peppers conceal little that is new. Amazingly, the patents claimed rights to the use of all known species of *Aframomum* for the specific concern of erectile dysfunction, premature ejaculation, or both. Moreover, the patents claimed embodiments of the drug in any number of vague forms, including capsules, pills, tablets, ointments, and possibly beverages. In fact, the patent's territory was so broad that it nearly canceled out the possibility of popular uses of *Aframomum* seeds in West Africa, the Caribbean, North America, or Europe.

Women's Health and Popular Knowledge

Many of those researching new pharmaceuticals from grains of paradise—whether for impotence or other ailments including HIV/AIDS—have close family ties to West Africa. This raises questions about how valuable information is transferred across lines of class and gender even within ethnic groups. In creating new drugs from grains of paradise, male scientists from the Republic of Congo privileged reproductive health treatments for men in their laboratory investigations. Knowledge on grains of paradise, although widespread, fluctuates within regions and social segments.

When I started studying recipes for grains of paradise in Ghana, I noticed that impotence remedies were missing. The African Union's *Traditional Medicine and Pharmacopeia* for Ghana published in 2000 listed a dozen recipes incorporating the pepper, including cures for boils, chest pains, cough, and fractures—but none for impotence. In fact, grains of paradise are women's business in Ghanaian society, given its status as a spice used mainly to prepare medicines for women and children. Ghanaian forester Albert Enti included it in his list of "rejuvenating plants for women" and explained that it was included "in almost every medicine, in enema preparations, decoctions, chewed raw, and [used] along with some rejuvenating plants. It is an important part of all 'bitters.' The seed is ground into a soft paste and this is applied into the vagina as a bactericide to treat various infections."[61]

It was after a conversation with my mother, a U.S. food sociologist who had noticed a surge of interest in grains of paradise in the United States as a rediscovered gourmet spice, that I realized that grains of paradise were so ubiquitous in my earlier research that they had basically become invisible. In returning to Ghana to do fieldwork, I began to ask people specifically about this plant. Suddenly, I noticed older woman strolling along the lanes of urban centers with trays of dried plants including *fam wisa*. The shriveled, darkened pods were at times encased in the film of a plastic bag. Other times they were loose, for the perusal of the buyer. To allow for closer inspection, the women would split the pods open with the edge of their fingernails and pop out the tiny reddish seeds that lay on pillows of fluff inside. Placing a few on my tongue promised to provide a bright, burning sensation and a hunt for something to drink. Alternatively, open air markets offered removed seeds stuffed into repurposed glass bottles. They were sold alongside jars of black peppercorns, *soro wisa* (*Capsicum annuum*). I bought bottles of each, and packets of the full pods, not exactly sure how I would use them but aware that they were going to be important to my research (see Figure 2.8).

But, what exactly were people using *fam wisa* for? And how widespread was popular knowledge of the pepper? I asked my husband's grandmother over tea. She was by then in her eighties and had outlived her husband, both of my grandparents, and my husband's paternal set of

Figure 2.8 Fam wisa, freshly dried. (Photo by author.)

grandparents as well. By that point, I had interviewed close to a hundred healers in Ghana, many of them in their seventies and beyond. But for some reason I had failed to see the significance of the plant and never included it in my questions. "Da, what is the purpose of this plant?" She seemed coy and looked away from my husband and me, sipping from her cup. "I'm forgetting many things." She fingered the pods, splitting them open to reveal the seeds and offering them to us for a taste. "I'm forgetting many things. Ask Monica when she comes back for lunch." I was surprised that Da didn't know what the peppers were for; she seemed to be hiding her knowledge. Did they remind her of the "fetish" priests, the witch-finders of the village from which she had carefully shielded her children? After all, according to a psychic healer in her hometown, my husband's mother had been slated to be an apprentice at a shrine, not a journalist. We assumed Grandma was reluctant to divulge her under-standing of a pepper closely associated in her mind with occult practices.

When her nurse returned from her classes to prepare the midday meal, I asked her if she knew what the peppers were for. She grabbed a few and pulled me into the kitchen. "Why, are you sick?" She looked concerned and started to giggle nervously. "No, I just want to know how these are used for my research." She looked into the hall to make sure my husband was safely seated beside his grandmother. She offered a possible recipe for their use, explaining afterward that she did not do it anymore, as she was trying to stop using alcohol in keeping with the directions of her pastor. "When you have your menses, grind a small handful of the seeds and steep in a bottle of gin. Then take shots of it, and you will sleep and your pains will disappear. The gin will become very, very red." Later, the family cook offered to buy me some *fam wisa* seeds so that I could use them in my recipes back in the United States. I asked her if she had ever used them in medicines. "No, but they are good for spicing meats. Well, you can add them to a lot of dishes, and grind them with other spices."

I was surprised to learn that the peppers were important in treating "women's sickness," as at this point I had mostly read of their centrality to curing impotence. One of the few studies of popular knowledge of herbal remedies in Ghana, conducted during the 1970s, found wide-spread knowledge of plants women used to stimulate miscarriages when a pregnancy was unwanted. Grains of paradise were among the medici-nal plants included in an extensive list of more than fifty abortifacients. During a focus-group discussion in a suburb of Accra, with women food merchants and their clients, I learned more.

Akosua Agyewaa, a fruit and vegetable seller in Kwabenya, a suburb of Accra, explained that *fam wisa* is cultivated from seeds: "Farmers grow it

Figure 2.9 Plant seller's signboard, Ghana. (Photo by Linda Nadeau.)

for three to four months. And then, they dig holes and put the seedlings inside. Until the plants are mature, if you taste the seeds, they aren't hot, they don't burn. But when it is fully matured and you taste the seeds of a *fam wisa*, they are very, very, very hot. Like a [chili] pepper." She sold dried pods at her popular food stand, explaining that people bought them to heal stomach aches and to make enemas and various medicinal drinks, including the alcoholic beverage *akpeteshi*.[62]

I asked a bystander, Michael Kofi Yeboah, if it was used for male medicines for man power (*obarima aduru*) (see Figure 2.9). "Oh, no. It is only for illnesses of the chest. Men crush it and add it to gin and drink it." I confirmed, "So, it's not man power? In the United States, they use it for Viagra." He got quite interested then and seemed genuinely surprised, "Really? Here, if you go to work, and you feel pains in your chest, you can drink the gin with *fam wisa*." I then mentioned other places where people used it for impotence. Pointing to my chest, I verified again, "So in Ghana, it is mostly for this part, it's not for this part, down here. In Nigeria it is for the down, in Ghana for the up." He nodded in agreement.[63]

A number of Ghanaian botanists have examined the trade in grains of paradise in the country, finding that it is controlled by women farmers and merchants in the Ashanti, Brong Ahafo, and Western regions.[64] I traveled to one area where women cultivated grains of paradise for their own use in the shade of fruit trees, with a small surplus to sell. In Pebase, a small enclave near Asankrangwa in the Western Region near the Ivory Coast border, families had migrated from across Ghana to work in the colonial forestry system as loggers. Now, a cosmopolitan assemblage of Akan, Ga, and Nzima speakers farmed cocoa, oranges, kola, rubber, and occasionally grains of paradise in forest enclaves. On the compound of a village chief just off the highway to Asankrangwa, I engaged his several wives in conversation on the many uses of the peppers. They confirmed that they favored the dried pepper, rather than the fresh red pods, and that they mostly used grains of paradise to treat themselves and their children. Chief Armah explained, "We cultivate it ourselves. If you use the seeds, it will be at least six months before they germinate. So instead, we use the roots, the rhizomes. In that case, you have new plants within two weeks." Several women on the compound interjected that "anybody can plant it, but normally it is the women who plant it. For us, it is an additional source of income, together with ginger."[65]

The women provided me with several recipes, including techniques to grind the seeds of dried grains of paradise to treat boils, vaginal problems, body aches, and runny noses. In contrast, Armah emphasized how he would use the peppers in combination with other dried barks and herbs to treat a variety of ailments, including impotence:

> Without *fam wisa*, none of these other herbal treatments can work. These herbs here are for hernia, but unless you strengthen them with *fam wisa*, you can't drink them. These are for waist pains, and these for low sperm counts or for those whose erection can't go up. To finish the preparation, you strengthen it with *fam wisa*. We call *fam wisa* chief; without it, all these other herbs can't work.

Quarm, a healer and environmentalist, facilitated our conversations and confirmed that sick patients frequently visited the Chief's compound for care. They were attended to by women, with Quarm observing, "Anything we want the patient to take, we tell the women and give it to them. So the women tend to know more about the medicines used to cure. They nurse the men and gather more knowledge." Men and women sustained different recipes for treating diseases with grains of paradise, including therapies for male and female reproductive concerns (see Figure 2.10).

Herbal knowledge in contemporary West Africa is a rich terrain open to extensive investigation. There are differences in personal knowledge based on class, age, occupation, and gender. And there seems to be some regional variation, as indicated in the maps of geographic distribution. Conversations with ten individuals in Ghana pointed to a variety of popular uses for grains of paradise. This account of discussions with family, friends, and informants may seem casual and ad hoc. But it mimics the practices of African scientists who discussed their bioprospecting activities with me. Ivan Addae-Mensah, a leading biochemist in Ghana, received explanations on plant uses from his mother. A biochemist in Madagascar discussed possible plants for experimentation with his

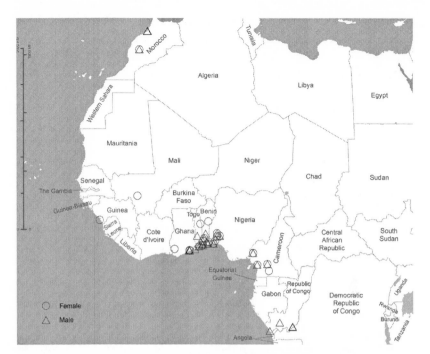

Figure 2.10 Collection points in Africa of medicinal recipes using grains of paradise to improve reproductive health. Key: circle = recipes for women; triangle = recipes for men. (Source: Author's research.)

brother, who had started using an herbal preparation to treat his diabetes. These are informal conversations where individuals go person to person sourcing data from those with different levels of knowledge. There is not always such a strong divide in African bioprospecting between the one embodying the knowledge and the researcher.[66]

Perhaps it was partly my status as a married woman or an investigator with foreign ties that prevented me from hearing personal recipes for treating impotence in Ghanaian men. Maybe it was so common in impotence tonics that plant specialists had not emphasized its importance for the AU survey in Ghana. On more than one occasion an elderly man would take one of my research assistants (either male or female) off to the side to describe a recipe that I was not allowed to hear or record. When I showed the pepper pods to my brother, an architect practicing in Ghana, he laughed. "I've seen those. After a long day's work, the builders on my site in Tema like to go to a small bar near the main road. In the back of the shop, there are always bottles with those peppers steeped in strong alcohol. The liquid is a dark red. I've never tried it, but the guys always wink and say it will make you strong if you take it."

Historically, expertise in African healing plants has depended on reticence. In the Akan language the symbol for wisdom ⚯ implies the proverb *"Mate masie,"* meaning, "I have heard and kept it."[67] Healers, uncles, parents, teachers, and friends strategically concealed information on plants, depending on the situation. A chief with efficacious tonics might display pulverized barks, rather than describe all the ingredients. A grandmother who once knew about grains of paradise, and perhaps still did, would be reluctant to discuss delicate matters like reproductive health in front of her grandson and his wife. A persistent student of herbal medicine would be able to gradually accumulate knowledge on plants, perhaps mounting plants on pieces of paper to help her identify them in the forests, as the herbal seller Juliette Yeboah in Accra once expressed to me. Indeed, even in the face of centuries of documentation on grains of paradise, some have still managed to hide valuable insights, and others are still experimenting to find more.

LIGHTWEIGHT EXCHANGES

Grains of paradise are light and easy to carry. The dried peppers and related information are simple to pass along without papers. When I was traveling to London for a Wellcome Institute conference to discuss my research, the pods slipped easily into my hand luggage, whereas the hefty bottles of cloudy bitters would never pass airport security. On several occasions, the small brown pods filled with seeds passed quietly from me to another person. These lightweight exchanges remind us that it is impossible to know exactly how plants and knowledge circulate over time.

In Johannesburg, the plants provided a lucky indication of my commitment to herbal knowledge. At the Oprah Winfrey Leadership Academy for Girls, the pepper, unknown in South Africa, remained a signifier of the potential of traditional herbs to cure many diseases, with no mention of sexual medicaments for my audience of young ladies. But, in an interview gone bad at the Faraday Market, grains of paradise assumed their attributes as a treatment for impotence. The pods, forgotten in my purse, emerged at just the right moment when my substitute translator for the day turned out to be a former police officer with whom the plant seller we were interviewing had had a previous altercation. My possession of grains of paradise suggested that I was in fact not working for the government in disguise, but actually deeply invested in herbal medicine in Ghana. Once the pods fell from my palms into the hands of my interviewee, with promises that they might actually grow in South Africa and provide a new stimulant for male customers, smiles returned and smooth conversations continued as in past weeks.

In New Haven, a South Asian doctor sat in the back of a lecture I gave on intellectual property rights and plants at Yale Law School. As I slipped out during a lull on the panel to use the restroom, he brushed my arm. Looking up from his chair along the aisle, he whispered, "Can I have a couple of the pods?" Laughing hesitantly, I asked softly, "For your personal use? Sure." He blinked shyly, "Of course." He passed me his card, a researcher at the medical school, while I slipped him some of my extra peppers. Are the seeds under a microscope, or a tongue? Who can know in the game of discreet bioprospecting?

Why did everyone always try to take my "herbal Viagras"? At a workshop at a small college in the United States, the Togolese professor motioned for me to pass him the small box of packaged Ghanaian "Man Power Powder" under the table at a meal, so that his French wife would not see. I obliged, and I do not think she noticed.

At the University of Pennsylvania, a Nigerian friend looked startled when she met me after my talk. She noticed I had a couple of alligator peppers in a plastic bag with my laptop and papers, the perennial handout to keep people interested. "My father used to chew those! I haven't seen them for ages," she laughed. My friend heard me out on how frustrating it was to take questions from an audience who knew so little about Africa. She shook her head, saying "I'm going back home to help the continent. You can stay here to continue to explain Africa to Americans." Yet, we assumed ignorance and misunderstanding of traditional African herbs, even when the pepper at hand had long been a part of the American story.

CONCLUSION: LOCAL ROOTS AND GLOBAL SEEDS

The movement of plant materials has long been undocumented. This is especially true in West Africa, where commerce, medicine, religion, and science have been widely practiced and recorded through oral communication and public witnessing rather than written documentation. Grains of paradise are indigenous to tropical West Africa and have been widely cultivated and traded. It is extremely challenging to identify shifts in how people living in West Africa have used the medicinal peppers over time to establish a history of traditional knowledge. In contrast to the story of *Strophanthus* recounted in Chapter 3, I was unable to find a cache of documents on how treatments for grains of paradise arose in a specific location in Africa at a specific time.

Thus, the scientist in Africa is in a bind when it comes to recent calls to share benefits from research on local plants and local knowledge. There is the potential for many, many conflicting claims to grains of paradise across geographic areas. Widely dispersed healing plants, backed by a range of popular and esoteric knowledge, are difficult to associate with specific communities or specific countries for the purposes of benefit sharing. If anything, my research suggests a gendered component to the knowledge appropriated, as African scientists have produced research experiments with a bias toward "proving" the efficacy of grains of paradise therapies aimed at ailments of men. Historian of medicine Londa Schiebinger reminds us that sometimes women have concealed information on herbal remedies, privileging male authority in botanical compendiums.[68] At the same time, scientists who basically adapt popular recipes into patented pharmaceuticals run the risk of stealing the innovations of many others.

There is a precedent for conflict when a patent for a plant-based chemical process is centered on widely held information, especially if the product proves extremely profitable. In that sense, the story of the widespread grains of paradise and efforts to make new pharmaceuticals from the pepper has much in common with the well-known neem tree controversy. Neem grows throughout East Africa, the Middle East, and South Asia, where it has been used for a variety of medicinal therapies as well as in agriculture for centuries. The Indian activist Vandana Shiva, who has promoted use of the term "biopiracy," participated in efforts to overturn European patents related to Neemix, an insecticide manufactured by the U.S. company W. R. Grace from the neem tree. The neem controversy led to uprisings of Indian farmers. It became a political case for the Indian government not because of the limited knowledge of the plant in a single community but rather because of its cultural currency across the nation. However, the plant is not commonly used just in South Asia; its popularity extends beyond the subcontinent to East Africa and the Middle East. Had the government pursued compensation from W. R. Grace, as the San did in the case of hoodia, it would have been quite challenging to determine who would benefit.

It is a huge problem to identify exactly who was first to use grains of paradise, as with the case of rosy periwinkle and calls for retroactive benefit sharing from Eli Lilly. The original owners of healing knowledge are often a moving target, making the rhetoric of documenting indigenous botanical resources to establish benefit sharing a complex proposition after centuries of relatively open access. Indeed, I have been approached by healers and environmental activists in African countries who seek advice on how to show who "stole" recipes from their ancestors through archival evidence. But, it is unclear how exactly knowledge of herbal medicines spread through tropical West Africa, the Sahel, Sahara, and Mediterranean to Europe and the Americas. Where possible, I have identified the slim number of references for African contexts to approximate the domain of healing knowledge for grains of paradise. Recipes must have spread from person to person, often in family settings but also at trading stands, ports, and on ships. Nonetheless, the signs of this transfer of knowledge remain with us globally today. People move, seeds spread, and although roots establish themselves in particular places, knowledge of how to use them can diffuse overtime to an infinite number of sites.

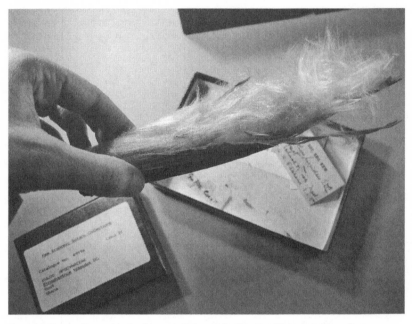

Figure 3.1 Strophanthus pod in the Economic Botany Collection, Royal Botanic Gardens, Kew. (Photo by author.)

CHAPTER 3

Take Arrow Poisons for the Heart

In nearly every narrative of exploration in uncivilised tropical regions accounts are given of poisonous substances, which in many instances are stated to possess remarkable properties.

—Thomas Fraser, *"Strophanthus hispidus"*

The fluffy seeds spring out of the tidy black boxes as I carefully lift the lids. Each brown kernel clings to a feathery parachute, leading me to wonder if there is any danger of inhaling a noxious residue as they launch themselves into the air (see Figure 3.1). Skittish seeds from East Africa led someone before me to scribble in red ink "TAKE CARE—FLYAWAY SEEDS." Separated from their downy tails, other seeds nestle together in large glass bottles. They come with fluorescent orange stickers of a skull and crossbones with the words "VERY TOXIC." Yet, in their various stages of entombment and decay, none of the seeds seem particularly harmful as I hold and examine them without gloves or a mask. Most have been segregated by region: *Strophanthus kombe* from Central Africa in one jar; *Strophanthus hispidus* from Ghana in another. The seeds harbor historical clues on their journey from Africa on tiny slips of yellow paper tucked into their boxes and bottles. From an expedition in Niger in April 1894, reddish *S. hispidus* seeds traveled to London. In January 1901, five pockets of *Strophanthus* seed and nine bales of pods lay in wait at the Drug Warehouse at the London Docks. A pod full of *S. hispidus* seeds from the

Gold Coast made its way to the Liverpool School of Tropical Medicine by 1915. On October 20, 1947, a cluster of gingerlike roots of *S. hispidus* left the forest on Bana Hill in the Gold Coast with a scribbled message, "The natives use the root against syphilis."[1]

For most of the nineteenth century, these seeds were best known as the crucial ingredient in arrow poisons. By the century's close, however, investigations on African poisoned arrows in Europe led to the creation of a new heart medication called "strophanthin." The *Strophanthus* case allows us to probe the question of how plants became pharmaceuticals under colonial occupation (see Figure 3.2). As with other plants examined in this book, researchers have engaged in a complex and interlocked set of activities, including the production of new kinds of records, experiments, explanations, products, and harvests. Colonial investigators sought ways to "civilize" African plant knowledge, much as missionaries and government officials hoped to industrialize and Christianize their colonies. Initially, scientists in Britain, Germany, and France commandeered African plant expertise through networks of European physicians and botanists visiting new colonies. Then, they placed seeds and arrows

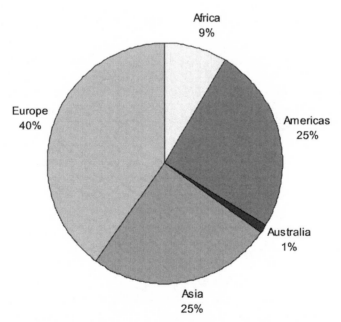

Figure 3.2 Plants included in 1885 British *Pharmacopoeia*, with approximate continent of origin (175 plants in all). (Source: General Medical Council, *British Pharmacopoeia*.)

under microscopes and injected laboratory animals with *Strophanthus* extracts. They created chemical and physiological explanations for why *Strophanthus* seeds increased the heart rates of animals, developed tidy tablets from strophanthin, and organized *Strophanthus* plants in walled gardens.

Although it has received little historical attention, the search for new drugs was a central component of colonial expansion in Africa during the late nineteenth century. European botanists appropriated plants and information, expanding diasporas of healing plants established through Early Modern African networks of trade and kinship. On the Gold Coast, the British overran the capital of the Asante Kingdom in 1874 and established a permanent capital for their self-declared protectorate among the Ga people in Accra by 1877. The colonial governor and his administrators were tasked with finding ways to make the emerging Gold Coast colony pay for itself, including the discovery of any medicinal plants of value. European appropriation of poisoned-arrow technology for an export trade in the valuable *Strophanthus* seeds—seeds critical to both the weapon and the drug—represented one of the earliest episodes in transnational drug prospecting in African colonies. It is significant that the 1884–1885 Berlin Conference that paved the way for European colonization in Africa demanded safe passage for scientific parties exploring plants along the Congo and Niger Rivers.[2] In the Gold Coast, a botanist from the Royal Botanic Gardens at Kew was dispatched to set up an experimental agricultural garden in the southern mountain town of Aburi. The governor also appointed a group of interested men to a Commission for the Promotion of Agriculture in the Gold Coast colony.[3] A close reading of pharmacological history suggests that the search for new drugs was also a factor in the scramble for Africa.

This chapter examines how new colonial subjects on the Northern frontier of the British Gold Coast resisted these efforts to appropriate plant information, ban arrow poisons, and harvest *Strophanthus* for emerging pharmaceutical companies like Burroughs Wellcome and E. R. Squibb and Sons. Fundamentally, it probes how populations in the Gold Coast confronted efforts of colonial officials to create new databases and gardens open to different European parties but closed to African subjects. Further, through the process of appropriation of *Strophanthus,* colonial researchers obscured the role of African plant experts whose prior experimentation was critical to the transformation of the plant into a drug. The first section considers how colonial officials attempted to take control of the use of poisoned arrows as a technology of warfare and resistance on the West African Sahel from 1885 to 1922. The second section examines the same period through the story of the conversion of the weapon into a drug, and the third explores the subsequent export scheme in the Gold Coast.

Read together, these contemporaneous narratives recover the social and global milieus in which pharmaceuticals continue to be embedded. The overlapping stories reveal the messiness of not only drug discovery from plants, but the rise of pharmaceutical science more broadly. It implicates scientists like Thomas Fraser, the creator and promoter of strophanthin at a laboratory in Scotland, in the coercive practices that marked rule on a colonial frontier. When understood as both a drug and a weapon, *Strophanthus* holds the ironic position of functioning as a tool for fending off the very empires that would have their children take chocolate-covered strophanthin pills. While here it might be tempting to draw easy lines between bad colonists, rotten African chiefs, and resisting subjects, the starker difference is between those who could set themselves up as scientific experts, whether white or black. Determining lines of ownership and rights to both plants and related traditional medicines continues to represent a political dilemma that often sacrifices those with less access to the language of the laboratory.

TAKING POISONED ARROWS

Poisoned arrows were critical weapons in the arsenal of African populations resisting invaders, slave raiders, and distrusted voyagers. The party of the Portuguese explorer Nuno Tristao made the earliest European report of poisoned arrows in West Africa after a failed attempt to land in Gambia in 1447.[4] Poisoned arrows also killed a large number in the party of the British navigator Richard Hakluyt as he approached Cape Verde in 1567.[5] Poisoned-arrow technology became synonymous with the mysterious dangers of the African continent. For several centuries, details on their manufacture were unavailable to Europeans at coastal forts. In 1673, the Danish missionary Wilhelm Müller wrote of Fetu poisoned arrows purportedly laced with crocodile bile on the Gold Coast, but until the late nineteenth century further information eluded inquisitive Europeans.[6]

During the 1890s, the British Colonial War Office and Royal Botanic Gardens at Kew witnessed a spike in correspondence surrounding arrow wounds as wars were waged to secure territory in Africa.[7] Earlier emissaries of the empire, including David Livingstone, had collected poisoned arrows for European museums. Army surgeons and military intelligence were now concerned with their actual deployment as weapons. Much has been made of the devastating impact of the repeating rifle in Africa after its introduction to British troops in 1889; as the satirist Hilaire Belloc famously quipped, "Whatever happens, we have got the Maxim gun, and they have not."[8] Regardless of their military advantage, however, British colonial authorities voiced continued apprehension. Indeed, Britain's considerable success was followed in the emerging West African colonies in newspaper reports on fatalities and advances. In June 1892, the *Gold Coast Chronicle* published a fight song describing the war in nearby Nigeria (sung to the tune of Bonnie Dundee): "Away with the Jebus! / Bring out the Rockets / The Maxims, the Gatlings / Give it 'em Hausas, Ibadans and Scouts / Finish 'em, Finish 'em, Finish 'em now!"[9] However, apprehension surrounding poisoned arrows and other African armaments shaped a surprising footnote to the official narratives of colonial conquest.

Africans capitalized on their superior understanding of local plants to challenge efforts to occupy their territory. Among Gurunshi communities living along the Sahel in West Africa, secret recipes for "red-tipped arrows" were disclosed to religious and healing figures such as the land priest or *tiindana* who prepared the poison in secluded parts of the bush ahead of offensives (see Figure 3.3). In the face of colonial espionage in the 1890s, those privy to poisoned-arrow production supplied incomplete details:

> The seeds are reduced to a powder by grinding *and the other parts added.* And to this a small quantity of water is added and stirred. The mixture is then boiled for some time until it becomes of a thick consistency. It is then allowed to cool and the arrows are subsequently smeared with the thick brown coloured resinous looking residue.[10]

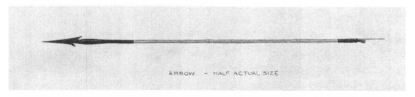

ARROW – HALF ACTUAL SIZE

Figure 3.3 Illustration of poisoned arrow from Albert Chalmers's "A Further Report of Experiments upon the Fra Fra Arrow Poison" (1899), in "Miscellaneous Gold Coast Cultural Products (I–W)," 1888–1906. (Source: Archives of the Royal Botanic Gardens, Kew. Image reproduced with the kind permission of the Board of Trustees of the Royal Botanic Gardens, Kew.)

Arrow poisons represented a form of African experimentation and innovation. Communities adapted inherited recipes to create secretive, localized formulas.[11] Parts and amounts of plants, time of collection, additional ingredients, and heating processes were committed to memory and varied from community to community. These diverse formulas combined available toxins—including potent plants, snakes, fermented urine, and scorpions—to devastating effect. Northcott reported that poisons included "ali," a water insect, and preparations from Wa used snake venom mixed with "yao" plants, possibly suggesting the Mole word for *S. hispidus*, "yoagba."[12]

Official colonial anxiety over poisoned arrows was based in ignorance of weapon manufacture. Without knowing what made the arrows so deadly, surgeons floundered in their attempts to treat the resulting wounds. Consider the experiences of Igala Grunshi, a Ghanaian Sergeant serving with the British forces in the northeastern area of the territory. In August 1899, his shoulder was pierced by an arrow dipped in poison by a warrior of a Frafra community.[13] Grunshi's men immediately ripped out the arrow and conveyed him to the army surgeon. The surgeon offered a solution of potassium permanganate, but Grunshi begged him instead to allow the application of an indigenous antidote. Most likely a Gur-speaker fighter on the colonial side, Grunshi would have been familiar with the effects of arrow poison (capable of killing within the hour), and may have been inoculated against the toxin.[14] British surgeon P. J. Garland reported that fifteen minutes after he administered his drugs, further treatment was provided by men who "made small incisions on [Grunshi's] back and placed the antidote in the incisions."[15] Within a week, Grunshi was well enough to return to the battlefield. Admitting his ignorance on the matter, Garland dispatched the arrow and antidote to London for identification.

Grunshi's experience on the battlefield illustrated a larger phenomenon of European unfamiliarity with indigenous weaponry and therapies by the nineteenth century. Further, from European traders along the coast, northern populations sustained knowledge of poisoned-arrow production alongside imported weaponry. In a rare document written in Hausa (using Arabic script), Abu Mallam described the Zabarma conquest in the Sahel in the nineteenth century. Mallam noted use of both guns and arrow poisons in battles for trade and slaves. Zabarma's leader, Babatu, fought against local groups, sometimes "gun against gun, man against man." Of the Guni, it was noted that "their poison was not a thing to play with."[16] Other examples of hybrid weaponry included Dane guns used by the Dagomba, who made their own iron bullets.[17]

Colonial physicians did their best to create written documentation of poisoned-arrow wounds, possible ingredients, and antidotes on the battlefields. To surgeons like Garland, arrow wounds presented a medical quandary. Garland reported that "it was impossible to carry out very accurate observations as the column was in motion and men were hit from time to time."[18] He advised his men and officers "that in the event of their being hit by arrows they were to immediately have the arrow pulled out without waiting for my arrival on the scene."[19] Once Garland or one of his dressers caught up with the wounded party, a solution of potassium permanganate was syringed into the wound. Brandy was also on offer, and in extreme cases a weak solution of cocaine. Like Grunshi, not all soldiers were confident in Garland's procedure to neutralize the poison. Limited evidence suggests that although French and German military convoys were more likely to use indigenous antidotes, British-led troops were encouraged to avoid using them.[20] In an atmosphere marked by war and conflict, misinformation ran rampant. A surgeon working with Garland noted, "It is highly improbable that natives against whom we were fighting would give any information with regards to such a secret and virulent poison."[21] Reports on poisoned arrows were marked "confidential" and labeled "not intended for publication."[22] Considering that soldiers in West African regiments were from neighboring communities, it is certain that, like Grunshi, they offered their own opinions and experiences, much of which unfortunately remains outside historical records.

From the battlefield, military officials carried available poison and plant samples. They submitted these artifacts of African military expertise to botanists tasked with creating new databases of valuable plants in the emerging colonies. The botanical ingredients of the parcel Garland sent to the Army Medical Department in London in 1899 could not be fully identified at Royal Botanic Gardens, Kew. A researcher there stated the following:

a. '[P]oison plant' is a species of *Strophanthus*. I cannot make it exactly; but it must be very near to *S. hispidus*. There are no twigs with the pods as stated in the letter.

b. The 'antidote' is a mixture of 4 or 5 diff. Plants or plant fragments, namely 2 very diff. kinds of Branches. They are like the second set of leaves [indeterminable] without a close anatomical examination.[23]

Although the antidote remained a mystery, efforts were made to compare research on the species and related poisons in other parts of Africa, including observations made in Uganda the previous year.[24] Partial identification of key ingredients in the arrow toxins submitted by Garland allowed the Army Medical Department greater control over the "nuisance" of African resistance by banning their production and limiting access to *Strophanthus*. *Strophanthus*, from the Greek *"strophe"* for twisting and *"anthus"* for flower, was a genus name first employed by the French botanist Augustin Pyramus de Candolle in his description of the hierarchy of flora in 1800. The botanists working at Kew made every effort to link the leaves and available plant parts within this global schemata and confirmed the existence of the much-sought-after *Strophanthus* in the northern reaches of the Gold Coast.

Some colonial physicians also attempted basic experiments in Africa. Albert Chalmers, a British surgeon based in the northern reaches of the Gold Coast during the Frafra expedition, carried a limited quantity of Frafra arrow poison back to the capital Accra to conduct a few laboratory investigations. His research agenda indicates the concerns of a battlefield surgeon:

1. How does a man or animal die when hit by one of these poisoned arrows?
2. What treatment should be adopted in order to preserve life?[25]

Chalmers's research was limited to available resources in the Accra laboratory. He had three possible plants acquired by three military officials during separate expeditions against the Frafra, a letter from Kew suggesting that Garland's sample was "a Strophanthin," part of Woodhead's report on Ugandan arrow poisons, his own notes, and two military reports on arrow poisons in West Africa. With little more than a few "leaves and a pod with only 3 seeds in it (the others having been all lost)," he managed to boil down a solution of *Strophanthus* and inject it into a guinea pig with a hypodermic needle. Since the guinea pig's heart failed and it died after about twenty minutes, Chalmers determined that it did not shiver in the same way as the wounded men he had observed: "There was a difference between the symptoms of the strophanthus poison and that of the arrow poison, and I consider that they are different poisons."[26] As we shall see, Thomas Fraser had, by this point, already been working on the chemistry of the new drug strophanthin for ten years, but it is not clear whether Chalmers knew of this work.

In addition to making careful records on *Strophanthus* and West African armaments, colonists in the Gold Coast sought to control access to arrow poisons among new subjects. By 1902, British-led troops secured control over the Asante kingdom and declared a protectorate over the

Northern Territories. Arrow poisons continued to be of concern in the colony as administrators replaced soldiers and military physicians, and limiting access to weaponry was a central government policy. Efforts in October 1905 to secure Navarro (now Navrongo), the last municipality before the Ghanaian border with Burkina Faso at Paga, were indicative of these attempts at control.[27] Lieutenant P. J. Partridge was sent with 150 men to establish the post. He was informed that "tribes in these parts settle all disputes by immediate recourse to poisoned bows and arrows and only within the last few weeks one village has fought among themselves over a trivial dispute resulting in the loss of several lives."[28] After a meeting with town leaders in which he informed them of the ban on poisoned arrows, Partridge reported the suggested "prohibitions met with great applause and clapping of hands. . . . I conclude they have as great an objection to the poisoned arrow as we have."[29] To confirm his mandate, Partridge displayed a large Maxim gun with 2,000 rounds of ammunition in the center of Navarro. He executed two criminals in neighboring towns and donated a "silk dress to the chief of Navarro" in return for his supply of "free labour." Within two days, Navarro's citizens mobilized the collection of 500 loads of grass and 200 loads of sticks to start building a jail, government store, courthouse, and military barrack. According to Partridge, the Navarro leader understood that the European presence "would make him into a big Chief and he and his people were prepared to do everything they could to help."[30]

Following the paternalistic policies of the colonial administration, Partridge seized arrow poisons and inflicted offenders with heavy fines, including the collection of herds of sheep and cattle. Outlawing poisoned arrows and possession of poison extended the Gold Coast laws of 1892 to the Northern Territories. Section 46 of Criminal Code 12 made not only murder, but evidence of intention to murder, a punishable offence with up to ten years' imprisonment: "Every person who prepares or supplies, or has in his possession, custody, or control . . . any instruments, materials, or means . . . by which life is likely to be endangered . . . shall be liable to punishment in like manner as if he had attempted to commit that crime."[31] Household and military weapons, including bows and arrows, were outlawed within colonial jurisdiction: "The information should be widely circulated that the carrying of poisoned arrows and the possession of the poison in their compounds will be considered an offence."[32] These new policies would have criminalized a wide spectrum of individuals—including hunters, war medicine specialists, rebels, and warring chiefs.

As colonial occupation solidified, banning arrows served the interests of both chiefs and colonial officers hoping to minimize resistance. "Frafra" was the name assigned to one of the five kingdoms of the Northeast province (now Upper East Region in Ghana) that the colonial administration christened in 1911.[33] To maintain control in this and other kingdoms, British authorities continued to give gifts to "friendly chiefs" and regularly propped up preferred leaders.[34] After the death of several chiefs in Zouaragu in 1918, regional commissioners confided on a list of possible replacements.[35] Given the close alliance of political authorities, those who chose to resist the colonial apparatus aimed their arrows indiscriminately at both European and African officials. In November 1917, Akanyele, a man residing in Bolgatanga, threatened to shoot arrows at both his king and the colonial police sent by the District Commissioner to arrest him. After several months in hiding, Akanyele reappeared the following April, prompting a standoff between village royals, mounted police officers, and the new acting District Commissioner, A. W. Cardinall. Akanyele stationed himself on the roof of his thatched house, armed with his bow and arrows. After one arrow grazed Cardinall and a second pierced the shin of Cardinall's interpreter, the Commissioner opted to fire a gun toward the house, prompting a flurry of activity and confusion that ended with a fatal bullet wound to Akanyele.[36] An official statement was sent from the capital to "please inform Cardinall that Governor considers he acted with courage in a critical situation and that the shooting of Akanyele was justified."[37]

In 1920, Cardinall capitalized on his experiences in the publication *The Natives of the Northern Territories of the Gold Coast, Their Customs, Religion and Folklore*. The Akanyele affair or a similar incident formed an ethnographic morsel for the colonial official to digest at his leisure:

Once my interpreter was hit by a poisoned arrow. The local *liri-tina* [herbalist] would not come. He was too afraid of a general fight, since the war-cry had been raised. He supplied the antidote, however, *but I could not learn of what it was composed*. The procedure was as follows. The wound was in the left leg just below the knee-cap, *the poison Strophanthus*. The arrow had pierced in about three-quarters of an inch and took me several seconds to extract. The man was made to sit down. His neck was cut in three places, but not so as to draw blood, and the skin between the fingers was treated likewise. The wound was then beaten with the flat of a knife, and after a little blood had flowed the medicine—a black sort of paste—was applied and a draught of some concoction given. . . . The man lived.[38]

In contrast to earlier encounters with poisoned arrows, like that of Grunshi in 1899, Cardinall's experience occurred at a moment of increased colonial authority. Cardinall escaped brushes with "primitive" warfare unharmed and was celebrated for his courage under fire. Whether he could be certain the poison was from *Strophanthus* plants when he had no clear indication of the composition of the antidote was unclear. But by then *Strophanthus* was almost expected in a colonial tale. Even Arthur Conan Doyle's *Lost World* described a mythical land where South American *Strophanthus* arrows were aimed at dinosaurs: "But where the conical explosive bullets of the twentieth century were of no avail, the poisoned arrows of the natives, dipped in the juice of *Strophanthus* and steeped afterwards in decayed carrion, could succeed."[39]

In the Northern Territories of the Gold Coast, arrows were officially outlawed, although they continued to be used to resist both European administrators and African leaders until at least 1918. Yet even as poisoned arrows were met with prompt gunfire, colonial administrators continued to make toxic armaments the object of scientific inquiry. Researchers in London pursued simultaneous efforts to remake arrow poisons into the drug strophanthin. Thus, exporting *Strophanthus* seeds meant gaining full access to a plant whose alternative uses in weaponry were forbidden.

MAKING STROPHANTHIN PILLS

Strophanthus species are found throughout Africa and Asia, including parts of China and the Philippines, with no overlap between species in Africa and those indigenous to Asia.[40] A climbing vine with dramatic spotted flowers, *Strophanthus* plants can be found growing south of the Sahel and throughout eastern and southern Africa in the underbrush of woody grassland and among treetop vines in more forested areas. Linguistic evidence suggests that a range of West, East, and southern African communities had terms for the genus.[41] Some populations protected it within communal gardens, perhaps implying popular knowledge of its uses beyond military, hunting, or medicoreligious sects (see Figure 3.4).[42]

African healers have long recognized the medicinal value of the genus, incorporating different species into treatments for muscular aches, open sores, constipation, food poisoning, venereal ailments, and heart failure.[43] In the Gold Coast, preparations of the plant variously called *"omaatwa," "yoagbe,"* and *"ajokuma"* relied on alcoholic decoctions made by steeping roots in a fermented, alcoholic beverage, a therapeutic method common to the region. The resulting bitter-tasting solution would be

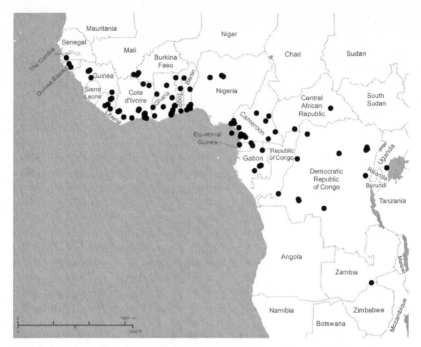

Figure 3.4 Collection points of 188 *Strophanthus hispidus* samples in herbaria, accessed through the online database of the Global Biodiversity Information Facility: Cameroon National Herbarium; Real Jardin Botanico Vascular Plant Herbarium (Madrid); Royal Botanic Gardens (Kew); Museum National d'Histoire Naturelle et Réseau des Herbiers de France; Botanic Garden and Botanical Museum Berlin-Dahlem; National Herbarium of the Netherlands; Missouri Botanical Garden; Conserve International; Herbarium Senckenberg (France); Phanerogamic Botanical Collections (Sweden); Herbaria of the University of Zurich. (Source: Global Biodiversity Information Facility Data Portal.)

taken in small sips over a period of days or weeks.[44] Local healing specialists throughout West Africa seemed to have closely monitored its use: "The stems are mashed and boiled and the liquid drunk, the dose being carefully regulated by the native doctor, any error easily producing poisoning."[45] A potent plant with both healing and highly toxic capabilities, *Strophanthus* was also integral to traditional legal regimes relying on ordeal trials. In such court proceedings, the physical response of the accused to a strong drink prepared with *Strophanthus* or another potentially toxic ingredient determined the verdict.[46] Among the secret women's societies of Sierra Leone, *S. gratus* represented a form of female knowledge in its use during closed ordeal trials.[47]

Even before *S. hispidus* was identified as a possible ingredient in Frafra arrow poison, the colonial administration in the British Gold Coast was keen to find signs of the plant. In the wake of the landmark 1884–1885 conference of Berlin that firmly secured European interests in the interior of Africa, the Gold Coast Governor appointed a group of interested men to a Commission for the Promotion of Agriculture on the Gold Coast Colony. Significantly—although mention was made of regional crops including rice, corn, yams, kola, and the newly introduced cocoa— *Strophanthus* was the only item listed under the subheading "Drugs" in the Commission's initial forty-page memorandum:

> A large number of plants are used medicinally by the natives, but too little is know of this department to allow of particularization. No doubt examination of the flora by botanists will lead to the discovery of many plants as valuable as the *Strophanthus*, to which attention has been lately directed.[48]

The enthusiasm for *Strophanthus* represented in the 1889 report of the Commission stemmed from the rising popularity in Britain of a novel treatment for circulation promoted by the Scottish physician Thomas Fraser. His studies on a series of poisoned arrows led to the isolation of purified strophanthin in the late 1880s, creating a demand for *S. hispidus* seeds from Africa.[49] He researched *Strophanthus* after corresponding with explorer David Livingstone and his companion, the physician John Kirk. Kirk kept his collection of poisoned arrows in the same bag as his toothbrush. Noting a throbbing sensation on his gums one morning, he thought the arrow tips might contain a potent stimulant.[50] Livingstone secured information on the source of the poison from arrows collected in Kombé during an expedition along the Zambezi River.[51]

Fraser pieced together available written information on African arrow poisons and assembled eight sample poisoned arrows on which he inflicted a battery of tests, eventually leading him to the elusive drug. Although Fraser's research did not rely on later arrows acquired during the Frafra expedition, the subsequent export scheme from the Gold Coast stemmed from his investigations. To make the drug, Fraser first ground and dried samples of seed obtained from his colleagues, which he surmised to be *S. hispidus*. His investigations on poisoned arrows reads like a crime investigation, with all available evidence excavated from letters dispatched from across the British Empire. Descriptions for his sample arrows, labeled exhibits A through J, reveal the circuitous routes of imperial science as objects collected in East, West, and Southern Africa recirculated in Europe. To some extent, he provided geographic and cultural information on the sites where people produced the collected armaments, noting, for example, that Arrow C came from a district "75 miles N.N.W. of Zanzibar," and that Arrow H, said to come from "one of the Manyuema tribes on the west side of Lake Tanganyika," was used only for the killing of game.[52]

Fraser mixed combinations of powdered *Strophanthus* seed with alcohol or water until he was able to create a concentrated form of the seeds that, under the lens of his microscope, revealed suspended crystals.[53] These crystals, described as "intensely bitter," were mixed with water and tannin to produce "the active principle." Then the tannin was converted to tannate through the introduction of lead oxide. Carbonic acid was passed through the remaining solution for several days. A solution of the dried residue could then precipitate strophanthin through the introduction of ether.[54] The rewards of this long process were "beautiful stellar groups of colourless and transparent crystals."[55] It was not until the early twentieth century that Fraser's experiments were corrected to reveal that the plant in question was actually *S. kombe*.[56] In response, the British *Pharmacopoeia* approved only strophanthin made from the *S. kombe*—a specification that affected the export schemes described in the following paragraphs.[57]

Strophanthin's effect on the heart and blood circulation was similar to that of digitalin, derived from common foxglove. As with strophanthin, in the classic story of the development of digitalin European men appropriated folk medicine, in this case the experiments of an elderly woman in Shropshire who kept a secret mix to cure dropsy.[58] In his publications, Fraser heralded strophanthin as being more potent than digitalin, with the added effect of increasing the action of the heart without raising blood pressure or causing indigestion. On frogs and rabbits of different weights, Fraser conducted further investigations on the effects of strophanthin to establish the minimal number of grains of the crystal required to induce death. For instance, with the frogs—after their twitching, slowing of breath, dullness of skin, and instability in the head area—he noted, "Jumping is effected with difficulty; sometimes delay is observed in the flexing of the limbs after a jump; then, each jump lands the frog on the back, and although, at first, the frog is able, after a little delay, to turn it afterwards it cannot do so, and remains motionless on its back."[59] Through his published experiments and explanations for the efficacy of strophanthin, Fraser's name supplanted those of the African informants who provided colonial collectors with valuable plants and arrows. In the absence of written documentation of their efforts, African plant specialists did not feature in Fraser's catalogue of previous investigations by French and British chemists.

By late 1905, clinical trials of intravenous use of strophathin for individuals suffering from weak hearts began in Germany under the direction of Albert Fraenkel in Strasburg.[60] During the next decade, German researchers perfected the use of strophanthin to combat poor circulation and came to dominate the trade in *Strophanthus* through their African colonies. Strophanthin was incorporated into official pharmacopoeias in the United States, Britain, France and Germany, with some dispensatories preferring *S. kombe* seeds.[61] Using both British and American standards, Burroughs Wellcome and Company prepared the "tabloid" brand Stropanthus Tincture from 0.01 grams of seed for each 0.1 gram dose.[62] The drug was prescribed to adults with heart murmurs; it was also masked in sweet syrup and given to children three times a day to alleviate many ailments, including "nervous asthma," typhoid, and pneumonia. For the American market, E. R. Squibb and Sons combined digitalis with strophanthin in a popular chocolate-covered tablet sold at 16 cents per 100 count. The recommended dosage for palpitation, "smoker's heart," and as a cardiac tonic was one tablet every three or four hours.[63]

During the early twentieth century, it does not appear that drug companies relied on patents to secure profits from the new drug strophanthin. Rather, they invested in branding and advertising of their *Strophanthus* products, creating distinctive labels and tablets (see Figure 3.5). Companies sought to create standard products that physicians and patients could use with predictable results. However, questions on the safety and purity of strophanthin plagued its initial acceptance. Physicians reported inconsistent results with strophanthin pills, preferring digitalis, or pills made with a mixture of both.[64] Although Fraenkel recorded no fatalities, his studies on *injecting* strophanthin were not widely accepted outside Germany.[65] What might have been behind the unreliable results reported with strophanthin? The white crystals of the extract suggested a pure, standard commodity, but accurate differentiation of various species of *Strophanthus* with their distinctive chemistry was difficult to conduct with dried seeds devoid of any other plant parts.[66] An early test used sulfuric acid to turn substances containing strophanthin green. However, even "pure" tinctures turned red or reddish-green, leaving chemists to "conclude that [they] . . . were prepared from a nonofficial seed."[67] The chemist alone with the delocalized seeds could not be confident that his drugs were pure.

HARVESTING *STROPHANTHUS* SEEDS

Efforts to export *Strophanthus* from the Gold Coast during World War I indicated a need for botanical expertise at the ground level. However,

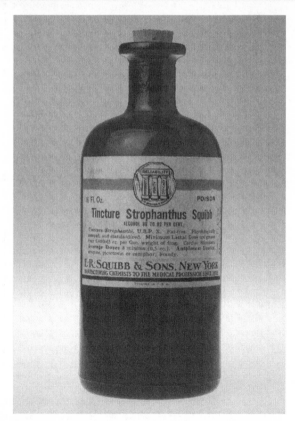

Figure 3.5 (a) Digital scan of label for Squibb Strophanthin, c. 1916. (b) Photograph of a bottle of Squibb Strophanthin tincture, filled in 1932. (Source: Archives of Bristol, Myers, Squibb.)

the very people who might have built on existing knowledge to accurately identify *Strophanthus* were banned from using it. For instance, Akan experts in the Gold Coast had their own schema for distinguishing categories in the genus: *S. hispidus* was categorized as "male," based on its amount of sap, in contrast to "female" *S. preussii*.[68] In late 1914, the British in the Gold Coast took advantage of instability during the World War I to invade neighboring German Togoland, entering an experimental *Strophanthus* garden in the northern town of Yendi. The motley harvest of "small dark brown seeds with yellowish hairy patches" triggered a short-lived export scheme.[69]

Initially, the Gold Coast administrators had high expectations for the *Strophanthus* plants. In April 1915, a high-ranking colonial official in the Northern Territories shipped off his cache of plundered seed to England for analysis. The Imperial Institute in London determined that they "differed widely, both in appearance and in their chemical reactions when tested," from the species approved for medical use, *S. kombe*.[70] Secondary identification at the Royal Botanic Gardens, Kew, posited that the seeds were of the species *S. hispidus*, the same type connected with Frafra arrow poisons in 1899 (see Figure 3.6). However, because current unrest in Nyasaland had interrupted the normal supply of *S. kombe*, importers in London would temporarily entertain purchase of West African *S. hispidus* seeds at a price of 1 shilling 3 pence per pound.[71]

In the Gold Coast, government agricultural scientists experimented with *S. hispidus* cultivation. The agricultural department, like all branches of the colonial administration, was weighted to the southern coastal areas and lacked experience with northern plants. At this time, the Director of Agriculture, W. S. D. Tudhope, was based at the government gardens at Aburi, where he focused on cocoa cultivation. Tudhope sent detailed explanations of how to prune the unfamiliar plants, adding that it would be best if he or his curator were to come in person: "It is difficult to advise on such a matter by letter."[72] In 1917, the harvest of *S. hispidus* seed did not seem promising. The abandoned German shrubs at Yendi would only provide "one or two hundred weights" and Tudhope was forced to admit to the Governor that "it is quite evident we are not yet in a position to enter into the commercial aspect of the subject."[73]

The initial enthusiasm for the export scheme soon proved unsustainable. Those appointed to manage the new acquisitions in Togoland were left to such tasks as the winnowing of poisonous seeds in difficult circumstances. The agricultural inspectors at Aburi were not sympathetic

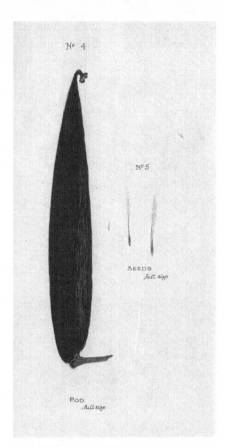

Figure 3.6 Illustration of *Strophanthus* pod and
seeds from Albert Chalmers's "A Further Report of
Experiments upon the Fra Fra Arrow Poison" (1899),
in "Miscellaneous Gold Coast Cultural Products
(I–W)," 1888–1906. (Source: Archives of the Royal
Botanic Gardens, Kew. Image reproduced with the
kind permission of the Board of Trustees of the Royal
Botanic Gardens, Kew.)

to their struggles. Defending himself in March 1918, the District Commissioner in Yendi explained to his supervisor in Tamale, "In my opinion [Tudhope's] criticism was uncalled for." He continued,

> With inadequate means at my disposal for beating out this seed,
> and then having to allow it to dry for days in the sun, that a small
> percentage of dust etc should be mixed with it is unavoidable. Added
> to which a too close contact in winnowing is dangerous to both

the eyesight and lungs as this poison is an extreme irritant. In my attempts to do so both my helpers and myself were caused the acutest discomfort.[74]

Given the difficulties of *Strophanthus* processing, colonial authorities turned to Africans for assistance. Tudhope presented a proposal to the Governor's office in 1917 "to encourage the Natives of the Northern Territories to collect the seed and bring it into the Agricultural Station, and to pay them say 3d [pence] per lb; and this may have the effect of stimulating production." In 1918, at least one community took him up on this proposal; the Chief of Lorha sent a delegation bearing seeds to the District Commissioner in Wa.[75] Further samples do not seem to have been gleaned from local residents.

The threat of poisoned arrows loomed large in discussions surrounding the *Strophanthus* export scheme. Although collection of the seed by Africans from wild sources was permissible, experiments in cultivation were left to Europeans. *S. hipsidus* was a favored ingredient in arrow poisons made in the northeastern and northwestern provinces and to a lesser extent in the southern one. The Chief Commissioner of the Northern Territories explained to Tudhope:

> There is no objection whatever to the cultivation of *Strophanthus* at Botanical or District Stations, but I do not advise that the natives should be encouraged to cultivate it; at any rate for the present.
>
> If we find that *Strophanthus* seed is worth the cultivation of the plant on a large scale, and that it would prove to be a profitable local industry, it will be time enough to encourage the Chiefs of the Province to establish plantations of *Strophanthus*.[76]

Efforts to grow *Strophathus* had limited success. In 1917, the Chief Commissioner of the Northern Territories reported that "the area planted with *Strophanthus* at the Tamale Agricultural Plantation is not yielding good results and few of the plants have reached a height of more than 6 inches."[77] Despite the difficulties of cultivation, difficulties that to this day have not been fully overcome, the Agricultural Department persisted in its efforts. Indeed, in the imagination of lobbyists in Britain at the time, "the forests of Africa . . . teem with useful drugs, prominent among which may be cited the kola nut and the *Strophanthus* seeds."[78]

By 1922, the export scheme of *Strophanthus* all but ceased in the Gold Coast and surviving plants in the agricultural gardens were dug up and burned. The failure of the export scheme was pinned on the "dangerous properties" of the plants and the untrustworthiness of people in "poison arrow districts." At Navarro, the Commissioner of the Northern Province received a letter from the regional office advising him to destroy all seedlings in early 1922:

> I note that you may have a crop of *Strophanthus* this year. In spite of the wishes of the Agricultural Department, I am not prepared to allow this crop to mature in a poison arrow district, unless you can guarantee that none of it will be stolen. Our first duty is to wean the natives from the use of this poison, and I think that it is a poor way to do it by actually growing it in our own experimental gardens right amongst them. There are other places where it can be experimentally grown without risk.[79]

This sentiment was echoed by other commissioners throughout the Northern Territories: "I beg to inform you that the D. C., Wa considers that it is quite impossible to guarantee that none of the crop will be stolen, and he has therefore had all the plants dug up and burnt."[80] By the beginning of July 1922, all commissioners were instructed to destroy remaining plants.

The difficulties represented in the Gold Coast export scheme suggest, in part, the circumstances behind impure, unreliable strophanthin stocks worldwide.[81] Commercially, *Strophanthus* buyers preferred pods or previously removed and cleaned seeds, which were ironically more difficult to identify by species and place of origin. German scientific opinion favored the readily defined, hairy seeds of *S. hispidus*, whereas "the British Pharmacopoeia, the French Codex and the Swiss Pharmacopoeia" recognized only *S. kombe* seeds for legal production of strophanthin.[82] By the 1930s intricate strategies were devised to differentiate between seeds originating in Mozambique and those from Sierra Leone or Upper Niger or the Zambesi delta, yet contamination was common and results unpredictable.[83] Although the African plant experts like the *tiindana* may have had greater fluency with plant identification, British agricultural officers, citing concern over "misuse" of the plants for poisoning arrows, did not formalize any basis for meaningful collaboration, and their investigations floundered. This was in contrast to the successful cocoa trade led by African farmers in the colony during the same period.[84]

Threat of potential abuse was sufficient grounds to abandon the initiative, despite an obvious lack of experience propagating and harvesting the plant on the part of colonial administrators.[85] As one commissioner rationalized, "*Strophanthus* is not hardly used as a drug in Europe" and scarcely warranted all the trouble required to cultivate it in the face of

African demand. Insufficient coordination with German scientists was also at play. Further, evidence of poisoned-arrow battles, such as that between Akanyele and Cardinall in 1917, suggested the instability of colonial authority as administrators and indirect rulers feared for their own safety.[86] The ancient proverb "They will beat their swords into plowshares and their spears into pruning hooks" foretold a time when weaponry might be put aside and transformed.[87] But—in this case, anyway—swords were not remade into plowshares, and poisoned arrows were not fully transformed to strophantin pills in the Gold Coast.

CONCLUSION: COLONIAL SCIENCE MASKED AFRICAN INNOVATION

The intersection of poisoned arrows, strophanthin medication, and *Strophanthus* export in the colonial Gold Coast allows us to interrogate the history of drug discovery in an increasingly interconnected world by the late nineteenth century. Colonialism was central to the elucidation of alkaloids and glycosides, the backbone of modern pharmacology.[88] Literary critics remind us that the colors and textures of Asia and Africa formed an important subtext to European novels of the nineteenth century. The cholera epidemics of China, romanticized ayahs of childhood, a lover gone abroad—all circulate in the subconscious of European characters of the imperial age and deserve closer scrutiny.[89] In the case of *Strophanthus*, a careful reading of laboratory reports constructed in Edinburgh suggested the residue of empire. Returning "colourless, opaque, and brittle" strophanthin crystals to an African setting provides the contours of the larger global economy in which "science" operated. The creation of standardized, white tablets hid from view the ways in which European scientists appropriated plant expertise from across Africa. Multiple contributors to innovation gave way to individual scientists like Fraser, who took credit for his discoveries even as other European researchers influenced his research. Colonial science became the mask through which we view centuries of African innovation on potent plants.

Second, the process of obscuring the roles of African plant experts led to shortcomings with the drug. In the case of the Gold Coast, the politics of who had rights to use and assign value to plants shaped strophanthin's trajectory. Before colonial occupation, people residing in the Sahel were free to incorporate local plants into weaponry and healing practices. Colonial scientists appropriated *Strophanthus*, grafting herbarium records onto a preexisting plant diaspora that spanned much of Africa. They expanded the number of potential claimants to *Strophanthus*. However, through the process of renaming a medicinal plant and reducing it to a standardized extract, European chemists and botanists transferred authority to science and the colonial state. Names of individual African informants did not feature in their reports and publications, as European scientists discounted the value of prior experimentation in arenas they deemed to be "primitive." Yet the colonial state was sometimes fragile, and its authority contested on all sides. In the Northern Territories of the Gold Coast, demand for arrow-poison plants intersected with aspirations of both Britain and Germany for an export trade. Power, literacy, and scientific authority separated the long-standing adaptations of arrow poisons among healers and hunters in African communities from the laboratory and herbarium investigations of European scientists. In the context of anxiety surrounding poisoned arrows, colonial authorities did not treat African plant experts as equals in the Northern Territories of the Gold Coast, undermining possible progress in the trade.[90] The *Strophanthus* transformation in colonial Ghana is echoed in current practices as multinational pharmaceutical companies continue to see "local politics" as a perceived drawback to using botanical sources in drug production.[91]

Third, the uneven process of transforming an orally transmitted indigenous technology, such as arrow-poison production, into the words and numbers of (European) science was part of an irrevocable shift in access to plant knowledge in Africa. British physicians and botanists participated in the production of "open access" databases of scientific information that were accessible to European researchers. Despite the shortcomings of the *Strophanthus* export initiative, the process of co-opting local knowledge systems through colonial violence was a precursor of what Arun Agrawal described in the 1990s as the troubling preservation of medical insights *"ex situ"* through "isolation, documentation, and storage of indigenous knowledge in international, regional and national archives; and . . . dissemination [of it] to other contexts and spaces—a strategy [associated with the rise of] western science."[92] Rereading pharmacological discoveries alongside colonial archives reveals the interplay between medical lineages and ultimately the multiple benefactors of our intellectual inheritance. Considering that nearly 10 percent of plants included in the 1885 British *Pharmacopoeia* were from Africa, what might the pharmacological industry owe descendents of those who suggested new therapies?

Finally, to justify this process of intellectual "appropriation," the colonial state commanded the language of poison and toxicity to demarcate access to promising plants. The potential of *Strophanthus* to cause harm and death in the "wrong" hands was the focus of government policies in the Gold Coast. From the outlawing of poisoned arrows, to the destruction of test gardens, a climate of fear and anxiety bred paternalistic laws. This preoccupation with poison was not necessarily the approach taken by healers; some recognized the medicinal potential of the plant. By the 1930s, although healers were better able to position themselves within the colonial state, government policies continued to emphasize the dangers of plant-based therapies in African hands.[93] European botanists and chemists used the specter of poisoning to justify colonial investigations, furthering the dissemination of regional plant knowledge beyond the Gold Coast.

Chapter 4 continues the story of drug development in the former Gold Coast, after its independence from British colonial rule. In postcolonial Ghana, a growing number of African scientists were able to seek credit for their contributions to the development of a new drug from the bitter roots of *Cryptolepis sanguinolenta*. And yet, as with colonial researchers before them, these Ghanaian scientists tended to obscure the role of traditional plant experts in the long process that led to a new pharmaceutical.

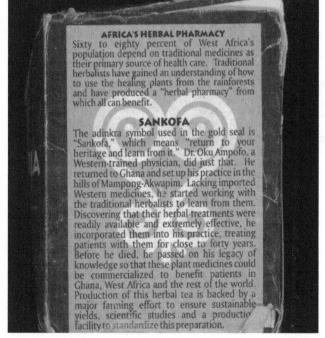

Figure 4.1(a) and 4.1(b) Phyto-Laria box for herbal teas to treat malaria. (Photo by author, used with the permission of Phyto-Riker Inc.)

CHAPTER 4

Take Bitter Roots for Malaria

The proposed Phyto-Laria packaging in my hands looks promising. The green cardboard features the Akan symbol for *sankofa*—"go back and take it" or "return to your roots," an admonishment to rediscover the past. The back provides an account of the late Ghanaian physician Oku Ampofo, who helped develop this herbal malaria treatment from old recipes. It details how Ampofo, one of the first Ghanaians to earn a medical degree in the United Kingdom, began to test plant medicines in rural communities during World War II. The boxes that will eventually be produced from this prototype will hold tea bags filled with dried flakes of the bitter roots of *Cryptolepis sanguinolenta,* the treatment Ampofo helped perfect for drug-resistant falciparum malaria. The packaging emerged from marketing research in the United States, but the products will be sold in West Africa, where malaria remains endemic. Through brilliant design, Phyto-Laria emphasizes partnerships between physicians and healers in Ghana. The flat sheets of paper, however, hide the necessary alliances that Ampofo built with North American scientists to make his dream for herbal drugs a reality.

An American biochemist, Diane Winn, first showed me the Phyto-Laria packaging mock-ups in 1998, shortly after the death of her friend Ampofo. Winn met Ampofo through a U.S. National Institutes of Health (NIH) research project in Ghana in 1962. When I met her at her office in Accra, she explained to me how Ampofo cured her of breast cancer using plants. Over the years, Winn developed a close friendship and reciprocal relationship with Ampofo. She provided vehicles and equipment for his laboratories at the Centre for Scientific Research into Plant Medicine (CS-RPM) in the mountainside town of Mampong that he cofounded with Ghana's first pharmacology PhD, Albert Nii Tackie. In turn, Ampofo explained that through Winn he met cancer researchers in the United States, supplying them regularly with plants for screening. When Ampofo grew blind and frail in his old age, Winn flew him to the United States to see physicians at Johns Hopkins. At his funeral, she announced that her company, Phyto-Riker, had secured access to Ghana's national pharmaceutical manufacturing facility. With the buyout of the drug wing of the Ghana Industrial Holding Corporation (GIHOC), Winn took advantage of a series of privatization measures instituted by the government in the late 1990s to jump-start ailing national industries.[1]

This chapter uses the story of *Cryptolepis* to examine how African scientists worked to transform plants into pharmaceuticals after the demise

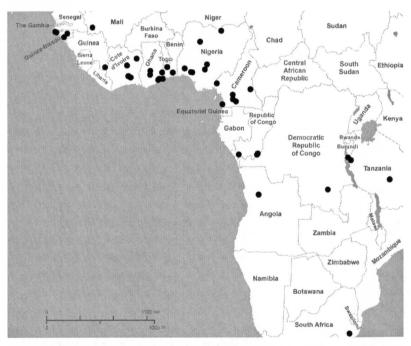

Figure 4.2 Collection points of sixty-eight *Cryptolepis sanguinolenta* samples in herbaria accessed through the online database of the Global Biodiversity Information Facility: Missouri Botanical Garden; Museum National d'Histoire Naturelle et Réseau des Herbiers de France; Herbarium Senckenberg; Royal Botanic Gardens, Kew; National Herbarium of the Netherlands; Missouri Botanical Garden, Real Jardin Botanico Vascular Plant Herbarium, Madrid; Ghana Herbarium. (Source: Global Biodiversity Information Facility Data Portal.)

of colonial occupation. As in Chapter 3, the account traces how scientists created records, experiments, explanations, products, and harvests in the process of transforming herbal medicines into marketable pharmaceuticals. However, this case emphasizes the role of scientists in postcolonial Ghana. Reclaiming the popular West African remedy they called "Ghana Quinine," a generation of scientists sought to replicate the success of the South American cinchona bark that contained the malaria treatment quinine.

The chapter begins in the last years of the colonial era, when Ampofo began to write down herbal recipes that healers used in the Akuapem mountains of Southeastern Ghana. It then examines how Ampofo's research merged with a nationalist project to document Ghanaian plant medicine after independence from Britain in 1957. The last two sections of the chapter show how Tackie and Ampofo depended on external support for their research into *Cryptolepis*, and how this led to patents being filed abroad. Along the way, the story traces the bitterness with which healers and scientists in Ghana remembered the partial transformation of *Cryptolepis* into a patented pharmaceutical, a dusty tea, and a stiff tonic during a turbulent time in the nation's history. Although many participated in the remaking of the plant into new medications, not everyone prospered from *Cryptolepis* (see Figure 4.2).[2]

RECORDING TRADITIONAL THERAPIES IN AKUAPEM

The long, winding road that led from the Ghanaian coast to the Akuapem highlands was once a footpath. A century ago, people walked for several days from their farms along the fertile mountain range to the ports, carrying merchandise on their heads. From the 1840s onward, the cooler highlands were popular with Swiss missionaries, who inspired their early converts to construct stone churches along the ridge. Then, British colonial officials and their wives spent the muggy rainy seasons at the Aburi hill station away from the fever-inducing miasmas of the sea marshes, traveling to the region in special hammocks balanced on the heads of four men. As a college student, I frequently traveled to the Akuapem town of Akropong to see family and friends. The drivers of the vans I rode knew all the twists and turns by heart, rounding bends at full tilt. Although the road has since been leveled and widened, at the time it was passable only until the former vacation home of the first President Kwame Nkrumah at Peduase. From there, the van drivers dodged pedestrians in the small mountain towns lining the narrow highway-cum-main street. Accidents were especially a problem on Fridays and

Saturdays, when people from the city returned to their hometowns to participate in family funerals. The vans whizzed by party after party of men and women adorned in fashionable combinations of white, brown, black, and red cloth that referred to the age of the deceased and preference of the family. Sometimes, I passed a cousin along the way, or a relative who drove vans between Accra and the Akuapem hills saw me as he honked his horn at our driver on his way down. In an era before mobile phones, my cousins in Akropong often knew I was en route before I reached the Kwabesco Hotel junction and walked the short distance to their maternity clinic.

These journeys to the highlands are how I first learned of Ampofo, an advocate of plant medicine and the founder of the Tetteh Quarshie Memorial hospital in his hometown of Mampong. My aunt and grandmother had both perished while delivering children in the Akuapem mountains, and I was keen to understand the role of private and public health care workers in alleviating maternal mortality. Because several of my cousins ran a maternity clinic, I found myself interviewing them and their colleagues in nine towns and villages along the ridge. I was particularly interested in the stories they narrated about nutritious herbal soups recommended for pregnant women, as well as the plants they used to stop bleeding or speed up delivery. Noting my interest in herbal medicines, the birth specialists with whom I spoke arranged for me to go visit Ampofo.

Ampofo, one of the first Ghanaians to receive medical training abroad, also worked to document plant recipes in the Akuapem region in the late colonial period. Ampofo's rediscovery of African healing emerged from his experience with colonial education. Ampofo recalled watching court proceedings at the feet of his father, Chief Kwesi Ampofo, in the southeastern town of Amanase before beginning primary school in 1916 at a Basel Mission school when he was seven. Unusual for the time, his mother had attended several years of primary school through the Basel Mission, whereas Ampofo's father had received limited formal schooling but learned to read and write in both English and the Ghanaian language Twi through the help of childhood friends. Boarding schools in Anum, Mfantsipim, and Achimota relocated Ampofo to institutionalized spaces geographically and linguistically distinct from his home community. He crammed for Cambridge exams in English, Latin, French, history of the British Empire, geography, art, mathematics, and hygiene—intellectually entering a literate space to which his parents never had full access. By 1931, Ampofo won one of the first government scholarships available to pursue medical education in Scotland at the

University of Edinburgh. When Ampofo struggled with the professional examinations in pathology and materia medica, the government decided to pull Ampofo's scholarship. He relied on financial assistance from his brothers while he prepared for the exams of Edinburgh's Royal Colleges of Surgeons and Physicians, from which he was to qualify in 1939.[3]

Although Ampofo maintained his Twi language skills and family ties, his circuitous path to a medical career abbreviated his experience with local healing methods and animist traditions. Ampofo's British education was critical to his rediscovery of Ghanaian art and healing.[4] As with many colonial scholars in Europe and America, university training abroad placed Ampofo in conversation with other African and Asian young adults grappling with the ironies of colonial rule. Ghanaian students, as was the case with their non-European compatriots, confronted acute racism and misconceptions about life in their home communities. Postcolonial memoirs cite difficulties renting apartments, animosity from fellow classmates, and racialized notions of their inferiority embedded in textbooks. These experiences solidified the desire to improve the image of non-Europeans on the world stage as they emerged as an integral component of the independence-era leadership.[5]

Rejected for a government medical post on his return to the Gold Coast in 1940, Ampofo found himself rediscovering the local plant remedies of his youth in the waning years of World War II. In his unpublished autobiography, Ampofo described his attempts to find employment with the medical service:

> The war had affected every business and even farmers were burning their cocoa. And in visiting the Director of Medical Services he laughed when I told him that I wanted to join the medical service. He said rather brusquely that there was no vacancy for African doctors yet. This upset me because I travelled with two British classmates on the same boat and they were admitted straight into the service, allocated hospitals, given cars and bungalows whereas another African classmate of mine was waiting for his appointment. In fact I took this statement of the Director as a challenge because I had originally intended to work among my own people in Akuapem. My only problem was how to find money.[6]

Ampofo's experience reflected a systematic program to limit the number of African doctors in the British colonial medical service. By the early 1950s, one Ghanaian doctor was thoroughly frustrated with colonial policies that favored white doctors in the colonial service:

Our numbers have been seriously restricted by the failure of past Governments to give more than an average of one medical scholarship a year. Even when we qualified, we were not sure of immediate employment as a result of a ratio policy whereby African medical officers were employed in a strict, and very small, proportion to the expatriate officers.[7]

Through a series of loans and support from his older siblings, Ampofo managed to establish a clinic in his hometown. He first used drug samples he had brought from England; when these had run out, he took public transport to the capital city several times weekly to restock. He described these early days as overwhelming, with people coming from across the Akuapem region to see if this new doctor could cure what the retired physician could not.[8] In contrast to European doctors in the colonial service keen to tackle specialized research in the handful of hospitals in urban centers, African doctors like Ampofo focused on providing care in rural villages.[9] With more cases than he could handle, the young doctor conscientiously tried to cure all he could, often taking gravely ill patients to the colonial hospital on the coast himself.

Malaria was the most prevalent concern among Ampofo's patients.[10] The disease is caused by parasites that enter the blood through mosquito bites. As the parasites mature, the body responds with cycles of high fevers and chills, which can be deadly if the parasites overrun the spleen or invade the brain. For centuries, people offered many causes for malaria fevers, confusing then with other feverish conditions like yellow and blackwater fever. The term "malaria," Italian for "bad air," comes from European interpretations that putrefying matter created toxic miasmas that created the fevers. By 1900, researchers isolated mosquitoes as the carriers of malaria parasites. The decision to focus on mosquitoes stemmed from previous studies on parasites borne through flies in China, as well as from indigenous theories of mosquitoes as a cause of malaria that European missionaries and explorers had collected in Africa and Asia over the years.[11] People who frequently come into contact with malaria parasites develop short-term immunities. This led colonialists in Africa to believe that their subjects were immune to malaria even though they harbored the parasites. We now know that, in some places, people also inherited partial immunity through genetic mutations that affected the shape of blood cells. Confident that malaria only killed whites, colonial administrators implemented systematic segregation in places like the Gold Coast and Sierra Leone to maintain "sanitary corridors" between white housing on higher ground and black housing in swampy lowlands. This was thought, erroneously, to prevent infected mosquitoes from flying into European homes.[12] It was not until the 1950s that European scientists began to acknowledge the huge problem that malaria posed for Africans, especially young children who had yet to develop any immunities.[13]

Ampofo, however, was keenly aware that malaria threatened the health of Akuapem communities during the colonial period. How did he treat malaria? Within the Gold Coast as a whole, people with whom I have spoken recalled how as children their parents would make them take bitter grains of quinine on Sundays as a prophylactic during the 1940s and 1950s. Quinine was derived from the bark of cinchona, a plant from South America that Portuguese and Spanish missionaries had appropriated as a treatment for fevers since the sixteenth century. As quinine came to be used around the world to prevent the growth cycles of malaria parasites, some stronger parasites survived and became resistant to quinine. In the 1940s and 1950s, Ampofo would have combated malaria cases with quinine as well as with tablets and injections of mepacrine, proguanil, and chloroquine. The colonial government also sponsored expeditions to drain swamps, cover household water tanks, and spray heavy doses of toxic chemicals like DDT to control the growth of mosquito larvae on open water. However, despite such efforts, reported cases of malaria in the Gold Coast surged from around 30,000 in 1930 to 130,000 by 1950. This reported increase could have been due in part to increased use of government hospitals and clinics where statistics were maintained, as well as increased resistance of malaria to drugs. Today, a less virulent malaria parasite, *Plasmodium vivax*, is immune to quinine in West Africa, and the more deadly malaria parasite, *Plasmodium falciparum*, has come to evade successive waves of drugs (see Figure 4.3).[14]

Over the years, Ampofo also found that people used herbal remedies to combat the fevers, nausea, headaches, and body aches that accompany malaria. Although colonial governments at the time did not acknowledge African malarial treatments, it was a rational decision to try plant remedies when quinine or other imported drugs failed to combat fevers. It is unclear when Ampofo learned of *Cryptolepis* as a useful traditional remedy against malaria. The specific species, *C. sanginolenta*, is a relatively scarce climbing shrub found in West African forested areas, including those in Ghana, Senegal, and Nigeria. The roots have been used for multiple purposes over the years, including treatments for chest ailments, fevers, dye for goat leather, and an aphrodisiac.[15] In an article published in 1983, Ampofo credited the healer Kwadwo Manu, who lived

Figure 4.3 Former site of Ampofo's *Obi Kyere* Herbal Center, precursor to the Center for Scientific Research into Plant Medicine. (Photo by author.)

in the nearby town of Apirade, with providing him with information on *Cryptolepis*.[16] Ampofo most likely documented this treatment sometime in the 1960s or 1970s, when government support allowed him to work part-time on herbal research and patients returned to traditional therapies to treat drug-resistant malaria. Ampofo explained that in his practice more than 80 percent of his patients also visited traditional healers to treat their illnesses. He began to spend time discussing local plant remedies with some of his competitors. He recalled working closely with Mr. Kwasi Adae, Mr. Kofi Addo, Mr. Kwame Awuku, Mr. Adu Darko, Mr. Kwaku Poku, Alhaji Sidi Biney, Mr. Nyampon, Mrs. Kyeiwa, and Mrs. Adae.[17]

In describing how his interest in local herbs was piqued, Ampofo explained in interviews with me in 1997 that while he was conducting his work as a physician, he was constantly negotiating with his patients about his treatments versus those of the herbalists. Laughing, he recalled that most of his patients, having already visited an herbalist, would tell him frankly the limitations they saw with his medicine. He said they would explain, " 'Well, I got ill and I've been treated by a herbalist but I've come to you just for injection.' "[18]

It soon became clear to Ampofo that he would have to assess the usefulness of the local plants that his government education and medical training had taught him to ignore. During the colonial period, there was little precedent for people living in the Gold Coast to write down recipes for the medicinal plants they used to treat diseases. One exception was the Fante herbalist J. A. Kwesi Aaba, from the city of Sekondi, who in 1934 was the first Ghanaian known to publish his herbal recipes. Aaba provided only the Fante names of herbs in his treatments, including *Adisikankyi* and *Kunkruma* plants, which could be used to treat *Konoruku* (malaria). Aaba related *Konoruku* to overconsumption of mangoes, miasmas, and mosquitoes:

> One of the symptoms of Konoruku is cough; and people, taking this to be common cough, are deceived by the cunning Konoruku until it poisons the whole blood and claims its victim. Whenever malarial signs are detected, take this Simple Formula:
>
> Ankama [Lemon] leaves. . . . . 1 part
> Kunkruma roots and leaves 1   "
> Adisikankyi roots or barks 1   "
> Dried pepper . . . . a sufficiency
>
> Boil with water for drinking and for injection by the rectum. The decoction must be used warm.[19]

It is unclear whether either of the two plants in Aaba's recipe may have been *Cryptolepis*. Given that Aaba lived on the more arid coast, it is less likely than if he had lived at the edge of the rainforest in Akwuapem.

Ampofo developed a system for documenting the herbal remedies of healers on the Akuapem ridge. Year by year, Ampofo amassed a personal archive of information gathered from these healers, which would later become a source of conflict between Ghanaian scientists at CSRPM and foreign visitors. After Ampofo's death, I sought out his old friends for their reminisces, including Jimi Moxson, the last colonial mayor of Accra, who had retired in Aburi after independence and presided as a local king. Weeks before his own death, Moxson spoke fondly of the late doctor and his wife. Moxson had married them while serving as District Commissioner in the Eastern Region. He requested that an assistant provide me with pages from his autobiography in progress, where he described how Ampofo honed his interest in plant medicine. Moxson explained how Ampofo documented information on local plants, which he "painstakingly recorded in his card index system together with the vernacular names (in three or four different dialects), the English names, and the botanical names." Moxson noted how Ampofo gained the trust of healers in the Akuapem area: "By patience and gaining their confidence and by a step-by-step approach he progressively succeeded in recording their treatment and cures and in classifying the herbs, roots, barks, leaves and flowers from which they cooked up their infusions."[20] He noted that Ampofo did not worry himself with spiritually based medicines, focusing only on herbs. In his own words, Ampofo explained to me that he had "no time to dance" when it came to the spiritual and fetish rituals integrated into plant medicine.[21] Rather, Ampofo concerned himself with the positivist project of classifying and documenting herbal remedies.

It was in this context of colonial disinterest in his medical expertise, limited drugs, and abundant competition from herbalists that Ampofo began an early project of "biological prospecting." He was disillusioned with the promises of the colonial regime and saw a real need for cheaper, more readily available medicine. Through his patients, Ampofo then found out the names of local healers and made it a habit of visiting them and learning about their work. The independent government then co-opted this research to create a national database of plants with the potential to create drugs.

TAKING PLANTS FOR THE NATION

After independence in 1957, Ghanaians positioned themselves to reclaim plant remedies, and exchanges between physicians, scientists, and healers intensified. By the late 1950s, Ampofo's investigations had gained the attention of independence leader Kwame Nkrumah and Ghana's emerging scientific community. Ampofo was asked to share his notes with pharmacists and chemists at the Kumasi College of Technology, and the University of Ghana engaged in the alkaloidal screening of local herbs. These researchers were known as the "Alkaloid Group." As plants passed from healers to botanists, chemists and physicians, the lines of credit became blurred.

Ghanaian Scientists Claim Alkaloids

The decision to focus on alkaloids was motivated in part by the accessibility of testing alkaloid-rich plants in Ghanaian laboratories. Alkaloids, the nitrogen-rich by-products of respiration in plants, had only recently begun to be systematically identified and studied. With around

98 percent of globally distributed plant species still to be analyzed for alkaloids by the late 1950s, workers in Ghana felt that they could more easily contribute to this aspect of drug development than they could to expensive synthetic approaches. The possibility of honing in on specific plants through the advice of local herbalists further amplified interest in this area. Located in the tropical rain-forest zone thought to be rich in plant alkaloids, Ghana, it seemed, was ideally positioned for an alkaloidal screening exercise. The clear crystalline structure of many alkaloids provided further appeal to the fledgling crystallography lab at Legon.

The original investigations of the Alkaloid Group in the early 1960s provided avenues for scientific collaboration within and outside Ghana. Although the Alkaloid Group gestured toward commercializing any findings, the primary focus was the expansion of scientific knowledge. Collaborating researchers worked to screen "a large number of both medicinal and poisonous herbs" and differentiate between "those containing alkaloidal active principles, and those containing non-alkaloidal active principles."[22]

Besides contributing to an increase in chemical data, the Alkaloid Group inspired national pride. After many years, black Africans were at the helm of political and intellectual leadership in the fledgling nation, and Ghana's participation in global scientific activity was in itself a cause for rejoicing. First President Nkrumah, at the laying of the foundation stone of the ill-fated atomic reactor in 1964, captured this sentiment as it related to his socialist agenda: "We must *ourselves* take part in the pursuit of scientific and technological research as a means of providing the basis for our socialist society. Socialism without science is void."[23]

The Alkaloid Group emerged from conversations among the handful of individuals with university training in science at Ghana's independence in 1957. Oppong-Boachie, former director of CSRPM, credited four Ghanaian scientists with heading studies initiated through the Alkaloid Group, which was a project of the National Research Council (NRC). These included Ampofo, pharmacist Albert Nii Tackie, and chemists J. A. K. Quartey and F. G. T. O'B Torto.[24] Tackie, the first Ghanaian to earn a doctorate in pharmacy, took a leading role in the research. Like Ampofo a decade before, Tackie was an Achimota graduate who had studied abroad. After completing bachelor's studies in pharmacy at London University, Tackie returned to the Gold Coast to serve as a principal pharmacist at the Department of Health.[25] By 1957 he joined his classmate Emmanuel Gyang and his former teacher Eric Allman as an instructor at the Pharmacy School of Kumasi College.[26]

The Alkaloid Group evolved from efforts of early pharmacists like Tackie and Gyang to secure control over drug distribution in the young nation. At the Korle Bu Dispensing School and later at the Kumasi College Pharmacy School, many pharmacy students became interested in studying the natural remedies that fell outside their training program but to which they had grown accustomed as children. During the 1950s, the Pharmacy School purchased samples of local remedies from markets in Kumasi—for instance, the purgative "Cawu," earth powders substituted for talcum, and barks important to divination rites. In collaboration with such institutions as the London University School of Pharmacy, initial pharmacological tests were conducted and efforts made to identify specific species and "active constituents."[27]

Trained to dispense imported patent medicines, Ghanaian pharmacists were keenly aware of other suppliers of drugs with whom they competed for business. On the one hand there were the "Licensed Sellers of Poisons." In contrast to the at most 300 registered pharmacists at the dawn of independence, over 3,000 individuals had obtained a license to sell poisons—for which there was no special training and only a £1 annual fee. Officially licensed to distribute chemicals not meant for human consumption, such as battery acid or insecticides, licensed sellers of poisons commonly sold patent medicines at "drug stores" masquerading as pharmacies. Alongside licensed sellers of poisons, herbalists and herb peddlers offered medicinal plants to ailing patients. Together, these two groups served to undermine the pharmacist's trade in imported drugs. Often cheaper and more familiar to potential buyers, indigenous medicinal plants were a clear challenge to the emerging pharmacy profession in Ghana.[28]

To coordinate these ad hoc studies within its broader development strategy, the executive board of the NRC established a special subcommittee on Native Herbs and Medicines in early 1959.[29] This subcommittee of the medical research initiative developed a collaborative system to test local Ghanaian plants traditionally used as therapeutics. Resources for laboratory examination of plants in the country were limited at this time to Kumasi College and the University College of the Gold Coast (now the University of Ghana at Legon) where chemists Torto and Quartey had begun attempts to elucidate the chemical composition of herbs. Using Ampofo's growing database and submissions from interested herbalists and the general public, a Local Herbs Unit at Kumasi College—in coordination with the NRC—orchestrated systematic testing of medicinal plants.[30] In 1960, both Tackie and the British pharmacist Allman headed to the United Kingdom to pursue doctoral studies in pharmacy and take advantage of facilities at the Chelsea College of Science and Technology to run tests impossible in Ghana. However, Allman quickly returned to lecture at Kumasi College, believing that he was too advanced in age to take up the course, leaving Tackie to emerge as the torchbearer for Ghanaian phytochemical studies.[31]

During 1960, the Local Herbs Unit at Kumasi was processing an average of six plants or eighteen plant parts every week, according to correspondence between Allman and A. H. Beckett, Tackie's supervisor at the Chelsea College of Science and Technology at the University of London. From available records, it does not appear that any of these were *Cryptolepis,* although many of the barks and roots tested could be used to combat fevers. By January 1964, the program had processed 314 separate species, focusing on those containing alkaloids for further tests. Identification of plants to process was closely tied to earlier colonial studies of Gold Coast flora and operations at the Forestry Department.[32] The majority of plants were brought into the laboratory from surrounding forests in the Ashanti region, often by Albert Enti, then keeper of the herbarium of the Kumasi office of the Forestry Department. Enti was a man steeped in folklore surrounding plants of tropical West Africa. Born in the Gold Coast when it was a British colony, he had been privy to a missionary education followed by training in botany through the Forestry School. After the Gold Coast's independence in 1957, Enti was one of the few Ghanaians poised to assume leadership positions at the University of Ghana, in his case as keeper of the university herbarium.[33]

Until Tackie's return in 1963, Allman, along with Tackie's classmate Gyang, oversaw preliminary alkaloidal screening of the plant samples. The Local Herbs Unit classified plants according to the weight of alkaloids they could hope to extract on a scale of a+ (at least 1 gram alkaloid for 10 kilograms of dried plant) to a++++ (more than 1 gram of alkaloid for 1 kilogram of dried plant). Although further examination of the reduced alkaloid was difficult to conduct in Ghana, the Kumasi department of chemistry was equipped to boil powdered plant material with alcohol to create a potentially alkaloidal residue. By dissolving the residue with ammonia and ether, a hydrochloric acid layer could be successively distilled and visually examined for alkaloids with Mayer's reagent (potassiomercuric iodide solution). The alkaloidal content procured was then contrasted with standard measurements of the alkaloids brucine or atropine methonitrate, poisonous substances originally derived from nux vomica seeds and belladonna plants, respectively.[34]

Tackie, as the first Ghanaian to receive a PhD in pharmacy, took a leading role in this research and was soon appointed professor and dean of the pharmacy school. However, the Ghanaian media emphasized the dependency of the Alkaloid Group on tests in Europe, failing to name Tackie in particular. In August 1963, the *Ghanaian Times* reported on *Mitragyna*, a key plant in the alkaloidal research, stating that "a team of British pharmacists" would soon be announcing their findings on this plant that was well known as a treatment for a variety of ailments, including malaria.[35] This publication created much unhappiness among the Ghanaian researchers, as Kumasi College Vice-Chancellor Robert Patrick Baffour explained in a memo to Allman:

> An article appeared in the local Dailies which paid tribute to the achievement of Chelsea and a group of British scientists in the study of a Ghanaian plant—the plant on which Dr. A. N. Tackie worked for his Doctorate Degree under Chelsea supervision. . . . The reaction that this unfortunate publication has created in the minds of academicians in Ghana is very unfavourable one and if indeed it emanates from Chelsea then very serious repercussions may likely follow.[36]

Allman, at that time still serving as head of the Pharmacy Department at Kumasi, was quick to smooth things with the Ghanaian "academicians"—noting the "mutually beneficial" arrangements between the two schools—that is, "they have trained Dr. Tackie in the very latest plant chemistry techniques," and "in return, we have provided tropical teaching experience for many of their staff."[37] In fact, Tackie's supervisors at the University of London had been careful to acknowledge cooperation between Kumasi and their institution in all publications.[38] A letter from Tackie's PhD supervisor Arnold Beckett, protesting the initial article published in the *Ghanaian Times*, explained that "some of the technical staff at the University in Kumasi have contributed significantly to this work in the collection of supplies and in extraction procedures."[39] The following June, a bold headline in the same paper proclaimed, "Ghana makes a big medical discovery," and included photographs of the newly returned Tackie and other members of staff at Kumasi "screening herbs in laboratory"[40] (see Figure 4.4).

Thus, Ghanaian scientists sought national and international recognition for the part they played in research on medicinal plants for new alkaloids. It was no small feat to make sure that colleagues abroad and newspapers at home recognized the efforts of individuals like Tackie. In a postcolonial context in which even Ghanaian journalists assumed Europeans held priority, members of the Alkaloid Group struggled to receive credit for their investigations.

Ghana makes a big medical discovery

GHANA recently made a very valuable contribution to medical science with the discovery of appreciable quantities of alkaloid in 40 Ghanaian herbs.

The discovery was made by the Local Herbs and Medical Service Unit of the Kwame Nkrumah University of Science and Technology, Kumasi.

This was the outcome of the screening of 300 local plants to discover their medical contents.

And it all started like this! Professor A. N. Tackie, Ghanaian head of the unit, had discovered more alkaloid from Ghanaian plants while preparing for his doctorate degree.

These alkaloid, it is believed, may be more effective and efficient for anaesthetics.

The search for more alkaloid started later at the Kumasi University.

University sources say that more plants are being collected both in Ghana and other Afro-Asian countries for examinations.

Pictured below are some of the scenes at the unit, which led to the discovery.

Screening herbs in laboratory

● Above: Professor Tackie (second from right), dean of the Faculty of Pharmacy and head of the Local Herbs and Medical Service Unit of the University, conducts a research in the laboratory.

His team of research workers include Dr Abraham Thomas (right), Dr T. Wood (third from right) and Dr D. Phollades.

Figure 4.4 Tackie's research in the *Ghanaian Times* (June 16, 1964). (Source: Ghana, PRAAD.)

Healers Claim Discoveries Too

For their part, healers involved in the research scheme also sought better recognition. The scientists encouraged individuals knowledgeable about plants to donate concoctions for study, at times promising to patent any cures in the participant's name. There are limited letters available from these exchanges, with none of the healers purporting to cure malaria. Lists of plants examined during the alkaloid screening exercises do not include *Cryptolepis*. One of the healers to contribute a sample for testing may have been Aaba, the herbalist who published his treatment for malaria in the 1930s. In this case, Aaba sent members of the local herbs committee two undisclosed combinations of herbs believed to stop heart palpitations in June 1961.[41] Tests conducted at Kumasi determined some effect of the mixture on frogs. Further samples were forwarded to Ampofo, perhaps linking two of Ghana's leading plant medicine advocates.

The same week that the Alkaloid Group received Aaba's treatments, Fosu, head teacher at the Catholic Middle School in Dwinyama in the Asante region, submitted a "preparation" that he promised would cure "infection." His status as a head teacher at a Catholic school meant that not all postcolonial healers were nonliterate or opposed to Christianity. Fosu noted that pregnant women should never use his medicine. In his own practice, two women to whom he had administered the drug "did not confess" that they were several months pregnant; during the first week of the use of the drug, the fetuses were aborted. Fosu submitted his preparation in two forms: "Apart from the roots I have enclosed in the parcel a small quantity of the powered form of the same *species*." Elsewhere he explained that the roots were accompanied by "a small quantity, about three table spoonfuls, which has been *winnowed (turned into powder)*." Fosu's use of the term "species" and his explanation of how the original roots were pulverized spoke of his desire to garner further legitimacy in the eyes of the scientists at the same time that he did not disclose details of his submitted plant preparation.[42]

In responding to Fosu, the executive secretary of the NRC explained that the government had in place "a research scheme for investigating such claims." He advised Fosu to forward one-pound samples of his "herbal medicine" to Allman, Quartey, and Ampofo for them to conduct "pharmacological extraction . . . chemical . . . and perhaps clinical" tests. Fosu was also informed that he had every right to withhold information on the source of his cures. The NRC secretary baited Fosu further by suggesting that in the event of "successful" tests, "the Government would continue to respect your claim and the cure might be patented in your name."[43]

In the postcolonial moment, freed from the threat of persecution as witch doctors, healers dared to seek credit for their contributions.[44] Media attention to the African scientists' work brought a deluge of plants to Kumasi for testing. However, the process of harvesting, drying, boiling off, and analyzing the many unidentified plant samples proved to be time consuming, and not without difficulty. Although the plants collected through the Forestry Service had been clearly labeled with Latin names, members of the public had their own methods for describing their plant expertise. Some, like Fosu, chose to withhold the plants' names altogether. These first collaborations between African scientists and healers were characterized by limited levels of transparency. Regarding plant identification, the NRC suggested that contributors might keep recipes for plant therapies secret in order to curry favor and augment their research base. For healers, participating in the national screening effort provided further avenues for substantiating their claims and securing clients.

Frequently, however, healers were sidestepped in the process of ascribing credit. The anonymity of healers in the process of research led to growing fault lines both among members of the newly formed Ghana Psychic and Traditional Healers Association (GPTHA) and between healer-scientist collaboration. In 1965, for example, the *Daily Graphic* proclaimed, "Healers find new drug: The Ghana Psychic and Traditional Healers Association has discovered a new herbal cure for high blood pressure and sun stroke." Yet, the credit was given to J. A. Nartey and N. A. B. Kofie, research scientists affiliated with a GPTHA hospital and clinic. Mensah-Dapaa, an academic and organizing secretary for GPTHA who attended an exhibition where "the herbal mixture 'Ahunanyankwa'" and other cures were on display, "urged the members to improve upon their methods of healing and appealed to them to refrain from charging exorbitant fees."[45] Despite these calls for professional and continental harmony and cooperation, the strains in collaboration were already at work. Indeed, the credit taken by Nartey for this and other GPTHA "discoveries" led to further schisms in the organization, a falling out with Ampofo, and further wariness of collaboration on the part of healers.

FINDING ALKALOIDS IN *CRYPTOLEPIS*

Ambivalence about the real intentions of scientists shaped later partnerships at the institute Ampofo and Tackie founded to study plant medicine at Mampong, Akuapem. In the years after independence, financial support for academic research waned, and Ghanaian scientists had to find ways to survive a grim economic climate. Interested biologists and chemists held out research to "develop" and better commercialize traditional medical remedies as one path to replenish research budgets and raise funds for family obligations. The plant screenings of the Alkaloid Group, along with other scientific projects, were stalled by the mid-1960s. Both Legon and Kumasi continued experimenting on medicinal plants, but without the intensity and success of the immediate independence years. As early as 1964, the expectations for alkaloidal screening through the Local Herbs Unit at Kumasi and the chemistry department at Legon far surpassed resources available through the Ghana Academy of Sciences. Despite growing media attention to the work of Tackie, the pharmacy department at Kumasi actually had to turn down supplies of plants from healers and botanists interested in supporting the initiative, "until the situation improves."[46] A decline in available funds at the universities was further compounded by political instability after Nkrumah was ousted in the nation's first coup d'état in 1966.[47]

Funding Plant Medicine Research

After several years of collaboration, the Alkaloid Group was reconfigured with a more defined economic thrust. In March 1964, the head of the chemistry department at Legon, F. G. Torto, floated a proposal to the Academy of Sciences and his collaborators at Kumasi to increase the number of research assistants available for alkaloidal screening and crystallography analysis. Noting the time-consuming, expensive nature of their investigations, Torto suggested that collaboration be initiated with "with one or more of the commercial concerns which have made approaches on this matter" to conduct further tests that "cannot be undertaken in this country." Torto argued that "such an agreement with a suitable [foreign] firm . . . would help to promote their commercial interests as well as the development of scientific activity in this country."[48]

Still frustrated with the pace of research, Ampofo and Tackie joined forces to establish CSRPM. Significantly, in their original proposal, Tackie and Ampofo enticed the government with the promise of a state-run herbal drug industry to be developed through CSRPM's investigations: "The purpose of the centre envisaged is not just for academic exercise as is done at present; but to . . . push the process further to the manufacturing stage locally. . . . The potential as a foreign exchange earner can be enormous."[49]

In contrast to Torto's initial suggestions, the proposal for the establishment of CSRPM did not explicitly mention cooperation with foreign firms, thus making a case for industrial prospects within Ghana. In their January 1972 proposal to the government, Tackie and Ampofo explained that "we are confident we can marshal every effort to provide a working unit to begin the project in a more rational way than has hitherto been undertaken by individuals and universities." The CSRPM proposal did not make explicit how any commercial results would be shared within wider Ghanaian society. Nor did it explicitly delineate the kinds of agreements that might be made with foreign scientists and firms. In making a case for large government financial support, Tackie and Ampofo drew the economic gains for plant research along broad, nationalized lines: "It will involve a great deal of money to embark on this project [in this country] even on a modest scale. . . . But it will nevertheless be a worthwhile investment." Ampofo and Tackie suggested three areas for immediate results, including isolation of "pure active principles" in popular remedies, synthesis of steroids from fruits of *Solanum torvum*, and purification of shea butter. The latter two areas were to be the focus of initial efforts "to divert the proceeds accruing here to meet some of the expenses in the third project [isolation of active principles]." Thus, with a suitable investment by government, they believed they might "capitalize immediately upon discoveries of promising pharmacological agents."[50]

When the National Redemption Council during the military regime of Colonel Acheampong approved the CSRPM proposal in June 1973, the authorized institutional structure pointed to this commercial and industrial agenda. A budget of 20 million cedis was approved for the first year of operation and CSRPM was placed within the financial care of the Ministry of Health, which supervised the disbursement of funds. Under government decree, the temporary advisory committee consisted of ten individuals representing all herbal medicine stakeholders. Most notably, the advisory council did not include a lawyer, as the legal implications of CSRPM's research were not a primary area of concern at this juncture. Along with Ampofo, who was designated CSRPM director, and Tackie, who was then chair of the Council for Scientific and Industrial Research, the advisory committee included "an educated herbalist," another advisor to the GPTHA, the chairman of the National Council for Higher Education, the Director of Medical Services, the Dean of Faculty of Pharmacy at Kumasi, and Blukoo-Allotey of the state pharmaceutical company of GIHOC. The CSRPM structure was novel in its attempt to create a collaborative environment between academic and government elites and representatives of Ghanaian healers, an approach that was to be mirrored in later government-sponsored initiatives on traditional medicine.

At the "outdooring" for CSRPM in November 1973, the Commissioner for Health, A. H. Selormey, explained that the new institute would project an air of openness to the world: "The Centre will pursue an open-door policy, welcoming contributing scientists and doctors as well as students of plant medicine."[51] Such altruistic language reflected the spirit of botanical research and exchange that had long been characteristic of herbariums and plant research institutes around the world. The university herbariums at Legon, Cape Coast, and Kumasi commonly fielded requests for specimens and identification from sister research institutes abroad, and Ghana's participation in these scientific exchanges was a point of pride. Yet, what would be the implications of a body like CSRPM, with the dual mandate to research and commercialize plants, openly sharing plant specimens and observations?

Within two weeks after CSRPM's well-publicized inauguration, private individuals knowledgeable about medicinal plants expressed interest in collaborating. However, they were eager to know precisely "what benefits" they would receive from sharing any information with the government-sponsored researchers. J. K. Apeagyei, an employee at the State Gold Mining Corporation, wrote, "Sir, I know some native herbs which can cure diseases. . . . I am prepared to send you samples of the herbs. *However, I would like to know what benefits I would derive if the research proves successful.*[52]

Obtaining Equipment to Study Plants

The traditional antimalarial *Cryptolepis* was one of many plants that CSRPM hoped to convert to a viable drug. Under Ampofo's leadership, *Cryptolepis* was transformed from a traditional plant with multiple uses to an internationally recognized treatment for the specific concern of malaria. However, the process by which this occurred did not necessarily go according to plan, and it led to a split in Ampofo and Tackie's strategies for survival in a disintegrating financial climate by the late 1990s. Tackie was forced to step down as director of CSRPM after he chose to market his own product from *Cryptolepis*. Ampofo opted to work closely with Winn and other U.S. researchers to make sure that *Cryptolepis* products would be produced in Ghana after his death.

Ghanaian researchers focused international attention on *Cryptolepis*. By 1974, *Cryptolepis* was included in the repertoire of plants Ampofo and collaborating healers and physicians at CSRPM used to combat fevers. CSRPM initially provided patients with aqueous decoctions of the plant as a natural alternative to synthetic malaria drugs. In 1978 Ampofo, together with physician Gilbert Boye, developed a two-year clinical study of *Cryptolepis* based on approximately 100 patients who were given either the synthetic drug cholorquine or an infusion of *Cryptolepis* made with dried plant roots and boiling water. They presented their findings at a first international conference on *Cryptolepis*, held in 1983 at the University of Science and Technology in Kumasi and sponsored by Ghanaian and Nigerian researchers.[53]

The Boye-Ampofo *Cryptolepis* paper was one of the few clinical studies to emerge from CSRPM, and it was published only in Ghana in a limited run of conference proceedings. The authors claimed dramatic success in reducing the parasitemia level in patients, eliminating the parasites entirely within six days.[54] However, the degree to which CSRPM conducted systematic clinical studies over the years has been open to interpretation. CSRPM affiliates have complained that new numbers were assigned to returning patients, undermining efforts to trace long-term side effects. As one chemist put it to me:

> Mamong [CSRPM] went in a different direction. It's now a center of Scientific Herbalism. Scientific Herbalism. You know, it's not particularly interested in determining active ingredients . . . not so much as the preparation of herbal potions of known efficacy. Yes, this is what they're doing [which is in a sense because] they need money.[55]

CSRPM, as well as other research bodies in Ghana, found it difficult to sustain high-level scientific activity during the 1970s and 1980s as political and economic instability continually disrupted funding streams. Identifying the chemistry behind the antimalarial effects of *Cryptolepis* proved even more challenging than clinical work, given the cost of laboratory equipment and dearth of funds. In 1972, Ignatius Kutu Acheampong's government was the fourth regime to take the reins of power in independent Ghana, the second to do so through a coup d'état. Despite a pledge to reverse the economic decline witnessed by previous governments, Acheampong's tenure as chairman of the National Redemption Council (1972–1975) and then the Supreme Military Council (1975–1978) were similarly marked by the "chronic problems of food shortages, aggravated debts, sectoral imbalances, and decline in production inherited from the Nkrumah years."[56] In this context of economic woes, the universities, let alone CSRPM, were in no position to afford laboratory instruments. By 1975, "it was decided that the Centre should confine itself to the gathering of data and documentation together with the clinical aspect of the work, while the analytical aspect of the work should be conducted at the higher institutions."[57]

Ironically, in their initial proposal for CSRPM, Ampofo and Tackie convinced the government of Colonel Acheampong that as part of an independent research center they would be better situated to investigate and commercialize medicinal plants than those operating within the university system. Reflecting on the founding of CSRPM, one of the university-based chemists who served as an advisor to the board explained:

> I was on their council, I was one of the first. . . . I told them that well, it would be very difficult for them to get properly established, because if it had been difficult for the universities to get proper funding for organic chemistry research, it would be even more difficult for them.[58]

Given the scarcity of its resources, CSRPM offered a two-pronged approach to enable laboratory work through university collaboration. First, along with the Faculty of Pharmacy at Kumasi, it requested 250,000 cedis to purchase three primary pieces of equipment: a mass spectrometer, a physiograph, and an automatic polarimeter in May 1975. The idea was that they "would be available for use by the whole country when received."[59] The mass spectrometer was seen as the most critical, and the researchers pointed to its nationalist value in researching *"our indigenous medicinal plants."*[60] Researchers at Legon also believed they ought to benefit, and later correspondence suggested that perhaps two mass spectrometers might be acquired. By November 1975, two spectrometers were made available in Ghana through the British Technical Assistance scheme and located at the Ghana Atomic Energy Commission. Ironically, they were "not suitable for the type of research envisaged by the [CSRPM] committee," and a further request was made to government for different versions of spectrometers.[61] A year after the original request, in May 1976, the government approved 460,000 cedis for the purchase of two mass spectrometers appropriate to biomedical research to be located at Legon and Kumasi.[62]

The second approach was to train postgraduate students specifically in medicinal plant research at affiliated chemistry laboratories at the University of Ghana at Legon, the University of Cape Coast, and the University of Ghana Medical School at Korle Bu. Reflecting on these early days at CSRPM, Tackie explained how they "thought it was going to be a wonderful place" but unfortunately, "the biggest drawback [was the] lack of requisite qualified people."[63] Nine selected individuals included N. Osei-Agyeman, C. B. Oseni, Twum Ampofo Ansah, and Paul Kofi, the latter a fine arts graduate who was to train in medical illustration in Mampong under S. K. Avumatsado. The postgraduate students were located at affiliated institutions through the Ghanaian National Service requirement, a scheme that in later years brought qualified scientists to serve at the CSRPM laboratories.[64]

What did this mean for *Cryptolepis* research? By 1978, the Ghanaian pharmacy professor Dwuma-Badu, who was working at Kumasi, had identified two alkaloids, quindoline and cryptolepine, from *Cryptolepis*. This led to enthusiasm for *Cryptolepis* studies in both Ghana and Nigeria, where parallel teams of researchers at the University of Nigeria, Lagos, also worked to understand the chemistry behind the plant. In 1982, the faculty of Pharmacy at Kumasi hosted a regional discussion of ongoing research on *Cryptolepis*. The chemical investigations on the alkaloidal content of the plant appeared alongside Boye and Ampofo's clinical studies.[65]

Yet, little was to come of this redoubled effort at scientific research. Further military coup d'états in the late 1970s and early 1980s, as well as financial distress and near-famine conditions in the 1980s, thwarted efforts to establish a stable research environment. CSRPM struggled to pay consultant herbalists and sought equipment and support from UNESCO and the World Health Organization to pay salaries and maintain the arboretum.[66] In the country as a whole, emigration reached an all-time high by 1984, particularly among physicians and university graduates. It was not until the late 1980s that further work on *Cryptolepis* was revitalized. Given the instability of the Ghanaian economy, university researchers fought primarily just to keep afloat, many opting to work in Nigeria and abroad. For instance, Tackie served as an external examiner at the University of Nsukka in Nigeria. The University of Cape Coast chemist Francis Tayman, after his wife nearly died from complications with gallstones in Ghana, weathered the early 1980s in Germany, where he made some money through textile import-export. On the other hand, some researchers found the 1980s to be their most productive period. A chemist who was later to head CSRPM, Archibald Sittie, was able to get a post at the Noguchi Memorial Institute for Medical Research during this period, where facilities for bioassays were available.[67]

During the difficult financial periods in the 1980s, Ampofo stayed in Ghana. He focused his energies on popularizing medicinal plants in nearby towns. Much as they had during World War II, people returned to healing plants in the face of drug scarcity and limited foreign exchange. In addition to running his practice at Tetteh Quarshie and CSRPM, Ampofo participated in the Ghana Rural Reconstruction Movement in collaboration with his brother, the obstetrician D. A. Ampofo, and lawyer Kwadwo Ohene-Ampofo. The Rural Reconstruction Movement hoped to bring the benefits of education and health to impoverished, rural settings, particularly in the Akuapem mountains. As part of this initiative, Ampofo published a guide to medicinal plants in 1983 to help local famers, schoolteachers, and other individuals treat simple ailments. The guide provided straightforward instructions in English and the names of plants in Twi. None of the diseases treated included malaria, however. This project indicated Ampofo's commitment to improving the life expectancies of everyday citizens in Ghana at a time when others with his international connections and training might have opted to leave the country.[68]

Despite these economic problems, *Cryptolepis* investigations continued at CSRPM and collaborating universities, and Ghanaian research teams were fairly certain of success. By the 1990s, it seemed that research on *Cryptolepis* might be reinvigorated through new possibilities in the field of biochemistry. In his inaugural lecture at the University of Ghana-Legon on April 18, 1991, chemistry Professor Ivan Addae-Mensah spoke of the great promise in antimalarial drug development with *Cryptolepis* extracts:

> Work on *Cryptolepis sanguinolenta* . . . is to me one of the best examples of how joint multidisciplinary, transnational research collaboration between sister African Universities can prove extremely fruitful and beneficial. . . . If my friends in Kumasi had not had faith in their own work, and trust in their preliminary results, they would have abandoned the work midstream. And who knows, someone else might have picked it up later in Europe or America and claimed credit for it. . . . If the clinical trials prove successful, this compound could well be a potential marketable antimalarial drug, developed mainly by researchers here in West Africa.[69]

Addae-Mensah had been active in alkaloidal chemistry research with Torto and B. A. Dadson since the 1960s, and in recent years had been encouraged to lend his support by developments in *Cryptolepis* studies. Aside from the plant's antimalarial properties, he and collaborators at Legon, Kumasi, and CSRPM conducted further investigations on the antimicrobial effects of *Cryptolepis*.

But commercializing plants alongside the "open-door policy" proved complex during CSRPM's first three decades. Increasingly, healers called on the state and its scientific operators to provide legal recognition and compensation for their research assistance at CSRPM and other institutes. Keen on scientific innovation and professional survival, scientists had to make a variety of moral choices to proceed with their investigations. The specific instance of the fight to market *Cryptolepis* reveals how Ghanaian scientists formed alliances with local healers and foreign investigators to further their research agendas, all the while competing with them for the monetary rewards of the end product.

As Ampofo's health failed in the 1990s, other scientific leaders sought control of CSRPM. Tackie assumed the role of director until a dispute about the future of *Cryptolepis* forced him to step down. Tackie made the decision to apply for a U.S. patent on cryptospirolepine, an alkaloid in the plant, in partnership with Paul Schiff, an American biochemist. Other Ghanaian researchers opted to focus on the production of a medicinal tea made from the plant, with the assistance of Diane Winn, Ampofo's longtime American collaborator. In both instances, scientists in the impoverished country relied on foreign contacts and investment to bring the dream of a Ghanaian pharmaceutical industry into reality. But, as patents soon expired because of failure to pay necessary fees, and plants died, scientists encountered challenges in transforming *Cryptolepis* into either a viable pharmaceutical or an herbal product.

A Patent and a Tonic

Tackie took over direction of CSRPM in the early 1990s. Ampofo's clinical studies in collaboration with Boye had provided evidence for the viability of *Cryptolepis* in treating malaria. Tackie, in contrast, focused on the chemical structure of the alkaloids responsible for effecting cure. Tackie had been awarded a 1970–1971 Fulbright professorship at Duquesne University in Pittsburgh in the United States. During his time in the States, he had met Paul Schiff, a pharmaceutical chemist at the University of Pittsburgh who was interested in medicinal plant research. Tackie explained that he lost touch with Schiff until the 1990s, when Schiff resurfaced and together they conducted collaborative work on the alkaloidal structure of *Cryptolepis*.[70] By 1991, their investigations—funded by a $100,000 grant from the NIH—led to the elucidation of the total proton and carbon nuclear magnetic resonance spectra of the alkaloid Cryptolepine.

As a result of this research, Tackie and Schiff filed a patent application in 1993 for a novel malaria drug formula from *Cryptolepis*. Tackie listed his affiliation at CSRPM; Schiff listed his at Health Search International (the precursor to Winn's company Phyto-Riker). U.S. Patent 5,362,726, approved the next year, might have served to establish a viable claim to malaria-related uses of *Cryptolepis*. The patent was a claim for a specific reformulation of *Cryptolepis* into cryptospirolepine, an "isolated, pure, active component," which the inventors described as an innovation on the "crude extract" from the plant.[71]

Schiff and Tackie, described as co-inventors, provided instructions for how the extract could be purified and reduced to a tablet form to be taken daily. The compound and method were specifically for treating falciparum malaria. But in the end, the patent did not lead to *Cryptolepis* being manufactured as described, and the patent expired in November 1998 because of failure to pay fees. When I asked Schiff why they decided to patent cryptospirolepine, he explained that at that time he and Tackie were working with Winn's company Health Search International and that a lawyer for the company encouraged them to file a patent. Regarding Winn, he explained, it "was great" once she "came on the scene. She is a real spark plug. She had some plans and hopes to market and produce [medicines] from *Cryptolepis*."[72]

Simultaneously, Tackie created Malaherb, an herbal preparation derived from *Cryptolepis*. He and his son Reggie Nii Tackie, manufactured the plant as a medicinal beverage in Ghana. Malaherb advertisements referred to investigations conducted by Ampofo and Boye at CSRPM in the early 1980s:

DESCRIPTION:
MALAHERB ["The Malaria Herb"] is the aqueous decoction of *Cryptolepis sanguinolenta*, a shrub indigenous to West Africa.

Cryptolepis contains the class of pharmacologically active compounds called alkaloids. The major alkaloid is cryptolepine. Malaherb is presented in 330 ml bottles.

EFFICACY:
Cryptolepine's efficacy has been well demonstrated and documented. In a series of inhibition experiments, the aqueous extracts inhibited significantly the growth of plasmodium falciprium strains, whatever their degree of resistance to cholorquine.

Data on the clinical use of *Cryptolepis* extract on file at the Centre for Scientific Research into Plant Medicine reports complete recovery from malaria symptoms within 3 days and negative blood films within 3 days.[73]

Printed with high-quality inks, one Malaherb poster titled "Fight Harder, Drink Malaherb," depicted a sick man in bed drinking red liquid from a spoon with a bottle of green-label Malaherb on his bedside table. A corresponding image showed a bottle of larger-than-life Malaherb with a face, legs, arms, and hands in red boxing gloves knocking out a huge, skinny mosquito, also in red boxing gloves and shoes suitable for the ring.

Tackie's private activities on *Cryptolepis* led to conflict at CSRPM and his resignation in 1997. Tackie's dilemma represented the financial and professional strategies necessary for survival in the Ghanaian scientific community during the late twentieth century. In Tackie's case, CSRPM did not have adequate resources for significant chemical tests. Although Tackie wanted to be perceived as a legitimate scientist, as an equal in the global scientific community, he was keenly aware that without dabbling in tonic production, he might not have sufficient funds at his retirement. In our conversations, Tackie explained to me that there was nothing wrong with his decision, because the plant was already in the public domain and commonly known. He was careful, however, to show me papers published by journals in the United States listing his name and the center as collaborators. This indicated that he strategically used patents, publication, chemical nomenclature, and trade secrets all in a bid to survive in Ghana's highly politicized scientific community.

International Partnerships and Rivalries

Ampofo, on the other hand, increasingly communicated his vision for plant medicine through Winn. This alliance made some scientists in Ghana wary. Ampofo originally met the Johns Hopkins biochemist in 1962 when she visited Ghana as part of a medical team from NIH. According to Winn, Ampofo introduced her to healers who cured her of cervical cancer in the 1960s and then breast cancer in the 1980s. In turn, Winn assisted CSRPM with purchases of equipment and medical assistance for Ampofo in the United States when he became blind and paralyzed. She also provided Ampofo with a novel sickle-cell anemia treatment for his daughter that had been developed by her former husband, Makio Kuruyama, in 1969. Winn became extremely interested in medicinal plants in Ghana and founded Health Search International, for which Tackie and Schiff initially consulted and filed the cyrptolepine patent. During the later years of Ampofo's life, Winn became very close to him, speaking to him almost daily while he lay on his death bed. Those who observed their relationship explained that Winn became "like a daughter" to Ampofo.

At Ampofo's funeral, Winn announced her plans to continue his research. She had reorganized Health Search International into Phyto-Riker Pharmaceuticals, a U.S.-based company (see Figure 4.5). In November 1997, Phyto-Riker officially bought out the pharmacy division of GIHOC. Originally established as the State Pharmaceutical Corporation in 1962, GIHOC's pharmaceutical division had met with struc-

Figure 4.5 Entrance to Phyto-Riker Pharmaceuticals in Ghana. (Photo by author.)

tural and financial difficulties in a politically turbulent postcolonial Ghana. Under the direction of Kennon A. Brennan, the African American cofounder who served as the company's first president, Phyto-Riker Pharmaceuticals hoped to manufacture low-cost drugs in Ghana to distribute to sixteen West African nations. While maintaining headquarters in New York, Phyto-Riker used a 1991 German-built plant in Ghana to produce its pharmaceuticals. Downplaying any possible negative interpretations of their takeover, Phyto-Riker put the nation of Ghana at the center of its plans, hoping to work closely with Ghanaian chemists and herbalists to design a number of drugs based on indigenous medicinals.

Phyto-Riker began operations with the production of generic drugs. I first toured the manufacturing plant at GIHOC in 1999 at a publicity event organized to attract African-American investors. We donned white paper shirts and shoe coverings to observe the shiny metal machines that would, in the future, produce pills. In its initial stages, Phyto-Riker described its research program on ethnomedicinals and their development for a global pharmaceutical industry with great enthusiasm and high hopes, referring to the potential for "new blockbuster drugs" for such diseases as HIV/AIDS and sickle-cell anemia. "Some day," a promotional handout proclaimed, "people the world over may owe their lives to Africa's Herbal Pharmacy!"[74]

Phyto-Riker introduced Phyto-Laria, a malaria treatment made from *Cryptolepis*, as their first product based on traditional knowledge. The company promoted Phyto-Laria as "a true herbal remedy," in contrast to the generic drugs: "The herbal tea based on *Cryptolepis* is a true herbal remedy containing the naturally occurring complex mixture of phytochemicals in a traditional dosage form with a long-established history of use, just like the old remedies from which the two aforementioned malaria drugs [*Cinchona* and *Artemisia*] were derived."

Marketing consultants based in the United States created an attractive preliminary package design for Phyto-Riker, taking Ampofo as the inspiration for the company. A stylized tropical green leaf design was overlaid with traditional Ghanaian symbols and a text detailing the story of Ampofo's rediscovery of traditional medicine. In fact, Winn, whose husband at the time was an international patent lawyer, secured exclusive rights to the use of Ampofo's name for Phyto-Riker.

Some CSRPM affiliates opted to work with Phyto-Riker, whereas others were more circumspect. Relationships between associates of the late physician and employees of Phyto-Riker Pharmaceuticals were often marked by deep-seated animosity and suspicion, as some Ghanaian chemists believed it was not in the national interest to cooperate with a foreign initiative. In the early period after Ampofo's death, tensions were so high that the CSRPM director requested military intervention to keep Winn and other Phyto-Riker employees off the center's premises. When asked about this period, one CSRPM affiliate explained:

> Dianne Winn was a daughter to Oku Ampofo, working with Dr. Boye and others. . . . But I think somewhere along the lines, when Phyto-Riker was established and wanted to annex this place, this place would be working for Phyto-Riker, wanted access to all the info here. . . . Prof. Tackie [then the director], Prof. Ayensu, Prof. Addae-Mensah, were all of the opinion that this was a national asset. . . . I think they [Phyto-Riker] also had certain terms that were not ideal.

When Ampofo was close to death and Winn broached the idea of buying GIHOC and establishing Phyto-Riker, it was a difficult period for CSRPM. For many Ghanaian researchers, it confirmed an ongoing problem with research disclosure at CSRPM. As evidence of similar problems, one scientist cited the work that Tackie had done on *Cryptolepis* with Schiff and his subsequent development of Malaherb:

The institution (CSRPM) must protect itself. . . . It is not foolproof. . . .
We've had a number of people who have worked here, [it became a]
matter of survival. . . . You are not bound to sign any oath of secrecy. . . .
Prof. Tackie, he left this place and left with a product. . . . All and all he
spent a lot of time, even when he was at Kumasi [on *Cryptolepis*]. Health
Search helped him isolate compounds, he decided to go commercial, no
rules really to stop someone. . . . [CSRPM] doors were open, [anybody]
could come in and go out. . . . The records were in the herbarium, in full
glare, anybody could just come in and have access. . . . Prof. Farnsworth
[of Shaman Pharmaceuticals] was invited sometime in 1991 or 1992.

In this discussion, the invocation of Farnsworth's group at Shaman
pharmaceuticals suggested the larger specter of bioprospectors. In fact,
Shaman pharmaceuticals also patented a form of *Cryptolepis* for use in
diabetes in 1997 and 1998, the former of which has since expired.

At the time of this writing, it is unclear how successful Phyto-Riker's
treatment Phytolaria will prove to be. As part of a bid to harvest roots
sustainably, Phyto-Riker contracted farmers to grow *Cryptolepis*. However, my informal conversations with traditional plant experts suggested
skepticism toward depleting plants for their bitter roots as a viable way
for farmers to secure a living. Others noted that Ghanaians were not accustomed to drinking their medicine in tea form, preferring alcoholic bitters or the universal injections publicized during colonial public health
campaigns. Indeed, CSRPM presented a competing product using *C.
sanguinolenta* in liquid form.

In addition, healers with whom I spoke in Ghana expressed continued grievances about CSRPM, confident that the institute was stealing
their herbal recipes. From the 1990s, Ghana's Food and Drugs Board gave
approval for CSRPM to test herbal products for toxicity. Healers were
certain that either CSRPM was itself copying their products or acting as
spies for other domestic producers. In its 2000 biannual report, CSRPM
explained that half of the institute's financing came from sales of plant-
based products.[75]

From the healer perspective, the interests of the CSPRM were confusing on several levels. As already described here, since the 1970s members
of traditional healing associations had been called on to submit herbal
recipes for alkaloidal screenings led by Ampofo, Tackie, and others.[76] Yet,
over time these scientists had collaborated with institutes and colleagues
abroad and the center had gained World Health Organization recognition. The liaisons between scientists, international researchers, and drug
firms caused alarm for healers. Healers' concerns about domestic piracy

were compounded by the circumstances of poverty that led many scientists to compromise their academic orientation. Although some may discount its fears as mere paranoia, CSPRM had indeed fallen on hard times by 2000, supplementing earnings with its own product line.

Conversations with medicinal plant experts revealed sophisticated understandings of the larger context of their practice. Tourists, rogue bioprospectors, and international journalists all showed interest in healer activities. Some viewed the greater publicity surrounding herbal manufacture with ambivalence. During an interview with prominent herbalist Sylvester Akadzah and his sons, I was shown a business card given to them by Chris Kilham, a notorious bioprospector who had done quite a bit of work on Kava in the South Pacific.[77] Surprised to hear that Kilham had been in Accra, I asked Akadzah if he knew what Kilham was up to. He and his sons laughed, pointing out that the card said quite explicitly "Chris Kilham, Medicine Hunter," and that, as had been their experience in the past with other interested foreigners, they had never heard back from him on the plant samples they had given to him.

CONCLUSION: POSTCOLONIAL SCIENTISTS TAKE CREDIT FOR HERBAL INNOVATIONS

Ghanaian scientists sought ways to privatize herbal medicine after independence in a bid to counteract colonial appropriation of plant data. Initially, key researchers—including Ampofo and Tackie—worked under a nationalist framework to circulate and share findings with the NRC. As funding for national research flagged by the end of the 1960s, Ampofo and Tackie established a semiprivate research center in the town of Mampong-Akuapem. Over time, Ghanaian scientists privatized their findings first through publications and chemical and plant names granted in their honor. Then, from scientific personalization they moved into commercial products and even patents. Ironically, in their efforts to establish a basis for personal prosperity, researchers like Ampofo and Tackie negotiated closer ties with expatriate researchers who shared resources and equipment.

The shift from a nationalized view of herbal knowledge to one with possibilities for private gain led to conflicts with scientists in other African countries as well as with healers. The case of the struggle to commercialize *Cryptolepis*, however, pointed to the complexities of the initial vision. It reflects general shifts in international systems of scientific exchange during the late twentieth century as the market came to influence patterns of biomedical research disclosure. Bruno Latour and Steve Woolgar, scholars of science studies, have convincingly argued that historically academic institutions were based not on the laws of the market but on systems of gifts and credit. Researchers gained prestige from rates of publication, academic honors, and personal networks in an atmosphere of relatively low pay for university professionals. However, the deepening bonds between industry and scientific research in fields like pharmaceutical discovery, biotechnology, and electrical engineering after World War II led to alterations in these historical regimes of academic credit.[78] Particularly in scientific fields where research might be patented for industrial applications, scientists no longer sought academic popularity through priority of publication but rather financial gain through timely patent applications. CSRPM, with a dual mandate to commercialize and research plants, found itself caught up in these transformations of research disclosure. Furthermore, the financial and political troubles in the country as a whole made the possibility of industrial rights both more alluring and more elusive to scientists determined to be seen as equals on the world stage.

The saga of Ghanaian efforts to commercialize *Cryptolepis* presents an alternative narrative and case study that deserves to be integrated into the discourse on contemporary bioprospecting. At the outset, the scientists in Ghana were trying to reverse the narrative of quinine, where colonists extracted fever barks from South America. Much like Mexicans interested in creating a drug industry from hormones in wild yams, the Ghanaians believed that they might develop an independent pharmaceutical industry through plant research.[79] CSRPM did not have the adequate resources for significant chemical tests that they believed were necessary for the realization of their dream of an herbal drug industry, forcing collaboration with foreign researchers. The takeover of the drug wing of GIHOC by Phyto-Riker, a U.S.-based firm keen to develop herbal drugs underscored the enduring difficulties Africa has had with managing and capitalizing on its commercial resources and was an ironic conclusion to efforts at CSRPM to make sure no one "in Europe or America claimed credit for" African investigations of *Cryptolepis*, as Addae-Mensah had hoped. Negotiating a complex terrain of professional ties, Ampofo and Tackie's dilemmas provide a poignant portrait of the financial and professional strategies necessary for survival in the Ghanaian scientific community during the late twentieth century.

But the story told here also points to grounds for healers' mistrust of scientists with greater access to international legal regimes. Healers were continually pressed to reveal plant medicine information in their interactions with scientists in Ghana and with members of internationally funded nongovernmental organizations. The scientists encouraged individuals knowledgeable about plants to donate concoctions for study, often making vague promises to give them credit or even to patent any cures in the participant's name. Given cases like that of *Cryptolepis*, why should healers trust someone who might transform a simple recipe for bitter roots into a drug protected with international patents and foreign trademarks?

Elsewhere in Africa, rural populations have been more successful in holding scientists accountable for their investigations into medicinal plants. Although Ghanaian healers sought priority alongside scientists, in other contexts priority went to communities as a whole, opening the way for joint benefit sharing. Chapter 5 examines the case of hoodia in South Africa, where the San share profits with pharmaceutical companies and research bodies. By the close of the twentieth century, a sea change in international policy on bioprospecting alongside shifting political dynamics in South Africa meant that scientists could no longer substitute their names for those of healers and disenfranchised groups who led them to new treatments.

Figure 5.1 Bottled hoodia in Economic Botany Collection, Royal Botanic Gardens, Kew. (Photo by author.)

CHAPTER 5

Take Kalahari Hoodia for Hunger

Apartheid had nutritional roots as well as consequences.
—Diana Wylie, *Starving on a Full Stomach*

If you take the green District line of the underground railway in London all the way west to Kew Gardens Station, follow the signs outside the train to the Royal Botanic Gardens, and look behind a small pond within the sweeping park grounds, you might catch a glimpse of the Sir Robert Banks building. Its underground structure emerges as a wall of glass above the water. Down below it contains a vast cellar, home to the economic botany collection that contributed to Kew Gardens' status as a world heritage site. It contains preserved specimens of all the major healing plants in the world, stacked along rows of shelves, stored in small drawers, and leaned against the walls. The cellar is a stark visual reminder of the wider complexities of sustaining not only medicinal plants in the environment but also the communities that used them and claim ownership to them today.

Here, on a damp autumn day, I held a bottled specimen of hoodia. Two pieces of the plant, bleached and bloated with age, floated inside a heavy glass cylinder with a flat top that I hoped would not open. Pickled around 1874, when the British governor of the Cape Colony, Sir Henry Barkly, donated it to science, the cactus-like plant hailed from Namaqualand, an arid region between South Africa and Namibia. It was one of many such specimens preserved in aging, amber liquids. Formerly, the collection of bottled plants—a set of odd curiosities into which school children would peer—had resided with the Pharmaceutical Society of Great Britain. But by the time I visited, these and other plant relics had moved into the temperature-controlled warehouse of the Banks Building to be perused by the occasional researcher. In the softly lit basement, ghosts of plants past mingled as I wondered what this plant's collector had known about it before dispatching it to England more than a century before. Were its prickles still sharp, or would they fall limp if the liquid spilled?[1]

The plant's name, hoodia, may sound familiar as a new miracle diet drug. At the Longs Drugs Store across from my house in Berkeley, California, I could purchase such products as Liquid Hoodia Extreme, DEX-L10 Hoodia Gordonii, Maximum Strength Hoodia Sure Rapid-Gels, or H57Hoodia, all of which claimed to contain *Hoodia gordonii* from South Africa. At Elephant Pharmacy, an alternative health store two blocks away, twenty bags of Bija Healing Teas' Hoodia Slimming lingered on the shelves. A pamphlet inserted in a box of Bija's Hoodia tea explained, "When Bushmen of Africa's Kalahari Desert went out on long hunting trips, they would munch on the stem of the Hoodia Cactus to stave off thirst and hunger. [We have] resurrected this thousands-of-years-old secret: curbing the appetite to help with weight loss."[2] The Bija box design incorporated photographs of cowrie shells, a South African stamp showing a rhinoceros, and an elephant bead. All of the hoodia packaging sported green, gold, and black colors, and several boxes contained copies of certificates from South African bodies guaranteeing rights to export barrels of powdered hoodia from the country.[3]

This chapter considers how hoodia migrated from its location in the Kalahari Desert region of South Africa, Botswana, and Namibia to my neighborhood pharmacy in the United States. It seeks to address the question of who benefits when plant medicines become drugs. In contrast to other plants in this book, the story of hoodia's transformation into a blockbuster drug and international nutraceutical in the late twentieth century came at a time when scientists were under pressure to acknowledge the rights of communities thought to have originally used medicinal plants—in this case, the Bushmen (or San) living in the Kalahari desert. The 1992 International Convention on Biological Diversity (CBD) recognized the contribution of First Peoples groups to drug discovery, and the 2004 South African Biodiversity Act provided a blueprint for how researchers were to compensate local communities in the country. As we shall see, however, the case of hoodia and the San rests uneasily as a model for profit-sharing herbals drugs.

Although it might be tempting to class this case as an apparently unique "success story" for benefit sharing, the history of hoodia suggests similarities to the trajectory of other medicinal plants in this book.[4] By the 1700s, use of hoodia among other African ethnic groups and settlers from Europe made the plant less San and more of a popular South African plant, with other communities staking a claim to knowledge. Additionally, other countries and other San communities outside South Africa sought benefits from the South African Council for Scientific and Industrial Research (CSIR), the research institution that patented the drug. Second, South African scientists (like researchers in Ghana, Republic of Congo, Madagascar, Canada, and the United States) also used patents to protect drug insights. Only as an afterword, history pushed them into benefit-sharing agreements structured to right wrongs from the apartheid period. It remained unclear whether CSIR initially obtained samples and information from the San in South Africa during the 1960s, although it was politically expedient to assume they did.[5] Finally, the pharmaceutical preparation from hoodia was as elusive as producing cryptospirolepine for malaria. In fact, the potential toxicity of a hoodia drug, as well as competing patents and unlicensed preparations, destabilized the CSIR's agreements with the San. The more I learned, the more this perfect case began to look like Madagascar and the rosy periwinkle, where important medicinal knowledge spread to Jamaica, the Philippines, Canada, and beyond.[6]

Thus, I began to wonder: What were the parameters of community and the period of original knowledge? How do we affirm and pay the codevelopers of knowledge without resorting to the racial and class logics of greedy (foreign) scientists versus poor indigenes? Philosophers have argued that scientific knowledge is a public good that cannot belong to a particular ethnic group.[7] Similarly, critics of expanding intellectual property regulations have argued that "ethnobiological knowledge should . . . remain where it belongs—in the global public domain."[8] And yet, the claims of patent holders and benefit sharers in South Africa showed how a society might construct lineages of scientific inheritance along class and ethnic lines.

MULTIPLE USERS OF GHAAP

The San were the first "bioprospectors" in Southern Africa. For at least 120,000 years, they have lived a nomadic lifestyle in the semiarid and arid regions skirting the Kalahari Desert.[9] To survive in areas with low rainfall, they experimented with different species of wild plants, identifying vegetation containing nontoxic sap and juices. To protect their skin from the harsh climate, San began to dry leaves of certain shrubs that they carried in small containers made from turtle shells. They mixed the dried leaves with animal fat, smearing it on their body as skin lotion. Their nearest neighbors, the pastoralist Khoi, named them in the Nama language after the bushes or *San* that literally adorned their bodies. For settlers from Europe, San use of herbal creams represented a sign of "filth" and grounds for discrimination.[10] Today, the more than 100,000 San residing across South Africa, Namibia, Zambia, Zimbabwe, Botswana, and Angola seek to claim rights to valuable plants associated with them, including the thirst-quenching hoodia and the antimicrobial bush leaves used for skin protectorants, *Pteronia onobromoides*.[11]

Historically, the San recorded the past through conversations, songs, and pictorial art on rocks. Until the late nineteenth century, they did not write down their language. The San's early twenty-first century lawyers therefore faced an evidentiary problem familiar from the other stories told in this book—namely, how to document the San's use of hoodia for the same purposes covered in the CSIR's patent. For evidence of the San's claim as a First Peoples group they could point to archaeological remains and distinctive murals throughout Southern Africa. Further descriptions of hoodia's use among the San came down through accounts of early explorers and settlers of European descent in South Africa. But those accounts also alluded to the ways in which the plant was used among other groups besides the San.

The earliest written references to consumption of desert succulents date to the 1700s, with increasing accounts from the nineteenth century made by European settlers, missionaries, and explorers as white and other African families displaced Bushmen communities. From the seventeenth century, Khoi-San lived near the Atlantic Coast of Southern Africa and provided information on hoodia to new migrants who came from Africa, Europe, and Asia. San included hoodia within the category of juicy, edible plants known as *"xhoba"* or *"ghoba,"* a term that migrated into the creole language Afrikaans as *"ghaap"* or *"ngaap."*[12] An early observer of ghaap on the Cape of South Africa in the 1770s noted that succulents were consumed not only by indigenous inhabitants that he called "Hottentots" (now seen to be a derogatory term replaced by "Khoi-San"), but also by Dutch farmers. The observer referred to several wild plants by their earlier botanical names and recalled in a memoir of his journey that "the *Stapelia incarnata*, a very branchy plant without leaves, was found in the vicinity of the mountains, though it was rather scarce;

the Hottentots ate it, after peeling off the edges and prickles."[13] He explained that Dutch settlers also harvested a species of *Stapelia*, possibly now known as *H. pilifera*, and preserved it in vinegar[14] (see Figure 5.2).

The term *ghaap* covers a number of stapeliads, a group of succulents, now considered part of the botanical family *Apocynaceae*.[15] Botanical names for hoodia varied over time and overlapped with related species of *Trichocaulon, Stapelia*, or hoodia. As scientists debated the botanical differences among the plants, not only did hoodia's assigned genus shift, but its many "discoverers," often amateur botanists, constantly affixed their names to new species. *Hoodia juttae*, for example, was "discovered by Mrs. Jutta Dinter on a botanical expedition to Klein Karas in October, 1913, and dedicated to her by her husband."[16] The genus *Hoodia* was renamed after one Mr. Hood in 1830, and *gordoni*—later *gordonii*—after colonel R. J. Gordon, who made or commissioned a drawing of the species on a journey along the Orange River in 1779.[17]

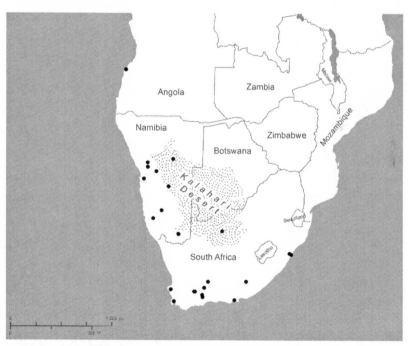

Figure 5.2 Collection points of forty-four *Hoodia gordonii, H. pilifera*, and *H. currorii* samples in herbaria accessed through the online database of the Global Biodiversity Information Facility: Southern Cape Herbarium and Bolus Herbarium, South Africa; Royal Botanic Gardens, Kew; Botanical Conservatory at University of California, Davis; Missouri Botanical Gardens; National Museum of Natural History; National Botanic Research Institute. (Source: Global Biodiversity Information Facility Data Portal.)

European settlers migrated to South Africa in large waves during the nineteenth century, helping to increase references to hoodia and ghaap. Dutch settlers competed with British colonists for jurisdiction over the Cape, with the discovery of diamonds and gold deposits pushing those of European descent further inland. European migrants looked to South Africa's unique flora for another avenue to acquire wealth. A physician who taught at South African College in Cape Town cited the usefulness of hoodia in a botanical survey he published in 1862:

> The stem of this plant, which grows in the dreary wastes of the Karroo, is fleshy and of the size and form of a cucumber. It has an insipid, yet cool and watery taste, and is eaten by the natives who call it *Guaap*, for the purpose of quenching their thirst. Infused with spirits, this plant is said to be a useful remedy in piles.[18]

As rail travel improved, missionaries, botanists, and settlers from European families moved further inland, adopting valuable plants like hoodia. Botanists at Kew Gardens in London received an early photograph of the plant around 1876, apparently sent because of the interesting shape of its branches. Barkly sent several living samples of hoodia back to Kew, in addition to the bottled specimen. I was surprised to open one herbarium page to see Barkly's thorny dried stalk with brown dust around it rather than a flat pressing or dried slice (see Figure 5.3). Someone had labeled it "Guap," suggesting wider awareness of its edible qualities by the late nineteenth century.[19]

In the 1870s, a Swedish botanist journeyed to Southern Africa, where he adopted a young San child whom he raised while in Africa and later Sweden. In his diary, he described how he ate a plant thought to be a species of *Hoodia*:

> I found a kind of cactus; it is called "kovapp" and is eaten by Hottentots, Bergdamara, and Bushmen. The sharp thorns are scraped away and then the whole plant is eaten as it is. It tastes somewhat bitter but leaves a sweet and pleasant after-taste as is often the case with a "bitter herb."[20]

This botanist sampled raw plants while his colonial contemporaries might have preserved edible succulents by making sugared hoodia jams, alcoholic decoctions using brandy, or dried hoodia. Afrikaaner communities of Dutch immigrants as well as British colonists circulated such recipes thought to cure upset stomachaches and hemorrhoids.[21]

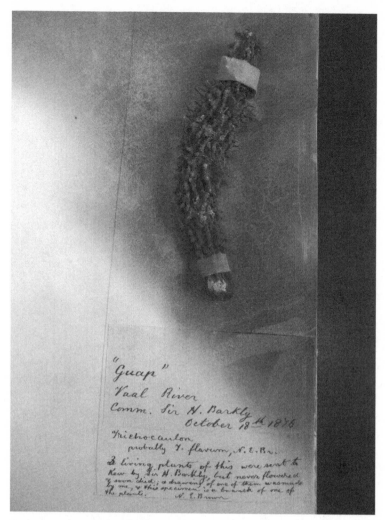

Figure 5.3 Unusual, full-stem Herbarium sample from 1876: "3 Living plants of this were sent to Kew by Sir H. Barkley, but never flowered and soon died; a drawing of one of them was made by me, and this specimen is a branch of one of the plants." The name "guap" suggests early awareness by botanists of its use as a food. (Source: Herbarium of the Royal Botanic Gardens, Kew.)

Knowledge of hoodia's multiple uses spread to Europe. At an evening meeting of the Pharmaceutical Society of Britain in November 1892, those assembled recalled the specimens of *Hoodia* sent to Kew on behalf of Barkly.[22] One of those assembled believed the sample before them could be a new species of *Stapelia*:

The two species previously known were . . . very rare, because the Hottentots [Khoi] and Bushmen [San] ate them whenever they were found, striping off the skin and eating the central part raw. He [Barkly] had never been able to get a plant so as to taste it himself in the fresh state, but he had tried other members of the genus. . . . When raw the plant was slimy and had a peculiar bitter flavor, which to him was not agreeable. When steeped in brandy the plant imparted a bitter flavor to it, and he thought made it more beneficial as an occasional restorative. . . . The name guap was apparently applied to [it].[23]

Still reeling from a devastating fire at the Wadsworth Laboratories of Burroughs, Welcome & Co. just two months prior, the Pharmaceutical Society members continued to discuss the economic potential of the wide array of plants pouring in from the colonies.[24] Bottled specimens, herbarium sheets, letters, and recorded discussions provide evidence of the circulation of Khoi-San plant knowledge within England by the late nineteenth century.

It is important that these early references indicate that ghaap consumption was linked not only to Khoi-San communities, but also to Dutch, British, and other groups of European and African descent who came to live among them. Early dictionaries of Afrikaans and South African English included the word "ghaap" as a term adopted from Khoi-San, although it is unclear if the word referred to just a thirst quencher or an appetite reducer as well.[25] Tense relations characterized early interactions between settlers from Europe and the preexisting communities, even as they shared knowledge about plants. Colonists used rawhide whips on Khoi-San workers bartered as slaves, and they kept brothels of Malaysian women forced to relocate to South Africa. Despite this checkered history, the Dutch settlers were the earliest group from Europe to survive several generations on the Cape and arguably emerged as an additional group in the mosaic of African ethnic groups. Through relationships shaped by coercion, hybrid cultures materialized, including the Afrikaans language—with words from Asia, Africa, and Europe—and apparently consumption of hoodia. In South Africa, "Cape coloreds," a racially mixed class descended from Khoi, Xhosa, Malaysian, and other groups that labored on farms and ranches, spoke Afrikaans and absorbed many of the San class through intermarriage.[26]

Despite—or perhaps because of—these shifting ethnic boundaries, by the nineteenth century, racial and scientific investigation had invested the San with radical cultural and biological difference as a means of separating them from other closely related groups vying for status as "coloreds" or "whites." By the early twentieth century, hoodia references noted the plant as peculiar to "Hottentot" or Bushmen diet, although people of European descent worked to cultivate it as an ornamental garden plant on the more humid coasts. They developed new strategies to cultivate hoodia in the 1920s and 1930s for the export market: "If the crown of the root is damaged in any way in the lifting from the ground, a *Hoodia* will not survive very long when replanted."[27] Thus, hoodia came to be a marker of difference to distinguish classes of South Africans as well as a marker of religious and cultural differences among the San in Southwest Africa (now Namibia) and Namaqualand. At the same time, some Khoi-San moved further inland to avoid white settlers, losing access to hoodia in the Cape areas (see Figure 5.2).

Given the bitter history of apartheid relations in South Africa, these historical indications that hoodia was a San, as well as Afrikaaner, plant destabilize clear lines of redistributive justice evident in the CSIR benefit-sharing agreements. The ethnobotanist Ben-Erik van Wyk has also argued that medicinal plants, particularly in the Cape, represented a shared heritage of Khoi, San, and Cape-Dutch (now Afrikaaner) citizens.[28] Given early descriptions of hoodia, other Southern African groups, including those of Dutch descent, could conceivably make wider claims to hoodia folk knowledge. The hybridizing of popular remedies continued in the twenty-first century. One of my initial interviews was with Patsy Vergeer, an Afrikaaner Sangoma (Zulu for healer). She explained how her family lived in the bush in the early twentieth century, learning about medicinal plants and trapping animals. She later spent time as an adult studying in the Netherlands, where she rediscovered Dutch healing traditions. On returning to South Africa, she became active in the promotion of herbal medicine, combining South African and international therapies, some of which she gleaned from a daughter-in-law of Chinese descent, into her healing practices. Participants in ongoing Afrikaaner folk healing might therefore be linked to hoodia, arguably with some claim to benefits from its marketing as a pharmaceutical.

TAKING VELD PLANTS

By the twentieth century, Khoi and San populations did not survive solely on wild plants. Some worked on European farms and lived in village settings with Tswana-speakers, with whom they intermarried. This allowed them better access to cultivated crops and even canned goods. Nevertheless, the San were subject to a number of state-run biopiracy efforts designed to obtain information on how their supposedly unique diet facilitated their survival in the difficult climatic conditions. In particular, the minority of San who sustained hunting and gathering lifestyles participated in research into their everyday eating practices funded through the South African government, the University of Witwatersrand, the U.S. National Institutes of Health (NIH), Harvard University, and the Smithsonian Institute, among other bodies. !Kung and Tshu-Khwe speakers in Southwest Africa (now Namibia) and Becuanaland Protectorate (now Botswana) helped establish conceptions of appropriate Bushmen diets for scientific researchers. They provided names, uses, and samples of plants. San living in the Kalahari Desert in Namibia, South Africa, and Botswana did not appear to consume hoodia, suggesting that knowledge of hoodia was sustained among multiethnic communities in the less arid Great Karoo where it thrived.

Research on San plant practices was part of an extensive legacy of ethnobotanical surveys in the region. Studies of plant use among other ethnic groups in Southern Africa also captivated botanists during the twentieth century. Investigations documented thousands of plants—some indigenous to Southern Africa, many imported—to detail their economic value. This work did not provide interviewees with informed consent waivers and frequently involved subterfuge.[29] Viewed retrospectively, a long history of biological takings allowed hoodia to serve as a catalyst to reassign benefits within South Africa.

Desert Survivors

During the 1920s and 1930s, San families survived droughts in the arid areas where white settlement forced them to find refuge. A small minority came to live inside the Kalahari Desert, outside hoodia's natural growing range. Dry even in the best of times, the Kalahari contained sandy grassland, with patches of green shrubs, occasional baobab trees, and the scarce water hole. The area held little appeal for people who farmed or herded in what are now South Africa, Namibia, and Botswana. But to those who had found ways to survive on liquid-filled wild roots and melons and to hunt wild wart hogs, wildebeest, porcupines, and other small animals, it represented a space of some respite from colonial authority and wage labor on the farms of European settlers.

Yet, even for the San, who managed to live a seminomadic existence in the Kalahari, successive dry seasons from 1926 to 1933 created challenges for food foraging. San discussed their lifestyle and diet with visiting researchers, including the usefulness of wild foods as well as processed foods purchased in shops. Scientists in the 1930s marveled at San strategies for survival at the same time that any physical differences served as grounds for ongoing prejudice. A rare reference suggests some San still had access to hoodia in the Great Karoo, on the outskirts of the Kalahari. In 1932, for example, a congenial, balding German pharmacist named Rudolf Marloth described his experiences among the San. He told how he had been traveling with a San guide during the severe droughts of the late 1920s. His companion suggested that they eat plants to get by. After sampling some, he explained that he was not hungry for the next two days and was able to survive until returning to camp. His description of what locals called "ghaap," *Trichocaulon piliferum*, now classified as *H. pilifera*, provided perhaps the first published indication that hoodia might thwart hunger:

> [The plant] consists of a number of erect, stout shoots two inches thick and up to two feet high. This is the real *ghaap* of the natives, who use it as a substitute for food and water when both are scarce. The sweet sap reminds one of liquorice and, when on one occasion thirst compelled me to follow the example of my Hottentot guide, it saved further suffering and removed the pangs of hunger so efficiently that I could not eat anything for a day after having reached the camp.[30]

He published this note in an extensive, multivolume study of plants with potential for pharmaceutical exploitation that was sold in Cape Town and London, having previously included entries on ghaap in a list of South Africa plants that was published in 1917.[31] At a time when South African nonwhites found their rights increasingly curtailed, such studies celebrated their plant knowledge and depended on their expertise. South African law forbade commerce in medicinal plants outside of Natal and restricted consultations with herbalists, even as scientists collected data with economic potential and white merchants sold herbal remedies.[32]

In the aftermath of successive droughts, San interacted with other researchers interested in documenting their lifestyle and plant use. In 1936, a San leader, whose name perhaps was Abraham, was approached by a prominent white South African sportsman with a proposition to obtain easy food for his friends and family. Abraham brought together seventy San to a point agreed upon by the end of May. The following month, they watched as three cars and a large truck, having driven along dry river beds, arrived to study them. They then spent several months with researchers from the University of Witwatersrand who subjected them to measurements of every imaginable part of their bodies. The San were placed inside complex implements for measuring their skulls, as the Witwatersrand team hoped to contribute to the field of phrenology that argued skull cavity (presumed to be smaller in nonwhites) indicated levels of intelligence and propensity for education.[33] Research subjects received food from the University as a gift for their time. Some sang songs for the lecturers, fascinating them with their unique clicking language and frustrating them by interspersing songs learned at missions. After all, to the San these researchers were just the latest in a string of white visitors, heralded with the ventures of missionaries like David Livingstone to the Kalahari in the previous century.

One Witswarsrand researcher led discussions of eating practices. He had hoped to find out more about medicinal plants used among the seventy individuals assembled, but "the situation of the camp in the arid sand dune country on the edge of the Kalahari Game Reserve near the junction of the Auob and !Nosop Rivers, did not offer much opportunity for the collective identification of medicinal or food plants."[34] He visually inspected teeth and gums, deciding that cavities were low because of effective cleaning with ash and blades of grass. He noted a general deterioration in diet, in part through "contact with civilised or semi-civilised races." He noted that San farmworkers still spent part of the year as nomads when they hunted wild vegetables, fruits, and game.[35]

Even as the researchers hoped to find the San in their "natural" habitat, the realities of their parallel existence as poorly paid agricultural workers who obtained coffee and sugar from storekeepers begged consideration. This particular lecturer received assistance from two individuals, Khanako, a middle-aged woman, and Molopo, a young man, who were most familiar with plants used to combat ill health. He noted that although they mentioned plants used to treat sores, stomachs, childbirth complications, and diarrhea, "Nowadays, of course, they have also become familiar with the European medicines which the farmers give them, and they have acquired great faith in Epsom salts as a general panacea."[36] Neither Khanako nor Molopo mentioned hoodia.

San health definitely deteriorated as the Witwatersrand study then compared eating habits in the artificial field conditions to ones constructed in town. In September, the assembled San filed into a University truck and traveled to Johannesburg, where they spent the remainder of the year until their return in January 1937. The discovery of diamonds in nearby mines had transformed this town in Gauteng province. But, these San were not bound for the mines: rather, their handlers placed them on display for Johannesburg families to come and observe. They stayed at Tweerivieren, the Witwatersrand University farm.[37] Several of the San did not make it through the close of the exhibition, including "an infant which died in hospital." Abraham, the eldest and apparently a leader who assembled his friends and family in the first place, was one of those who succumbed to "arterial degeneration."

Under South African authority, the San were to experience a slower death through cultural annihilation.[38] Ongoing questions circulated over whether San chose to live in the Kalahari and other remote areas by choice, or if Khoi, Tswana, and Herero families moved there with the onslaught of settlers encroaching on water and land resources, thereby being reclassed as San. Historically, Khoi groups herded livestock on the southwestern coast of Africa, providing useful husbandry expertise to European colonists. They referred to poorer members of their communities who lived in the wilderness without farming or herding as "Sonqua" or San, perhaps in reference to adornment with dried leaves.[39] When people from Germany settled along the southwest coast of Africa in the 1890s, Khoi-San refused to live and work on their farms. Some San families lived on reservations, but the conditions were very poor. After German men raped Khoi-San women in the early years of the twentieth century, their husbands and brothers attacked white farms, and settlers shot back, setting the stage for a government-sanctioned genocide of Khoi-San populations from 1903 to 1907. Some escaped into remote parts of the Kalahari Desert to avoid death. Later, San died during "Bushmen patrols" when European settlers in South Africa went on genocidal hunting expeditions.[40] The term "Bushmen" derived from the Dutch term *"Boschjesmans"* and a German term for bandits, *"bossiesman,"* pointing to their position in the history of resistance and colonial defiance.[41]

Even researchers in the 1920s and 1930s admitted that San moved into less desirable, vacant lands as a means of survival.[42] Contemporary scholars have detailed efforts to contain San within game reserves in the region and the extent to which the San are genetically and culturally related to other populations in Africa. Political scientist James C. Scott situates the San as an example of a multiethnic group with a lower class position who perfected "the art of not being governed."[43] By the late twentieth century, some advocates for indigenous rights argued that San life was a choice gladly made to live outside of European jurisdiction, and represented a several-thousand-year tradition only partially interrupted through migration patterns in the region. Archeologists sought evidence for early San settlements devoid of anything but stone tools. Elizabeth Marshall Thomas, described later in this chapter, would call this "The Old Way": a lifestyle based on simple pleasures and open air promenades that people who are consumed with possessions and technology might do well to emulate.[44]

Mining Workers for Plants

San families were not the only ones to provide fodder for South African botanists interested in documenting regional diets and edible plants. The cuisine of the majority populations in South Africa—including Zulu, Xhosa, and Sotho ethnic groups—also met scrutiny.[45] During the 1930s, in the wake of recurring droughts, families confronted the real possibility of food shortages in the region. Outside of South Africa, stories of the region's lack of precipitation invariably led to claims that people resorted to eating their own children to combat hunger. In fact, women scoured the countryside for edible plants, which they boiled and served with porridge or dried to eat during the winter months. They found ways to sustain their communities without meat, as millions of sheep died and cow herds were at risk. In particular, women provided names and details of how they cooked plants for researchers conducting a survey on the increase in tooth decay. However, women in remote locations were harder for the researchers to meet than men already recruited from across the Union of South Africa and from Portuguese and British colonies in Southern Africa.[46]

Around 1936, miners provided information on their typical diet for biochemists at the University of the Witwatersrand and the South African Institute for Medical Research (SAIMR), hoping to improve productivity in the context of limited food stocks.[47] These miners came to work in South Africa at the gold and diamond mines from as far as Mozambique, then Portuguese East Africa. In one study, more than 600 men sat down for interviews with an interpreter who spoke both English and their native languages. "The interpreter would gain the confidence of each case by explaining that the investigation was in their own interests, and would then ask the question, 'What foods do you eat in your kraal?'"[48] The miners expressed that back home, they ate meat, potatoes, mealie, milk, beer, bread, coffee, jam, tea, and corn. The miners also provided information on their consumption of edible plants. Over sixty miners recruited from their homes in the Transkei explained that they frequently ate wild roots, with more than seventy miners who migrated from Mozambique regularly consuming them. In contrast, the miners from Transkei were reluctant to admit that they ate "spinach," a term covering wild greens, for fear of appearing "effeminate."[49]

At SAIMR, investigators collected further data on the importance of these spinaches. Francis William Fox, a biochemist interested in the threat of malnutrition in South Africa and the prospect of increased scurvy, heralded the use of plants. He noted that in a good harvest, like that of 1925, South Africa could be assured of at least 25 million bags of the staple food corn, but in low rainfall years like 1933, that number had been drastically reduced to 8 million bags.[50] At the time, he made a distinction between the "Red Natives," who wore blankets and leather and adorned their bodies with earthen creams, and "Dressed Natives," who had acquired a taste for "European clothes." Fox noted that "Dressed" informants did not eat as many wild greens. Although he suggested further collection of edible wild plants among nonwhites, his overall conclusion was that "the main line of development here must no doubt be to encourage the growing of European vegetables."[51] Fox's fears that *"imifino"* use would die out proved incorrect, as a variety of greens continue to be popular in South Africa in the twenty-first century.[52]

When read closely, this research shows that both whites and blacks in South Africa were consuming wild greens by the 1930s. Fox's published reports acknowledge that "the more discriminating Europeans living in country districts not infrequently make use of some of these plants and occasionally even cultivate them in their gardens."[53] Studies also found that most of the plants were not solely original to South Africa but rather were "introduced weeds," like pigweed, thistle, and nightshade. Such statements complicated the racial logic at the time that only blacks would deign to eat wild plants as they had for centuries. Investigations showed that women sustained a culinary culture surrounding the plants, which were rarely eaten raw. Given the context of drought and forced migration within diverse ecozones, the research showed that people had a complex understanding of the plants and used them to survive droughts, although this knowledge was unevenly distributed throughout the population. Chemists at SAIMR conducted simple tests for the presence of moisture, protein, mineral salts, calcium, magnesium, phosphorus, iron, and ascorbic acid (vitamin C) in plants collected through the investigations.[54]

Subsequent research on veld plants at CSIR built on this early research on nutritional prospects in the country. Laboratory studies continued to assess major minerals and vitamins, with efforts aimed at improving the diet of the majority of South Africans. SAIMR received funding through CSIR by the 1950s, with investigators sharing research. The few whites with tertiary training in chemistry and biology tapped their personal networks of botanists and farmers and interrogated miners,

rural women, and prisoners about plants within an increasingly strati-
fied society.

Preserving the Old Way

During the 1950s and 1960s, San in Botswana and Namibia provided
further samples of edible plants for bioprospectors from the United States
and South Africa, although it is unclear if specimens included hoodia.
In addition to providing plants, !Kung families gave samples of their
blood, explained lineages, mapped out their migration patterns, and
allowed measurements of their blood pressure, cholesterol, urine, skin-
folds, height, and weight to more than twenty U.S. and South African
visitors during the 1960s as part of field studies of nutrition and behavior
among !Kung Bushmen.

In Namibia, !Kung communities worked with a U.S family from Mas-
sachusetts interested in filming, recording, and writing down their daily
activities. In 1951, they met Laurence Marshall, a conservative entrepre-
neur and founder of Raytheon, and his wife Lorna.[55] !Kung living in the
Kalahari Desert in what was then Southwest Africa, a protectorate of
South Africa, posed for photographs and described religious beliefs and
other practices for the Marshalls, who were collecting these materials for
Harvard University and the Smithsonian Institute. The images of Bushmen as living outside of modern life were carefully constructed, including their intake of wild plants versus prepared goods. A retrospective
analysis of photographs made during the Harvard-Smithsonian Kalahari
study led by the Marshall family revealed snapshots in their private collection of Bushmen informants leaning casually on cars clothed in trou-
sers, or taking food from cans. However, in photographs from the same
period published in *National Geographic,* Bushmen sat on the ground,
wearing traditional skins, using handmade tools. They received gifts of
food and tobacco from the Harvard Expedition as a form of compensation and watched their interlocutors consume tinned food.[56]

As part of the Harvard Expedition, !Kung also collected samples of the
wild plants they typically ate. In July 1952, they accompanied a research
team sent from Cape Town University to study plants in Southwest Africa. Then, from December 1952 to February 1953, they explained to the
Marshalls how they used the plants. Because the July expedition was
in winter and some of the plants were not available, the San reviewed
herbarium samples sent through the Marshalls to clarify the identification of plants. The names of San informants were not mentioned in the
two reports that were the outcome of these investigations. Nor did they

provide details on hoodia. Individuals living in the Kalahari explained that they did not differentiate between plants that the visiting botanists placed into different categories. As one author explained in a report,

> it would probably be best and simplest to have [the plants arranged] in botanical order. This would ensure that the things that are closely related and similar would be together—for example, Caralluma, Stapelia and Duvalia; which would be widely separated in alphabetical order but look so alike that even the Bushmen do not distinguish between them.[57]

!Kung plant experts provided details on 113 vegetables and fruits that they used for surviving dry conditions in the Kalahari. Their detailed understandings of plants impressed botanists involved in the project, who declared the outcome of the biological takings "a very useful contribution to the field of ethnobotany."[58]

The unpublished journals of the Marshall family provide further details on how !Kung speakers used moisture-rich plants, although they do not appear to mention hoodia directly.[59] In her diary, daughter Elizabeth Marshall marveled at the San's many uses of plant products even after they were sapped of all nutrients. In July 1955 one entry describes a trip with a young boy and several women named Tsetwe, K;xokwe, Gasi!na, and !Gai to a melon patch. The women smoothly loosened the valuable melons and balanced them in sacks on the walk home to camp, where they were carefully harvested to preserve the tough rinds for containers.

San across the border in Botswana similarly welcomed teams of researchers sent by Harvard and the Smithsonian into their midst. Between 1963 and 1969, !Kung who frequented a water source at Dobe near the Namibian border spent time speaking with Richard Borshay Lee, an anthropologist conducting research for his PhD dissertation at the University of California, Berkeley, who joined the Harvard Expedition. Around 360 !Kung and 300 Herero and Tswana lived near the water source, with seventy !Kung who tended to visit the area in the summer months. Fifty of the !Kung worked for Tswana and Herero families, or were married to them, eating primarily corn and dairy products that they produced on small farms.

At Dobe, the San shared meals of the wild foods they collected with Lee and other visiting researchers as well as members of other ethnic groups in the area. In 1963, they began to show plants to Lee and two interpreters who spoke Tswana/English and Tswana/!Kung, respectively. Gradually they worked more directly with Lee, who learned to ask them basic questions about plants. In 1964, they watched as he set aside his assortment of tinned goods to subsist solely on wild plants with them for six weeks. As part of the project, women displayed their daily collections on skins and leaves for the Harvard team to photograph; U.S. nutritionist Marjorie Grant Whiting weighed the nuts, berries, and fruits they brought back. Young men accompanied Lee and Graham Guy, curator of the Queen Victoria Museum, to Tsodilo Hills to identify and preserve plant samples in the field. By 1965, the !Kung, Tswana, and Herero families living within a forty-mile radius of Dobe and the nearby Tsodilo Hills helped the researchers document 190 plants, seventy of which were edible.

According to Lee's notes on this exercise, the San frequenting the water hole at Dobe divided wild plants into four categories. For plants that were still growing, the !Kung at Dobe called the above-ground parts "/um" and the below-ground parts "ci." For plants that were ripe, they called the fruits and leaves above ground "n/om" and the roots and tubers "tx'um." They indicated to Lee that they ate thirty-four ripe fruits or n/om plants, including baobab, custard apple, tsamma melons, and monogo nuts. The !Kung provided samples of twenty-three tx'um ripe underground roots that they ate, and eight ci developing underground tubers, many of which they valued for the liquid content. The people who spoke with Lee indicated seven !um pods, leaves, and stalks that they consumed while the plant was still growing. Finally, they included sixteen plants that they called "gum," for their resinous content. None of the eighty-five plants they provided in this exhaustive survey and census of !Kung diet included hoodia.[60]

The research on !Kung diet suggested a loose grouping of families of about sixteen each who were able to maintain their nutritional needs without farming, earning money, or purchasing goods in shops. The study coincided with a dramatic three-year drought in the British protectorate of Bechuanaland (now Botswana), when the majority of families experienced famine conditions and the World Food Program donated food to a third of the population. Lee's investigations, in particular, sought to determine how !Kung and their neighbors were able to avoid the famine through wild foods. Nor did the study of !Kung diet suggest that the affected individuals were overburdened with their search for wild foods: "In all, the adults of the Dobe camp worked about two and a half days a week. Since the average working day was about six hours long, the fact emerges that !kung Bushmen of Dobe, despite their harsh environment, devote from twelve to nineteen hours a week to getting food."[61]

Thus, at a time when poverty and famine prevailed, Lee emphasized Bushman happiness and even overconsumption. From his careful observation of the harvesting of "84 other species of edible food plants including 29 species of fruits, berries, and melons and 30 species of roots and bulbs," Lee determined that the !Kung were hardly starving, and in fact were arguably "affluent" when compared to either other populations in Botswana or even the United States.[62] Later studies made by the Japanese researcher Jiro Tanaka in an even drier region, the Kalahari Game Reserve, where approximately 1,000 San lived, counted seventy-nine edible plants. In contrast to Lee's findings, Tanaka's concluded that, given greater water scarcity, the ≠Kade communities there spent inordinate amounts of time sourcing water-filled fruits and roots, especially during September and October when "the trees and grasses wither away and every part of the Kalahari turns brown and barren."[63]

The San at Dobe outwardly resisted the gifts of tobacco and food that the researchers provided as a form of compensation for their incursions. !Kung seemed to resent to some extent Lee's attempt to offer a large cow for a Christmas day feast in 1967. In a classic anthropological article, "Eating Christmas in the Kalahari," Lee explored the insults the Bushmen leveled at a fat ox that he had offered as a holiday gift. He interpreted their taunting the supposed smallness of his cow not as the fantasy of a starving people, but a pastime of well-fed inhabitants of a remote setting who used their insults to show Lee's team that they were independent and superior to the visitors.[64]

In 1987, Lee and a South African team revisited Bushmen communities in the Dobe region of Botswana, finding "a society in transition from the hunting-gathering lifestyle . . . to that of pastoralism and dependency."[65] They argued that San children attending schools and provided with more food to eat were taller and heavier, suggesting that the small stature of the Bushmen was a result of low nutritive intake, not genetics. By the 1990s, scientists at the University of Witwatersrand estimated that the !Kung San, the subject of the Marshalls' investigations in Namibia, now numbered at around 6,000. By the 2000s, when the CSIR patented a chemical from hoodia as an appetite suppressant, the organization came under pressure to share profits with the roughly 4,000 San living in South Africa and the 80,000 San in Namibia and Botswana who all claimed hoodia as their own.[66] Ironically, given hoodia's wider distribution in less arid areas, forty years of intensive study of San foragers did not document wide use of hoodia among those living in the Kalahari in Botswana and Namibia.

Scientists at the CSIR in South Africa received plants from the Harvard Expedition studying San on the Botswana-Namibia border. CSIR researchers were involved in ongoing research into the dietary habits of nonwhites in South Africa in a bid to improve nutritional standards among workers. Beginning in the 1960s, they looked to reports of wild plants in a scheme to analyze foods from the veld. As part of this large survey of wild plants, CSIR scientists reportedly received samples of *Hoodia*, which they rejected as a viable plant for increasing the weight of undernourished black Africans. By chance, in the 1980s, biochemists at CSISR decided to start researching hoodia again with new equipment. In the process of transforming hoodia into the appetite suppressant P57, scientists at CSIR omitted the names of Khoi-San informants who may have helped shape their investigations.

The CSIR leadership changed over time to better reflect the racial diversity of South Africa. In 2007, the new CEO of CSIR, the black physicist Sibusiso Sibisi, introduced himself to me at a legal rights conference at Yale Law School. He reminded me of efforts at CSIR to find useful applications for the research that would benefit all people in South Africa. Specifically, he invited me to visit CSIR's offices in Pretoria to learn more about the success of the hoodia case.

Hoodia in the Laboratory

The Union of South Africa government approved a Scientific Research Council Act in October 1945. From its main campus in the capital, Pretoria, CSIR centralized previous medical, geological, and agricultural study at South African universities. Nutrition research emerged as a core area of study. In 1946, CSIR funded a Biochemical Nutrition Research Unit in Johannesburg and tapped British scientists to set up a National Chemical Research Laboratory (NCRL) in Pretoria. By the 1950s, concern over the health of mine workers and day laborers in urban townships led in part to the establishment of a National Department of Nutrition. In addition, the CSIR better coordinated study groups on nutrition and diet at the predominately white Universities of Pretoria, Johannesburg, and Cape Town, and in 1954 established a dedicated National Nutrition Research Institute (NNRI, later the National Food Research Institute, or NFRI).[67]

Scientists at CSIR sought ways to "improve" dietary habits in South Africa. Researchers at the NNRI examined eating practices among those at risk for cardiovascular disease from obesity, which at the time was more of a concern for whites. They also investigated malnutrition among black, colored, and Indian workers.[68] Such research led to the fortification of foods essential to everyday diet, including the addition of nutrients to corn or maize flour.[69] Veld plants entered the Food Chemistry division laboratories at NNRI during a research scheme from 1964–1982. Wild roots, leaves, and fruits promised avenues for new research on nutritious plants. Scientists at CSIR believed that black or "Bantu" workers in cities, and those on rural reservations, might have indigenous food patterns that were healthier than a European-style diet of grocery store goods. For instance, they surveyed foods that people living in Transkei, one of the official African homelands, or "Bantustans," ate on a daily basis. The researchers advocated promotion of the more than eighty easily accessible wild weeds, such as pigweed (*utyuthu, unomdlomboyi*, or *imbuya* in Xhosa), many of which were boiled and consumed as greens high in protein.[70]

Scientists obtained plants from around South Africa and surrounding areas with great difficulty. They first analyzed fruits found growing close to research facilities in Pretoria, branching out to include plants from different regions of South Africa and surrounding countries. However, they found that roots, berries, and leaves sourced farther from CSIR laboratories could easily spoil. The team at NNRI could not afford a dedicated truck to convey samples, relying on gifts from interested farmers, lay botanists, and collaborating researchers.[71] Regarding plants sent from Botswana and other locations, CSIR adopted a set of standards for preservation. Researchers urged contributors to scrub roots and tubers "thoroughly with a brush under tapwater, to remove all traces of sand and soil" before further cleansing "with distilled water."[72] Rather than using plastic bags, which promoted mold, CSIR advocated use of paper bags, or cotton wool and insulated boxes with "dry-ice packs" for more fragile items. At CSIR, researchers would freeze-dry samples to −40°C in hopes of being able to "determine the nutritive value of the products as normally eaten by the indigenous population."[73]

Available evidence does not indicate exactly where hoodia samples and information might have been obtained for CSIR research in the 1960s. In 1967 and 1968, archival reports from Harvard and CSIR indicate that the director of the NNRI Food Chemistry division, A.S. Wehmeyer, received "60 samples of veld foods eaten by the Bushmen" from Lee and his team at the Dobe site of the Harvard Expedition to the Kalahari.[74] There is evidence that eighteen or more wild plants from Namibia also came through channels at the State Museum in Windhoek. Wehmeyer's reports for CSIR, compiled by 1986, list 300 plants analyzed in the Bushveld study at the NNRI over two decades, including plants associated with San users. None of the references are to *Hoodia* or related stapeliads.[75] While the research was under way, it appears that CSIR scientists hit on nutritious plants like the mongongo nut and cucumbers popular among the San for possible cultivation and commercialization.[76] They promoted several other plants, including Tsi bean and the Sa plant, which had high levels of protein and water content.[77]

Wehmeyer's team continued to study veld plants through the 1980s. They tested hundreds of plants for vitamin C content, carotene, moisture, protein, fiber, carbohydrates, ash and minerals, phosphorus, thiamin and riboflavin, fats, and calories. Wehmeyer reported that once fresh materials were freeze-dried, they would be hand ground with a porcelain mortar and pestle. He and his colleagues then used small portions to conduct each of the tests for nutritive content. For example, he would burn plant parts at 550°C to obtain ash that was then mixed with hydrocholoric acid. He placed the resulting solution in a spectrophotometer to determine relative quantities of different minerals. To identify the amount of fiber in a sample, the researchers would mix the plants with sulfuric acid and sodium hydroxide and examine the residue left after boiling out the moisture. The Food Chemistry lab at NNRI cooperated with other laboratories at CSIR to conduct the chromatographic work to identify plant content like fatty acids and amino acids.[78]

In addition to research on plants for nutritive purposes, CSIR researchers at the NCRL identified alkaloids in plants for possible pharmaceutical applications. They received sponsorship through a number of bodies, including Smith, Kline and French, a drug firm based in the United States.[79] In 1966, NCRL tested samples of plants from South Africa for the presence of steroids as well as alkaloids. The researchers hoped to determine active ingredients so that they could then be synthesized within the South African pharmaceutical industry.[80] By 1967, NCRL believed that they might have found a local plant that could be used as chemotherapy for cancer patients.[81] NCRL obtained funding from the U.S. NIH for their research on alkaloids in South African barks with anticancer properties. The laboratory also received assistance from a French drug company to create a steroid from plants.[82] By 1975 the laboratory was able to obtain nuclear magnetic resonance (NMR) spectrum equipment, including a Varian CFT-20 NMR spectrometer capable of identifying "the carbon-13 isotope present in all organic compounds."[83] Thus, research at both the NCRL and NNRI led to further data collection on plants in Southern Africa.

Building on nutrition research to date, researchers at the NFRI at CSIR began to reexamine hoodia in 1983. Reportedly, a scientist there, P. S. Steyn, decided to begin tests on the genus using new techniques to identify the structure of plant chemicals. Researchers fed animals different extractions and studied their chemical structures to pinpoint the source of activity. One of the investigators on the project was Vinesh Maharaj. Maharaj's main area of responsibility was understanding microtoxins (secondary metabolites produced by fungi), but he had also been asked to explore possibilities for adding value to South African plants.[84]

In an interview, he elaborated on why the hoodia research was shelved for so many years, from the 1970s to 1980s. Working with previous surveys of South African plants, including Watt and Breyer-Brandwijk's 1932 study *The Medicinal and Poisonous Plants of Southern Africa,* trained botanists secured samples of plants for animal study, possibly from sites within South Africa or surrounding countries.[85] They may have supplemented their research with informal conversations in rural communities.[86] Maharaj explained that rather than focus on nutritive plants to increase the weight of workers, scientists now valued plants that made people lose weight:

> The [hoodia] project was mothballed for a long, long time. It wasn't until the early '80s, when the scientists, including myself, got involved in the project. And then the scientists then acquired technology. Namely, what you call nuclear magnetic resonance and the most sophisticated chromatographic techniques, to be able to separate the plant extracts. We, the scientists then relaunched the program, and identified and isolated the active ingredient in the plant that was mainly responsible for the anorectic effects that they observed in small laboratory animals.

Maharaj began to work at CSIR to study for his bachelor of science degree. He worked on the hoodia plant, in particular, to hone his skills in biochemistry. His family had ties to South Asia, allowing him better access to tertiary education under the apartheid regime. He continued to work at CSIR through his PhD studies, taking a leading role in natural products research by the 1990s.

CSIR did not release details of its research on hoodia until after the first patents were filed and licensed. Nor would scientists allow me to review internal reports at CSIR. Hoodia-related patents and published articles after 1997 provide some information on how the scientists identified P57, the active ingredient in *H. gordonii*. They knew that because *Hoodia* was a milkweed (*Asclepiadoideae*) it most likely would contain secondary metabolites, the types of chemicals which they had been examining in fungi. They examined the chemical structures of different compounds in extracts of *Hoodia* species. To make some of the extracts, they dried sliced *Hoodia* stalks from the "Pella district in the Northern Cape."[87] Next, they ground the desiccated hoodia to make a kind of flour, which was mixed with chemicals to create crystals that could be studied for their structures. In part of the research, they used animal models to see the effects on a dozen rats fed different concentrations of dried hoodia over a three-week period. The rats were "sacrificed" at different points in the trial, and their vital organs analyzed. The rats fed hoodia lost weight.[88] It is important that *H. gordonii,* the more bitter jackal's ghaap that Khoi-San and Afrikaaner users had rejected as too strong, was found actually to contain higher quantities of an appetite suppressing glycoside, $C_{47}H_{74}O_{15}$. Other species of *Hoodia*—for example *H. pilifera*— could also be processed to provide the chemical, which the researchers named P57.

Hoodia in the Market

In 1996, CSIR partnered with Phytopharm, a British company interested in natural products, to assist in the filing of international patents for their discovery of an active ingredient in hoodia that could be used to suppress the appetite. They first gained rights to a South African patent in 1997 for a twenty-year period. Maharaj, one of the scientists whose name appeared on the patents for hoodia's use as an appetite suppressant, explained that even though Phytopharm absorbed the costs of filing patents, the "intellectual property belongs to CSIR." Others listed on the hoodia appetite suppressant patents filed in South Africa, the United States, and the European Union included the CSIR scientists Fanie Retief Van Heerden, Robert Vleggaar, Roelof Marthinus Horak, Robin Alec Learmonth, and Rory Desmond Whittal.[89] CSIR viewed patents and intellectual property management as fundamental to its research strategy; the organization's 2005–2006 annual report reminded readers, "The number of patents produced by a research institution is a frequently used measure of research productivity."[90] Presumably, CSIR scientists found themselves under pressure to prove the commercial viability of their research. Maharaj recalled that it had actually been easier to conduct research at the CSIR during the apartheid era, when a smaller pool of researchers were competing for guaranteed funds:

And it's become much more competitive, now, compared to the early '80s. It was not that competitive. I think most of that time was much more relaxed. . . . Especially in terms of actually getting access to funding to do your research. You know, you're not the only applicant to any of these funding agencies, [they] are all global in these days.[91]

Promising financial rewards to CSIR, Phytopharm then licensed patents first to Pfizer, the transnational drug firm, and then to Unilever, makers of the diet beverage "Slim-Fast." Their aim was to find ways to produce a pharmaceutical preparation from hoodia, to be marketed as a diet drug. In 2000, Phytopharm paid CSIR $500,000 after they brokered the first agreement with Pfizer.[92] Unilever then purportedly invested over £20 million in hoodia research, cultivation, and chemical analysis before pulling out of the agreement with Phytopharm in late 2008 over fears about possible side effects.[93]

Meanwhile, many companies began operating outside of the patent and licensing agreements for P57. Scientists at CSIR described their disappointment at this turn of events. Threatened by the proliferation of hoodia products, Phytopharm published reports on the dangers of many supplements. Maharaj became visibly distraught when I explained that hoodia emails had become the spam of choice in the United States, even displacing Viagra advertisements. The U.S. television network CBS had aired a special on hoodia for *60 Minutes*, providing further ammunition for spammers. On the show, journalist Leslie Stahl traveled to the Kalahari and sampled hoodia from the wild, making the plant look accessible and safe.[94] Reflecting on this, Maharaj explained:

That's one of my biggest disappointments, as a scientist, in terms of how people have manipulated the innocent, and misled the public and actually made hoodia available to them in a nonscientifically validated product. And people have been so gullible, that they've actually fallen for it. . . . Hoodia has to be processed in a certain form. And it's only that form that actually works.[95]

First reports of fake hoodia in the market gave way to medical opinion that "several years back as little as 30 percent of the Hoodia products randomly tested contained adequate amounts of the botanical. However . . . in recent years, the percentage of products with adequate Hoodia has risen to 60 percent."[96] In Brazil, food regulators decided that cultivation and promotion of hoodia was too much of a health risk, banning hoodia commercialization there in 2007.[97]

Further setbacks with the release of hoodia pharmaceutical preparations plagued the commercialization of the CSIR's hoodia patent. First, there were delays in Phytopharm's release of a targeted diet drug. In 2003, Pfizer abandoned the research project, given the extreme cost of synthesizing P57 and the difficulties of obtaining raw material from the rare plant. This led to a downturn in Phytopharm stocks. Unilever's decision to terminate its collaboration with Phytopharm in 2008 sent the smaller firm's shares spiraling down by 43 percent in November of that year.[98] Initially, the company still hoped to marshal collaborators on hoodia, explaining that "the preclinical and clinical findings on the extract encourage its further study for obesity, as well as for pharmaceutical and veterinary applications."[99] By the close of 2010, Phytopharm abandoned further development of a hoodia diet drug, turning research back over to CSIR, with an agreement that profits from a future product developed there would be shared with the company.[100] During this period, inspired by CSIR research, Phytopharm laboratories also secured several patents for other uses of hoodia, including as an antidiabetic agent.[101]

REDISTRIBUTING BENEFITS

CSIR researchers were not the only ones with something at stake in the transformation of hoodia into a pharmaceutical. Relationships between scientists and healers had changed in South Africa by the mid-1990s. After the fall of apartheid, South African scientists recognized the role that San knowledge played in their discovery of P57 in hoodia. The San hoped that a portion of the profits generated through sales of a hoodia-based weight-loss drug could be paid into a trust fund as a form of compensation for their years of oppression under apartheid. This well-intended plan, however, would prove difficult to implement, as questions of safety and efficacy plagued the drug's developers. By 2010, with Phytopharm withdrawing support of a diet drug, CSIR continued to conduct investigations, with hopes of extending 6 percent of the profits to the San.

In a landmark settlement that characterized the new CSIR, the research body agreed to share financial benefits from hoodia patent licensing with San communities in South Africa. Roger Chennels, a South African lawyer retained by the South African San Council, threatened CSIR with legal action for theft of San intellectual property surrounding hoodia. Although hoodia was only one of the hundreds of plants described in the books that most likely influenced CSIR studies, the case had cultural currency in an atmosphere of post-apartheid reparations and redistributive justice. In the initial stages of the suit, Phytopharm argued that the San or "Bushmen" populations of the Kalahari Desert region were actually extinct and would not figure in any compensation due. Chennels successfully negotiated proceeds from sales of hoodia-related drugs for Khomani, !Xun, and Khwe communities after a protracted battle with the CSIR. Funds were placed in a trust held jointly with the South Africa San Council and the Working Group of Indigenous Minorities in South Africa (WIMSA).

Chennels claimed that distinctive San knowledge of the hunger-suppressing capabilities of the plant in South Africa spurred initial laboratory research at CSIR. Secondly, Chennels argued that the research, in contradiction of the CBD protocols, had not ensured informed consent among the San.[102] Given that the initial investigations were conducted under apartheid conditions when government policies regarded the San as the most primitive, inconsequential population in the region, the argument worked: the CSIR hastened to make amends for their self-described biopiracy.[103]

At the time of the settlement, San communities in South Africa were splintered and, essentially, culturally decimated. The San were eager for a flow of money and goods into areas that had been financially depressed for so long. The South African filmmaker Rehad Desai incorporated footage of the signing agreement at Andriesvale in March 2003 in his documentary on hoodia, *Bushman's Secrets*. Two spokespeople for the San—one in a suit, one in traditional animal skins—and Chennels flanked Sibisu, the CSIR CEO. One of the San representatives declared that he did not want any mistakes at the signing; for too long signatures on paper had negatively affected his people.[104] News reports cited the positive nature of the agreement:

> The resonance of the case for South Africa, with its history of dispossession of African people and devaluation of their culture, is huge.
>
> "We apologise to the San for having ignored them," said Dr Marthinus Horak, manager of CSIR's bioprospecting programme, speaking at a workshop on biopiracy held during the World Summit on Sustainable Development in Johannesburg last year. . . . The San plan to spend the money on education, skills development and create jobs for their people, who are among the most marginalised and poorest in the region. . . .

Ragel van Rooi walked aided by a stick painted with traditional San symbols. She wore a colourful flowered skirt. A pale blue scarf framed her wise eyes. Van Rooi did not know her age but neighbours estimated she must be about 70.

"I am happy that others can benefit from our plants," she said, when asked about the meaning of the day to her.

"Yes, but it would be wrong if fat white people overseas get slim thanks to us while our children go hungry and uneducated," replied Magdalena Kassie, 30, a community development facilitator with the South Africa San Institute in Upington, 225 km away.

"We lost our land and language, we were killed, driven out and demeaned," said Kxao Moses, WIMSA chair and a San from Namibia. "This agreement is a positive example, for once people are not exploiting us, as was the norm."

"It was the right thing to do," said Minister of Arts, Culture, Science and Technology, Ben Ngubane.[105]

As became the case with the other herbal medicines discussed here, the potential benefits to the community were overestimated, even as scientists hoped to use them as a means to right historical wrongs. Initial outlooks for hoodia were good, with the San Council receiving a first payment of R560,000 from CSIR in May 2005.[106] CSIR publicity featured the hoodia story as celebration of partnership in the New South Africa (see Figure 5.4). Fanie Retief van Heerden, another one of the scientists at CSIR whose name appeared on hoodia patents, was extremely enthusiastic about the promise of the commercialization of a Bushmen plant "to generate income for the people who have been the *custodians of the indigenous knowledge.*"[107]

San communities in neighboring Namibia and Botswana, however, were not content to garner income just from the cultivation of hoodia. They negotiated for a share in benefits from CSIR through Chennels. Meanwhile, concerns about the potential environmental costs of over-harvesting hoodia led to legislation regulating the export of the succulent. The plant had long been subject to animal grazing and required shade in order for young plants to survive. Forecasting massive efforts to collect the rare hoodia, Botswana, Namibia, and South Africa established the plant as an endangered species at the 2004 Bangkok Convention on International Trade in Endangered Species.[108] Hoodia plantations were developed in the Southern Africa region to encourage sustainable use of the plant. Unfortunately, the slow growth rate of the plant suggested continued potential for wild harvesting.[109]

Figure 5.4 "A New Partnership" featured in CSIR, "Annual Report (2005–06)." Original caption: "Discussing properties of the Hoodia plant at an experimental growth site, are Dr. Marthinus Horak, CSIR, Mr. Petrus Vaalbooi of the South African San Council and Dr. Vinesh Maharaj of the CSIR."

Knowledge Owners and Takers

The hoodia case in South Africa unfolded at a moment when the government was under pressure to make amends for the apartheid era through compensation and affirmative action. For instance, beginning in 2003, the South African government implemented a set of policies to Africanize industries under the rubric of Black Economic Empowerment. Best practices for affirmative action in employment and hiring affected not only black South Africans but also those formerly classified as colored and Indian, who had historically been underrepresented in business and education. Empowerment initiatives extended to the mining sector, where they shaped new forms of indigeneity, or what the sociologists John and Jean Comaroff have labeled "Ethnicity, Inc."[110] Similarly, anthropologist Rebecca Hardin has situated policies to grant South African communities extensive profits from mines located within their jurisdiction as a new form of "concessionary politics" reminiscent of colonial-era agreements between local chiefs and European mining agents.[111]

Redistributive justice along ethnic lines also affected efforts to extend benefits of biodiversity prospecting more widely. The hoodia case at CSIR represented a model for concessions made for ethnic minorities in nationalized research. The story of CSIR and hoodia percolated in reading materials for South African primary school students as educators and policy makers hoped to raise awareness of environmental preservation and community rights. As Chennels explained, "Benefit-sharing . . . improves resource distribution in the world, and thus contributes to global justice."[112]

In 2004, South Africa implemented a Biodiversity Act, which had been in planning since 1995 to comply with the CBD.[113] One of the main provisions of this act was that communities would be guaranteed potential benefits at the time interested foreign and domestic parties harvested plants for research. Article 83 outlined the parameters for benefit-sharing agreements. Such documents were required to indicate the form of "indigenous biological resources to which the relevant bioprospecting relates," "area or source," "quantity of indigenous biological resources," "traditional uses," and "present potential."[114] These agreements focused on access to plant harvests, even where information about the resource under consideration was already in the public domain given extensive colonial and apartheid-era investigations. The Biodiversity Act led to extreme instances of patrolling biopiracy as various provisions came into effect by 2008, and it immediately met calls of over-regulation from affected industries such as those for aloe cultivation and Amarula oil production.[115] Of particular concern were new kinds of permits that provided income for those tasked with managing the national environment. For instance, in 2010, regulators for the Act argued that anyone who had commercialized rooibos, a popular South African tea made from the leaves of a red bush, needed to apply for retrospective commercialization and export permits, which cost R5000 and R5200, respectively. Alternatively, they might face a prison stint of five years or hefty fines of R5,000,000. The Biodiversity Act also regulated permits for academic research and export of South African plants for a modest fee of R100.[116]

At CSIR, scientists described to me how additional forms of benefit-sharing agreements characterized research on other medicinal plants. Since 1996, traditional healers in South Africa had brought plants for testing to CSIR, with the understanding that they would be compensated in the long run. Tshidi Moroka, a nutrition expert at CSIR, discussed with me one possible route for providing benefits to local communities, not just along ethnic lines:

I think the system that we have in place now is running smoothly. But what the government wants to set up, like this trust fund, I really don't know how that's going to operate. But I think it will be better even for us because it will reduce the amount of time, and all this interaction, that we have to do with communities. If it's more than one community, it becomes a huge task. We'll have to go and identify these communities and make sure that you put them together. And they can then elect a representative who can go and represent them. So that can take years. Because they might not agree. . . . So it becomes a real huge problem. However, if we have a national trust fund, then it becomes easy because the government, from the province to the local structures, can handle that. So it will be easier for the researchers to be able to send in the money into that fund, which will then cater for the communities themselves.[117]

Moroka, a government worker tasked with actually implementing benefit-sharing agreements, confronted the realities of distributed knowledge. Critics of South Africa's emerging benefit-sharing policies around plant and genetic resources also noted the layers of complexity regarding the practicalities of future agreements. For instance, property disputes circulated around who owned land where plants were sourced, who used plants on a given territory, who held genetic rights, and arguably who held knowledge about their uses as well.[118]

CONCLUSION: SCIENTISTS RECOGNIZE INDIGENOUS INFORMANTS

From the offices of CSIR in Pretoria, to the garden of a white sangoma, to the economic botany vault at Kew, my investigations on the history of hoodia led me to a complex story of information exchange. Despite interviews with the scientists behind the CSIR patents, it proved difficult to determine how these scientists first learned about the plant's potential as an appetite suppressant. Early photographs in Kew, alongside the pickled bottles and the pages of pressed herbarium samples, proved long-standing circulation of information on hoodia by the 1960s, before CSIR research, on plants from the Kalahari and Western Cape. Scrutiny of early writings on the plant suggested multiple avenues for the dispersal of information. Hoodia emerged as not only a substance consumed within Bushmen communities but also as a product pickled on the colonial Dutch South African frontier and as a garden plant propagated for export at the beginning of the twentieth century. Available evidence also suggests that the CSIR may have sourced hoodia samples from Namibia or Botswana, as well as South Africa, further broadening the scope of community claimants. Although the path is difficult to trace, somehow scientists did appropriate this knowledge, and the eventual benefit-sharing agreement between CSIR and the San in South Africa was held up as a model for a more appropriate, and sustainable, relationship between local communities, national research bodies, and multinational pharmaceutical companies.

Yet, hoodia has not materialized as the blockbuster drug it was initially expected to be, partly because of concerns about long-term side effects. Nevertheless, the clear-cut dimensions of the story appeal to those advocating benefit sharing for "indigenous" communities.[119] Arguably, hoodia represents a novel case for two reasons. First, the plant grows in a small geographic location, roughly equivalent to the homelands claimed by the migrating San populations who have used it. In other words, historically, people harvested hoodia from the wild on the spot; it was difficult to cultivate elsewhere, and it was not part of a larger trade in medicinal plants. This minimized the number of wider claims to knowledge of the cactus-like plant outside of the arid regions of Southern Africa. Second, a specific set of scientists conducted research on hoodia without prior consent of a specific community associated with the plant. The researchers protected their chemical and clinical analyses with patents, before publishing results, and developed an agreement with the San as a retrospective gesture toward South African unity when the apartheid era closed in the 1990s.

This case provides a contrast to others in the book because scientists began to recognize multiple contributors to innovation through financial transactions at the dawn of the twenty-first century. Although Eli Lilly could secure research samples and harvests of periwinkle from Madagascar in the 1950s and 1960s without paying benefits to Malagasy citizens at Tolanaro, the tables had dramatically turned by the 1990s and 2000s. And although scientists in countries like Ghana emphasized their shared heritage with informants, South African scientists were happy to underscore their cultural and biological alterity to the San. The CBD, although not implemented in some countries like the United States, was appealing to emerging nations. South Africa, with a progressive new constitution and post-apartheid agenda of fairness and transparency, was well situated to experiment with implementing the CBD's provisions on a massive level. It was clear that CSIR saw patenting as fundamental to securing intellectual property rights, with no immediate plans of abandoning conventional legal regimes. At the same time, CSIR was confronted with a lobbyist in South Africa who wanted a piece of the pie. It remains to be seen how effective all the pending cases might be for communities eager to consolidate along ethnic lines for the purpose of profit sharing. Even as hoodia's potential subsided, Chennels, lawyer to the San, negotiated a new agreement with the South African company HGH Pharmaceuticals to extend benefits from research on a "San Prozac," *Sceletium tortuosum,* with San communities in Paulshoek and Nourivier at South Africa's Western Cape.[120]

In a 1971 film *Bitter Melons*, anthropologist and filmmaker John Marshall showcased music created by Ukxone, a blind San man whose songs carefully described the sensory side of life in the Botswana Kalahari where he lived with ten others. On a bow balanced on an upside-down, rusted metal dish, Ukxone brought forth complex melodies. Spliced in the footage of his playing were shots of a man munching on a sharply prickled plant. John narrated: "One of the songs that Ukxone learned from Jhuro was about ghoba, the small cucumbers with sharp spikes that ripen late and retained their water for a month or longer after the tsama melons have dried." As the melancholy melody continued, the man polished off the ghaap, then a wide shot showed him walking off into the distance. Later in the film, John continued, "'Bitter Melons' was probably Ukxone's favorite composition. Planting a melon land was an exceptional event among the people living at !Takaho. Their livelihood was gathering, an activity in which everyone participated except Ukxone who was too blind to see root vines in the grass." Scenes showed women planting seeds, and others expertly harvesting the tsama and slipping them into rough sacks.

It would be easy to think that John had simply set up his camera as idyllic scenes unfolded. Yet, notes for the film were already in preparation in his sister's notebooks from the field expeditions, each scene laid out in sequence in a period when expensive film equipment meant every second, every reel counted. From anthropologists, to botanists, to lawyers, a preoccupation to "preserve" the original culture of the San, a culture of survival and resistance sustained the San in unexpected ways. Ukxone's music, the tunes of a disenfranchised blind man, played for anthropology students around the world decades later.[121] In time, this lifestyle of gathering and collecting in the Kalahari brought unintended rewards, with San continuing the strategic performance of traditional ways at the ceremony for their rights to a hoodia patent. Whether the money would be as much as anticipated, and whether San would need to don suits or skins to survive, remained to be seen. Regardless, at Kew, the bottles of hoodia lingered, and the dried fingers of ghaap, rich in genetic possibilities, pressed up between herbarium pages, waiting for the next chapter of their rediscovery.

Toward Bioprosperity

The Western belief in an African culture that needs preservation confuses rainforests with people.
—David Hecht and Abdou Maliqalim Simone, *Invisible Governance*

This investigation of the transformation of healing plants into new medicines has probed the challenges behind assigning owners to plants and related chemicals. A historical view of past efforts to create plant-based pharmaceuticals points to the wide distribution of plants and healing knowledge. This leads to a multiplicity of stakeholders across ethnic groups, nations, and continents. At the same time, African scientists, healers, and communities have asserted themselves as unique owners and discoverers of plants and chemicals. This has been in keeping with general trends in both global scientific practice and environmental policy, which emphasize priority in product research and benefit sharing. Chemists in Madagascar, the Republic of Congo, South Africa, and Ghana have applied for patents related to their research on phytochemicals. San communities and Malagasy citizens have sought to assert first rights to potent plants. The dilemma remains as to how to encourage innovation in drug research while acknowledging the wide range of contributors to drug discovery. How do we extend prosperity from biological resources more broadly?

The book has tested several approaches to the histories of healing plants. The search for knowledge about these medicinal and potentially lucrative plants has been at once contested and cooperative, given the multiple actors involved. It is a search that has seen different protagonists and approaches over time, a search that has followed the geographic journeys and biological idiosyncrasies of the six different plants. To understand different routes to new drugs, the book has considered both plants that are widely distributed and those with a more limited geographic domain. It has addressed drug discovery from plants under colonial occupation and postcolonial rule. In constructing narratives about these bitter roots, we need to pay attention to the social interplay among curious scientists, careful healers, single-minded drug prospectors, and the phytochemicals they have sought to master. Throughout, we must consider who discovers the healing properties, who owns the knowledge, and with whom they may choose to share it. Given the multiplicity of researchers in multiple locations, the answer may be that many have discovered healing properties, and many presumably claim ownership of the plants and related knowledge.

Our modern framework of bioprospecting and the discourse of benefit sharing are not wholly adequate for capturing the complexity of the relationships among all the actors in the search for healing plants. An emphasis on priority and patents, national jurisdiction, and profit sharing with first peoples discounts the simultaneous trajectories of plant medicine in different places. This book has sought to tease out and untangle the rich strands of motivation, competition, and cooperation that underlie human science and culture. In this way we can move toward a renewed understanding of medical history that allows for synchronous, multivocal accounts.

Important shifts have occurred in our efforts to find ways to compensate and acknowledge multiple contributors to drug innovation. In African contexts, colonial occupation allowed for the obfuscation of the names of plant medicine experts who advised visiting botanists and chemists. By the 1950s, African nationalist scientists sought to be added to the roster of discoverers, assigning their names to patents, papers, and products. However, they often continued to omit the names of traditional healers or family members who had less access to the language of the laboratory but who had assisted them in their research.

It is a complex endeavor to broaden the definitions of intellectual property to better include the heritage of nonscientists. By the end of the twentieth century, nations with biologically diverse environments demanded fair trade in plants and information. The 1992 United Nations

(UN) Convention on Biological Diversity (CBD) prompted governments to implement new policies to ensure better documentation of companies harvesting natural resources for commercial gains. In Africa, for instance, the African Union worked with member states to create a Model Law for controlling access to genetic and biological resources, including seeds and medicinal plants. Kenya instituted a set of laws, including Article 53 of the 1999 Environmental and Coordination Act, "Access to genetic resources of Kenya." In 2004, South Africa established the Biological Diversity Act, which demanded that researchers compensate specific individuals and communities where knowledge and plants are sourced. Recently, the UN sponsored the creation of the Nagoya Protocol on Access to Genetic Resources and the Fair and Equitable Sharing of Benefits Arising from their Utilization to the Convention on Biological Diversity (ABS). The ABS sought to reconcile the demands of scientists, particularly in wealthier countries, for wide access to genetic materials with the demands of less wealthy countries for direct compensation through benefits.[1]

In essence, these broad policy changes over the past two decades have been in reaction to several centuries of open access to biological property for those with political power, especially under colonial regimes when there was little accountability to communities where plants were harvested.[2] The famous case of Eli Lilly's development of cancer medications from Madagascar's periwinkle led to calls for better compensation of host communities to promote environmental preservation. This book has documented how colonial networks allowed for the circulation of herbarium materials and information on their potential uses. French harvests of periwinkle and pennywort, Scottish documentation on Strophanthus, or Dutch collections of hoodia all relocated both plants and early recipes on their uses from African contexts to European ones.

It is not, however, a simple matter to monitor plants within a closed, national context. Take the case of grains of paradise. Even before the colonial period, the vagaries of Early Modern trade, trans-Atlantic slavery, and long-standing markets from tropical West Africa to the Middle East led to a wide dispersal of the spicy pepper and a range of recipes for its uses. When environments and ethnic groups overlap, and plants and people move over time, who is to say that Country X somehow owns Plant Y? Policy shifts at the international level also raise questions about the extent to which national scientists will need to acknowledge local claims to plant data within African countries in their quest for patents.

The questions raised in this book may be applied to a number of other arenas, including seeds, synthetic pharmaceuticals, genetic research, and other emerging fields in the life sciences. The commercialization of biological materials and treatments for disease bring together environmental settings, farmers, patients, scientists, and products that all shape rights to profit. Viewing them as globally and historically integrated complicates unilateral ties among specific communities, companies, and individuals. On one hand, we have the reality of a shared planet, where all of humanity deserves access to plants and information that might sustain life. On the other hand, we have the struggles for human survival that have established hierarchies to access, often within national boundaries. Our attempts to distribute the gains from these resources equitably must balance national and communal claims to autonomy with the free movement of plants and information.

I suggest a new term, "bioprosperity," to identify an approach for sustaining plants, pharmaceuticals, and people within relationships of equity. I take bioprosperity to be an amalgamation of the biological with the desire for economic and environmental sustainability and prosperity in settings outside Europe and North America. In contrast, biopiracy implicates scientists with political power and economic wealth in the theft of natural resources from poor people, or impoverished countries. To some extent, bioprospecting introduces a measure of balance, with companies pressed to afford benefits to affected populations. Bioprosperity recognizes that nations in Africa, Latin America, or Asia seek to profit from healing plants through their own initiatives. Bioprosperity acknowledges the multiple levels of exchange and distribution of plants within and across national territories in both the past and the present. Thus, bioprosperity is not a matter of simply "catching" invading biopirates to document export of plants and information. Indeed, strict regulations in one country would merely encourage interested parties to harvest overlapping species from weaker states. Additionally, substantial data on plants, including storehouses of genetic material encoded in seeds and herbariums samples, has long since migrated to locations with political power. Bioprosperity is about governments and companies investing in national research programs that cooperate within regional alliances of shared biological resources.

In what follows, I offer several suggestions for how scientists and herbalists in places like Ghana or South Africa might achieve bioprosperity given the history of actual practices surrounding plant drug discovery in African contexts. My concern is with limited definitions of priority, location, scientists, and indigeneity that proliferate in everyday discussions among stakeholders in African countries and within international discussions on managing access to genetic resources. I consider the challenges of assigning owners to pharmacological data if we take into account the history of simultaneous plant use and drug discovery across geographic boundaries. I address ways to resolve open and closed models in the search for healing plants, situating them within temporal and political contexts. And finally, I show how my initial four questions—who was first to use a plant, was traditional medicine local, how did plants become pharmaceuticals, and who benefited—might be applied to new cases, including that of bioprospecting for oil-bearing plants.

FIRST PEOPLES, FIRST SCIENTISTS

First, efforts to assign patents and benefits in plant-based drug discovery depend on constructions of priority that may be untenable when viewed historically. This book has shown how the quest for patents spread globally as African scientists participated in international standards for claiming discoveries.[3] From the 1950s, African scientists incorporated standards for scientific research that included the development of esoteric nomenclature, published reports on research, and increasingly, patents. The stories of how Ratsimamanga, Ibea, Tackie, or Maharaj secured rights to their research highlight the value of priority to scientists in Africa. The 1990s saw scientists based at African universities and research institutes adopting patents as a sign of their innovations on traditional recipes. Importantly, patents frequently were filed with the assistance of European and North American research partners.

For historians of science, the patent can be seen as an artifact of culture, socially constructed to lend credence to the idea that there are individual inventors who are first to claim discoveries. The patent is a symptom of scientific culture that privileges the fiction of the genius. Philosophically, then, the proliferation of patents even in African settings by the late twentieth century can seem like social farce. Each patent depends on a long chain of researchers, including unsung assistants, that belies the priority of those listed as the official claimants. For African scientists, however, the patent can be a way to recognize at least their contribution to research in a postcolonial context when they are still overly dependent on foreign assistance to complete laboratory investigations. Not long before he died, Tackie, the first Ghanaian pharmacy PhD recipient, proudly showed me offprints of his publications describing the new alkaloid from *Cryptolepis* named after him, cryptotackiene. Tackie was at the center of a dispute about whether he had stolen the research of other Ghanaians also invested in *Cryptolepis*. For Tackie personally, then, patents, chemicals, and publications were signs of his membership in the world community of scientists, a position he had struggled to secure throughout economic and political upheaval in a postcolonial context. Even if we situate these signs within a web of social fictions dependent on greater access to external support and education, Tackie emerged as a leading scientist within Ghana who inspired younger generations and prompted envy among his peers.

Similarly, the idea of "First Peoples" with greater claim to medicinal plants can be seen as more fantasy than reality. The fiction of groups of individuals, somehow outside of historical time, who sustain plant expertise while the rest of the world carries on with modernization has been central to recent discourse on bioprospecting and benefit sharing. I am reminded of glib discussions of "custodians of the rain forest" that I read in the 1990s that bore no resemblance to the desires of my indigenous Guan family members seeking passage away from impoverished, albeit tropical, environments. There is something deeply troubling about equating people with environments and bribing them to stay there whether they want to or not for the benefit of drug companies. As with the idea of genius, the concept of a primordial society is a symptom of our desire to limit and contain the diversity of human experience. This is not to say that there are not some populations who have tended to reside in a particular location for centuries, honing specific local expertise. However, even in the case of the culturally and physically distinctive San, we do not find a closed society that has evolved completely in isolation from the rest of the Southern Africa subregion. In particular, the plant at stake, hoodia, was not theirs exclusively, as written records demonstrate from at least the 1700s.

Yet, as with the idea of the African genius, a claim to plant priority through ethnic heritage has real value for those involved. And again, it is complicated to just dismiss indigeinity as a cultural imaginary. For the San, who received tangible financial rewards from hoodia research and other emerging plant-related drugs, asserting their rights as first inhabitants and first plant specialists of the region holds real economic value. The argument might be made that if Ghana's southern Guan populations were credited with discovering the antimalarial properties of *Cryptolepis*, they might want to stay on the Akuapem ridge sustaining biological diversity, too. Ambivalence marks the terrain between policy and historical evidence on First Peoples and first scientists.

Conflicts around plants like *Cryptolepis* and hoodia point to the challenges that remain in extending benefits from products protected with patents along national, private, or ethnic lines. There is little precedent for benefit sharing from plant resources in African settings, which might be viewed as reestablishing North-South dependencies. In colonial periods, scientists based in Europe and North America appropriated plant information without making provisions for benefit sharing. Since independence, African scientists competed with drug companies for rights to plant chemicals without acknowledging the rights of healers or communities to recipes. More recently, steps have been taken to provide access to profits from research to specific rural communities and ethnic groups, as the hoodia case explored. The historical cases show that there are continuities in colonial and postcolonial scientific investigations, that African researchers prefer patents even when pills are elusive, and that healers successfully market herbal therapies based on regional recipes.

LOCAL SEEDS, GLOBAL ROOTS

Second, the wide geographic range and historical circulation of medicinal plants like periwinkle, pennywort, and grains of paradise complicate claims to unique local knowledge. Benefit sharing as a model for redistributive rights has generally overemphasized local priority and failed to acknowledge the widespread distribution of herbal knowledge. A move toward bioprosperity must come to terms with the real temporal geography of plants and knowledge. It must reconcile local concerns with the legacy of colonialism and the open access to resources that it fostered.

Through the biography of each plant, we can locate trends in how people in African settings shared valuable information. Periwinkle, pennywort, and grains of paradise have a relatively wide distribution, and there are many competing popular medicinal recipes in wide circulation within open networks of exchange. This is why it was difficult for activists to get Eli Lilly to redistribute benefits from periwinkle drugs, or even La Roche and Ratsimamanga to pass on profits with respect to pennywort. In contrast, *Strophanthus* and *Cryptolepis* allowed for less open networks of exchange from the perspective of African actors. These relatively closed healing plant Diasporas depended on the military applications of the former and the nationalist model for interpreting "Ghana's Quinine" in the latter case, but they did not lead to a discourse on benefit sharing along ethnic lines in Ghana. Hoodia shows how postapartheid researchers hoped to close access to information on plants, even though South African flora had widely circulated in the past.

Beyond the relative distribution of information for each individual plant, it is possible to identify wider trends in how people seek to gain access to plants and information in African countries. These models for open access to shared plant resources are embedded in colonial power structures. Plant drug discovery has fluctuated between an open model of plant access and one of more restricted, or closed, access. Recently, especially for wealthier countries, a shared planet has become the symbol of environmental efforts to control climate change. Global governance structures would supersede national regulations for the good of all humanity. It is the colonial aspect of a shared planet that leads to conflict on bioprospecting regulation, as well as to the more recent move to national jurisdiction.

Increasingly, digitized photographs of herbarium sheets are widely available online. The conditions under which much of the material was collected have been acknowledged as unfair and unequal, hence the reluctance to keep herbarium information in a closed environment only available to elite researchers with financial resources to travel. In the context of plant resources, the question remains who will have access to harvests, genetic materials, and ethnobotanical records. A global, but colonial, view suggests that medicinal plants should be available to all, for the benefit of all. Resistance to open access stems from concerns over who will actually prosper from plant data. Even within global access structures, herbaria maintain rights to the use of their data, claiming ownership of plant parts and information frequently pirated during colonial occupations.

This "God's-eye view" of plant medicine relies on an imbalance of political power. The image of a fragile planet floating in space harkens back to the earliest photographs taken from the Apollo 8 spacecraft and remains a sign of U.S. hegemony in the middle of the twentieth century. Indeed, the most celebrated photograph shows a cloudy Africa, virginal and ripe for the taking. The outlook from space extends the longer trajectory of European-led efforts to dominate the planets, first through ocean voyages on Earth, now through interplanetary ventures. But a global perspective does present a vision of shared resources for those left living on our planet. National boundaries disappear, with apparently natural landscapes consuming our view. Africa's tropical rain-forest belts ring the continent; the Kalahari and Sahara Deserts lap against multiple countries. Biological resources emerge as a delicate phenomenon to be managed on a global scale.[4]

How are countries that have already relinquished so much biodata meant to prosper? At the international level, there is a conflict between those who want plant data to remain open and those who want it be closed. Even when nations hope to secure their "own" plants, there is greater awareness of the possibility of genetic resources crossing borders, if not in the past, then in the future. Article 11 of the 2011 Nagoya ABS Protocol points to a need to reconcile rights to biological property in more than one country or community:

1. In instances where the same genetic resources are found in situ within the territory of more than one Party, those Parties shall endeavour to cooperate, as appropriate, with the involvement of indigenous and local communities concerned, where applicable, with a view to implementing this Protocol.
2. Where the same traditional knowledge associated with genetic resources is shared by one or more indigenous and local communities in several Parties, those Parties shall endeavour to cooperate, as appropriate, with the involvement of the indigenous and local communities concerned, with a view to implementing the objective of this Protocol.[5]

It can be difficult to ascertain who really owns traditional knowledge, as well as the extent to which information has circulated across communities, nations, and even continents. Although it is of great importance to acknowledge the many individuals and groups who shape medical knowledge and biological discovery, it remains to be seen how all parties may be fully compensated for their roles. Take the case of grains of paradise, which promises several innovative pharmaceuticals. Are we to compensate governments or populations from Brazil to Trinidad, from Italy to Cameroon, from Mali to Morocco at a fractional level for their stake in related recipes for grains of paradise? Or do we close our eyes to the long history of a potent plant, and pretend that one or two informants in the Republic of Congo deserve to be compensated along with their villages? These are questions that still must be addressed, and they may have no clear resolution. Signatories to the Cartagena Accord—Bolivia, Ecuador, Colombia, Venezuela, and Peru—pioneered a regional alliance to monitor bioprospecting through the Andean Pact. But with the withdrawal of Venezuela, even this laudable effort came under threat. The fact that plants and people move across oceans further complicates regional legislation.[6]

This study provides new strategies for identifying the multiple contributions of different populations to plant drug discovery. In the absence of substantial written records for the history of traditional healing, it is possible to closely examine accounts of research scientists to see where they appropriated ethnobotanical information. Eli Lilly looked to the Philippines for herbal remedies for periwinkle, Peya Biotech used impotence cures from the Republic of Congo, and Phyto-Riker took up Ghanaian healing therapies for *Cryptolepis*. At the same time, these scattered healing therapies depended on larger networks of exchange, with similar treatments sustained beyond the source nation. Through herbarium records, recent surveys of herbal pharmacopeia, and interviews with plant specialists, it is possible to map the wider distribution of herbal knowledge. Although it is a time-consuming and imperfect process, such a strategy can allow for the relative distribution of benefits through biodiversity trust funds in multiple nations, in keeping with current approaches in extending benefits through class-action suits against corporations.[7]

Yet, not all healing plants lead to lucrative pharmaceuticals managed through wealthy companies. Hoodia may live on as a nutraceutical, now that information on P57 is in the public domain. And *Cryptolepis* found initial success not as the pill patented through Health Search International but as a tonic made using traditional Ghanaian recipes. As indicated in the ABS, benefit-sharing agreements place the onus of innovation on foreign commercial interests and research scientists. Importantly, I would argue that in African contexts, for instance, plant sellers are themselves heavily invested in producing tonics for regional markets. African scientists similarly make herbal products, and increasingly Western companies profit from unregulated plant-based teas, syrups, and capsules. African plant sellers present testimonies of discovery and priority, much as scientists and drug companies do. Herbalists like Assiamah even contract a network of sellers to sell their branded products. Yet, as the survey of recipes for grains of paradise showed, regional recipes show many continuities. Plant-based therapies rely on multiple contributors over time.

One possibility for managing healer autonomy over herbal remedies with similar ingredients might be through the cooperative use of geographic indications, after the French model of *appellation d'origine*. This unique approach to guaranteeing wine quality and authenticity through strict laws for the use of geographic names grew out of centuries of empirical observations. It was formerly brought under state jurisdiction in the late nineteenth century after a landmark ruling on the marketing of Champagne in 1894 whereby local foodstuffs such as wine and cheese were subsequently tied to the names of the towns and regions where they were produced. Throughout the twentieth century, the system of geographic indications was elaborated to suggest that the particular bacteria, grapes, variety of wood in wine presses, and other aspects of production created distinctive flavors not replicable elsewhere. The producers themselves monitored the quality and distinctiveness of regional products through rules on varieties of plants used, propagation, and production.[8] The *appelation d'origine* system drew wider attention in the early 1990s when the European Community adopted strict codes for geographic indications along the lines of the French system. This peculiarly local practice was globalized for trade-related reasons as the former Soviet bloc, the neo-Europes (New Zealand, Canada, United States, and Australia), and Latin American nations adopted comparable systems.[9]

Using geographic indications for the emerging herbal industry in Africa would acknowledge the viability of long-standing plant therapies with local demand. Already, herbal manufacturers exhibit a keen sense of protecting trademarks and assuring quality to maintain a customer base. Branding regional plant medicine favorites as geographically imbedded systems with long-standing techniques and distinctive flavors suggests one approach to placing more authority over ingredients and composition in the hands of regional producers. Indeed, some African countries have already pioneered this system for managing cultural products like textiles and folk music, as well as traditional foodstuffs and key global commodities like coffee.[10] Of course, the geographic indications approach is not without drawbacks. Critics of the system have questioned the scientific basis for geographic indications—in the face of genetic innovation and industrialized technique, to what extent did geography affect actual taste and quality?[11] The resurgence of interest in geographic indications in Europe in the face of widespread immigration and a perceived assault on "indigenous" European cultural practices also opens the door to romanticized, highly codified nostalgia.[12] Nonetheless, it suggests a further avenue for exploration in the intellectual property debate in Africa that allows for the regional distribution of plant knowledge and greater respect for producer autonomy. It changes the debate from one of foreign control of global resources to one of indigenous management of regional property.

Third, achieving bioprosperity will require a better understanding of the globalization of scientific research within African national frameworks. The scientific appropriation of herbal remedies is a historical process that implicates African researchers as well as the foreign drug companies classically identified with biopiracy. Approaches to drug discovery from plants may be placed within closed and open frameworks for the sharing of genetic information. Several past studies of bioprospecting have analyzed recent benefit-sharing agreements between, for instance, government bodies in the United States, U.S. research teams, and universities in Latin America. In contrast, this study has emphasized historical precedent in Africa for research on plant medicine before the establishment of such bioprospecting agreements. In this way, the book assesses the historical dynamics of scientific research, where plants have moved with and without documentation across borders.

The nationalist approach to controlling plants and information in African countries after independence can be viewed as a model of increasingly closed access. However, like the colonial model, the nationalist model incorporated private claims to pharmaceutical manufacture from plants by the end of the twentieth century. This brought nationalist scientists into conflict with citizens of their own and neighboring countries. These conflicts bore similarities to biopiracy led by foreign scientists and companies. Thus, the nationalist and colonial models for sharing and hiding plant information have both led to conflicts.

The nationalist model was part of a process of asserting control of resources within a country's borders. For African scientists, then, waiting for foreign parties to "discover" new chemicals and kick back rewards constitutes a naïve, dependent relationship with little historical precedent. Since the 1960s, emerging African countries have invested in efforts to document healing plants on both a national and regional basis. However, by the 1970s and 1980s, economic downturn and flailing government support led African scientists to seek foreign sponsorship for their research. This complicated the urge for independent control of plants within a national framework.

Further, scientists in the emerging nations of the world saw themselves in direct competition with foreign interests who similarly sought new chemicals from plants for commercial gain. The last decades have seen policy makers in places like Ethiopia, India, and Brazil keen to assert their rights to control national assets, including healing plants. But as chemist Ivan Addae-Mensah reminds us, "The Organization of

African Unity (OAU) has regulations that forbid export of medicinal plants in commercial quantities without the explicit permission of the host government. Research cooperation is likewise supposed to be regulated and monitored. But except for a few countries, these regulations are totally ignored."[13] Even when export papers were in order, African scientists were reluctant to send harvests abroad. Here, I am reminded of a teary conversation with another scientist in Ghana who for many years supplied barrels of plants to researchers in North America, sidestepping documentation. He claimed he needed the limited money he got for each kilogram of dried materials to help build his family home, pay his children's school fees, and survive a difficult academic environment in which university lecturers received less than US$100 a month. At another warehouse, I watched as workers assembled bales of dried and frozen plants for shipment to Bodyshop in the United Kingdom. Or, recall the biologists in Madagascar who managed a storeroom of dried specimens for interested researchers. At the same time that they expressed interest in the results of research on local plants abroad, these researchers described their frustrations at not having access to better research equipment to run sophisticated tests themselves.

Efforts to control national plant resources, when so much had already been documented and culled through colonial piracy, have been discouraging for African scientists. Observing the two decades since the UN's CBD, the Canadian environmentalist Pat Mooney remarked:

> While claiming to establish national authority over the biodiversity within national borders and creating a modest (albeit welcome) space for the participation of indigenous and local communities, the de facto impact of the CBD was to establish that all of the biodiversity (genes and species) pirated by colonial powers prior to 1992 and kept in zoos, herbaria, botanical gardens or gene banks instantly became the legal property of the coloniser. In one myopic moment, all the biodiversity that had been collected (and studied and considered to have value) became the heritage of the thieves; leaving to indigenous peoples and post-colonial governments all the remaining biodiversity not collected and not known to have value. This was presented as a great victory for the people.
>
> In the intervening 18 years, indigenous peoples and the governments of the global South have been fighting an uphill battle to gain acceptance of some kind of 'access and benefit sharing' deal that would be both fair and financially beneficial. Some now believe the goal is in sight while others fear it is slipping away—overridden by new tactics and technologies.[14]

It is the history of this uphill battle, before and on the tails of the CBD, that I have sought to unearth. If we still have hope in the future of African science and traditional medicine, which I do, then there are some obvious concerns. Africanization of bioprospecting at national universities and research institutes requires critical levels of capital investment, as well as capacity building. The cases in this book show how scientists based in Ghana or the Republic of Congo relied on colleagues in North America or how scientists based in Madagascar negotiated contracts with French parties. The key issue here is to find ways for such collaboration to be in the interests of African scientists and to identify channels for funding within Africa. Frequently, researchers at African universities confessed that they did not receive promised goods from their foreign partners.

Further, the cases in this book have shown how a national approach to controlling plant access must acknowledge the multiplicity of priority in drug discovery. The wide distribution of healing plants and extensive records on their uses complicates claims to national sovereignty over herbal remedies. One response to this paradox might be to derail the use of patents in plant drug discovery. African scientists (as well as their Asian or Latin American counterparts) continue to seek exclusive rights to pharmaceuticals through international patents. Policy must recognize the proliferation of claims to priority among scientists and communities, including patents in the global innovations arms race. As one African scientist explained to me, "If we do not claim our own knowledge and file patents, who will?"[15]

Frequently, government officials in African countries wanted to know the policy implications of my research and the comparative findings elsewhere. Economic stress has led academics in African countries to frequently apply their scholarship through consultancies with nongovernmental organizations (NGOs) and government bodies. When health concerns are omnipresent, and economic insecurity common, it seems myopic and unethical to conduct histories of medicine and science without any societal value.

For instance, an environmental activist in Ghana presented me with the story of an oil-bearing plant that he believed to be first used in a particular area. In his estimation, a foreign company had learned of its value through a Ghanaian cocoa buyer who traveled frequently to the town. The activist asked if the company owed money to the town. The company had presented the cocoa buyer with a car, apparently to thank him for the information, but it was now in disrepair. Turning to my analytical framework, I wondered who was first to identify the oil in the plant, was this knowledge limited to a single ethnic group, how did companies extract and modify the oil, and who ought to benefit. It turned out that colonial records suggested the plant was used across West Africa and had been relocated to German test gardens in what was then Tanganyika. Beyond the geographic dispersal of the plant, scientists in Europe and Africa had long appropriated the uses of the plant, leaving a string of patents, publications, and NGO consultancies in their wake. We visited the area where people were growing the plant for export, which turned out to be a cosmopolitan mix of families from across Ghana, complicating any ethnic-based claims. The company provided glossy descriptions of how cultivation of the plant was helping farmers across Africa. At the Kew Herbarium, I saw posters on the wall describing the potential of this plant. Yet, the question remained how to actually share real profits across a spectrum of communities and countries. My contact emphasized the need to press the foreign company to provide better compensation to his town, besides any efforts to extend wealth to farmers growing the plant in East Africa. Another possibility might be to encourage Ghanaian scientists to conduct further research and file patents, the nationalist approach that African universities still struggle to assert in their fight to control biological resources. Indeed, young people in the town told me how they were training to be science teachers and sought better access to technology to create their own products.

I hope that this book has demonstrated how highly personal experiences can interface with timely policy questions about rights to information and mechanisms for sharing wealth. For me this search began in Akuapem, near my father's hometown, in evening chats with relatives who used medicinal plants to combat infant and maternal mortality. Some of the healers assumed I wanted to document their practice closely, and they showed me leaves and recipes for me to record. I looked up letters of early missionaries and colonial authorities to the area, hoping to show how traditional medicine interfaced with religious and state regulations. I theorized that in Ghana, a discourse of toxicity shaped how colonial and postcolonial scientists and politicians gained control over medicinal plants. But then, when I traveled to other places I found that the narrative of toxicity was not universal. Instead, I realized that what interested me most was how widespread "local" knowledge really was. I wanted to know how the recipes people shared with me mutated into laboratory experiments and patent applications. I was curious about how people at different socioeconomic levels had sought control over medicinal plant medication in a bid for survival.

The parallel stories from different settings—Ghana, Madagascar, South Africa, Cameroon, and the Republic of Congo—remind us that traditional medicine is at times widely distributed across societies and at other times esoteric and highly guarded. The stories show us that botanical knowledge has crossed borders through colonial maneuverings, and that nationalist regimes are desperate to secure data for their own uses. The different stories I have told indicate that it may be almost impossible to prove who discovered something, even as we validate patents, and even harder to prove who owns traditional knowledge, even as we seek to provide profits more broadly. It is my hope that these cases will provide fodder for further research and debate as the search for healing plants continues into the future.

Many people contributed to the research that led to this book. Above all, I thank the many individuals in Ghana, South Africa, and Madagascar who allowed me to spend time speaking with them about our shared interests in plant medicine. Most wished to be remembered by their real names, which appear in the list of of Persons Consulted at the back of this book.

This project began with a conversation I had with the late Ghanaian physician Oku Ampofo in Ghana in August 1997. I am grateful to my first African history professor at Harvard University, the late Leroy Vail, who made invaluable comments on my initial attempt at a biographical essay on Ampofo, and to Fiona Hill, the undergraduate fellowships advisor in Cabot Hall who helped me write a successful Fulbright proposal to learn more. When I returned to Harvard, Mario Biagioli introduced me to debates around intellectual property rights, providing me with welcome inspiration. I thank my mentors for their support and encouragement as I traveled much uncharted territory to synthesize one of the first studies of the history of science in postcolonial Africa. Allan M. Brandt and Bridie Andrews expertly guided me through the history of medicine literature, while Emmanuel Akyeampong and Anthony Appiah taught me to better appreciate the nuances of Ghanaian history, culture, and philosophy.

I first described my plans to expand my earlier work to consider other cases from around Africa to the fellowships committee of the Harvard

Academy for International and Area Studies. It is fortunate that they believed in me, and I thank the director Jorge I. Dominguez for his mentorship and support during my two-year fellowship there. I benefited greatly from interactions with senior scholars Robert Bates and Roger Owens as well as with junior scholars, including Lisa Blaydes, Rebecca Hardin, Nahomi Ichino, Devra Coren Moehler, Harris Mylonas, Suresh Naidu, and Jun Uchida. I thank Emmanuel Akyeampong, Rosemary Coombe, Calestous Juma, Randall Packard, Charles Rosenberg, John Swann, and Larry Winnie for participating in an early review of my draft manuscript at the Academy.

I also thank colleagues at the University of California, Berkeley, for providing me with my first academic appointment and a lovely office in which to think and write. I appreciated David Hollinger's willingness to craft a space for me in both African History and the History of Science. My book is stronger given the conversations I have had with those in the history department who share my interest in Africana studies, science, medicine, and education, including Mark Brilliant, Cathryn Carson, Thomas Dandelet, Tabitha Kanogo, Thomas Laqueur, Jack Lesch, Waldo Martin, Maria Mavroudi, Massimo Mazzotti, Michael Nylan, Mark Petersen, Tyler Stovall, and Wen-hsin Yeh.

I thank colleagues in the Bay Area for their warm welcome into communities of science studies scholars and historians of medicine. At the University of California, San Francisco, I have learned much from Brian Dolan, Dorothy Porter, and Elizabeth Watkins, and their students who made the journey across the Bay to my seminar on the world history of drugs. At Berkeley, the Center for Science, Technology, Medicine and Society stretched me to consider themes in environmental studies, gender, and law. I especially thank Cori Hayden, Alastair Iles, Charis Thompson, and David Winickoff for their helpful suggestions on my work. Jenna Burrell shares my passion for science and culture in Ghana. I also thank Gilian Hart, Denise Herd, Daniel Hoffman, Amy Kapczynski, Carolyn Merchant, Hélène Mialet, Nancy Peluso, Paul Rabinow, Angelica Stacy, Kim TallBear, and Michael Wintroub for their encouragement.

As a transplant to California, I have appreciated the warmth and hospitality of colleagues who share my interest in Africa and the Diaspora. My thanks to Kwame Braun, Lane Clark, Catherine Cole, Mariane Ferme, Trevor Getz, Amma Ghartey-Tagoe Kootin, Percy Hintzen, Allyson Hobbs, Stephan Miescher, Mahasin Mujahid, Ugo Nwokeji, Leigh Raiford, and many others for providing intellectual community.

This project depended on generous support from the Institute of International Education the Fulbright Foundation, the Ford Foundation, and the Dibner Institute for the History of Science and Technology. At Harvard, I received grants from the Radcliffe Institute, Sheldon Traveling Fund, Weatherhead Center for International and Area Studies, and Jennifer Cole Fund of the Center for African Studies. Further funding to travel to the United Kingdom, South Africa, and Madagascar came from F. Warren Hellman through a fellowship for junior faculty at Berkeley.

I have conducted interviews in Ghana since 1997 and incurred innumerable debts. I especially thank Osofo K. Quarm of the Ghana Ethnomedical Foundation, Daniel Acquaye of Agribusiness in Sustainable Development in African Plant Products, and Kofi Adusei and Peter Arhin of the Ministry of Health for providing a wealth of experience and access to research reports and contacts. I thank the staff at the Ghana Public Records and Archives Administration offices, for their tireless efforts to help me unearth archival evidence. My gratitude goes to Solomon Ansah, Gustav Komalga, Ralph Kuagbenu, and Barbara Wulff for their assistance translating interviews. A number of scientists provided me with important insights and documents. I thank K. Francis Oppong-Boachie and Archibald Sittie, successive directors of the Center for Scientific Research into Plant Medicine; Ivan Addae-Mensah, chemist and former vice-chancellor at the University of Ghana; and David Ofori-Adjei, former director of the Noguchi Memorial Medical Research Institute. My thanks goes as well to Anthony Normeshie and other leaders of the Ghana Federation of Traditional Medicine.

The research in South Africa would not have been possible without the support of Vinesh Maharaj, Tshidi Moroka, Sibusiso Sibisi, and their colleagues at the Council for Scientific and Industrial Research in Pretoria. I also thank Shurnell Andersson, Thembelihle Malatji, and Jade Sasser for their assistance coordinating visits to Johannesburg and Cape Town. I am grateful to Sonya Anderson for the opportunity to share my research at the Oprah Winfrey Leadership Academy for Girls.

In Madagascar, I thank Jean Joseph Andriamanalintsoa, Lucie Rakotomalala, Hélène Ralimanana, and Isabelle Ramonta for their advice and support. For their assistance with coordinating interviews, I thank Christopher Golden, Elvie Razanami, and Charles Donné Zafidahy.

My appreciation goes to Craig Brough, Stuart Cable, Michele Losse, and their colleagues at the Herbarium, Library, and Archives of the Royal Botanic Gardens, Kew in the United Kingdom. I thank generous staff at the Public Record Office, Commonwealth Library at Oxford, Archives of the School of Oriental and African Studies, Wellcome Trust Library and Archives, London School of Tropical Medicine and Hygiene, and Museum of Natural History Archives. Patricia Kervick at the Archives of the

Peabody Museum and staff at the Economic Botany Library at Harvard were extremely helpful. I am grateful to the staff of the Interlibrary Loan Services at Berkeley.

Many people offered their thoughts at early presentations I gave on my research. For the opportunity to participate in formative conferences at the University of Edinburgh, Wellcome Institute in London, Bryn Mawr College, and Yale Law School, I thank Lawrence Dristas, Hal Cook, Paula Viterbo, and Ethan Katz, respectively. Jean Allman, Warwick Anderson, David Barnes, Adam Biggs, Daniela Bleichmar, Stephen Casper, Pratik Chakrabarti, Steven Feierman, Duana Fullwiley, Tamara Giles-Vernick, Evelyn Hammonds, Kenji Ito, Sheila Jasanoff, Dennis Laumann, Julie Livingstone, Gregg Mitman, Alondra Nelson, Naomi Rogers, Londa Schiebinger, Gabriela Soto Laveaga, Abha Sur, Helen Tilley, Eric Worby, and many others provided important feedback at annual meetings of the African Studies Association, American Association for the History of Medicine, History of Science Society, and Society for the Social Studies of Science. Since our early days at Harvard, I have benefited tremendously from conversations with Jimena Canales, Deborah Coen, Barrington Edwards, Jason Glenn, Jeremy Greene, Orit Halpern, David S. Jones, Nicholas King, Deborah Levine, Daniel Margocsy, John Mathew, Mara Mills, Sharrona Pearl, Alisha Rankin, Hanna Shell, Elly Truitt, and many others.

The material in chapter 3 first appeared in the journal *Social History of Medicine* published by Oxford University Press in 2008 as "Bioprospecting and Resistance: Transforming Poisoned Arrows into Strophanthin Pills in Colonial Gold Coast, 1885–1922," vol. 21, no. 2, pages 269–290.

I wish to thank my editor at the University of Chicago Press, Karen Merikangas Darling, for her early support of this project. Audra Wolfe provided invaluable feedback. Kevin Koy and Mark O'Connor at Berkeley's Geospatial Facility prepared the wonderful maps. At Harvard and Berkeley, I enjoyed working with student assistants to compile the geographic data, including Erica Lee, Diana Gergel, Jonathan Cole, Will Moine, Anouska Bhattacharyya, and Coral Martin. I thank Deborah Kerlogen, Ellen Thompson, Kathleen Hoover, and Clare Putnam for their administrative support at Berkeley and Harvard.

For their warm hospitality in Ghana over the years, my gratitude goes to Elizabeth Ohene and George Ofosu-Amaah and their families, Gloria Mensah, Agnes and Afua Ampofo, Timothy and Emily Barnes, Albert and Dora Barnafo, Barbara Baeta, Laura McGough, Olaf Kula, and Akosua Perbi. During our forays from Osu to Adukrom, I have gained so much from my friendships with Henrietta Asare, Sandra Ahwireng, and Ato Wilburforce. Riley Bove, Naunihal Singh, Kwame Zulu Shabbaz, Warigia Bowman, Dorinna Bekoe, Jonathan Roberts, Glenn Adams, Maame Ewusi-Mensah, Ijeoma Ikoku, Tammy Brown, Mark Ardayfio, Kofi Anku, Regine Jean-Charles, Aba Cecile McHardy, and Chris Wheat moved between Africa and North America with me. In London, I am forever beholden to Alice Peters. My friends Solita Alexander, Christine Folch, Anne Charity Hudley, and Meriel Baines Tulante, have all helped see this project through.

My brother and sister, Dk and Masi, share my soul. My cousins, Ernest and Sam, have kept me honest as they made their way between Ghana and Pennsylvania. My parents, Fran and Kwadwo, have always given me the resources to reach my potential. My husband, Koranteng Ofosu-Amaah, weathered my frequent trips abroad and kept me digitally aware and conversationally sated from the beginning to end of this project. Our children Danso and Kumiwah encourage us to sustain our healing plants for the coming generations.

Regarding Plant Medicine in Ghana

Abaitey, Alfred. Pharmacologist, Pharmacology Department, Kwame Nkrumah University of Science and Technology, Kumasi, April 3, 2003.

Abbiw, Daniel K. Ethnobotanist, University of Ghana Herbarium, Legon, October 3, 2002.

Acquaye, Daniel. Director, Agribusiness in Sustainable Development in African Plant Products, Accra, October 15, 2002, and July 2004.

Addae-Mensah, Ivan. Former Vice-Chancellor, Biochemist, Department of Chemistry, University of Ghana, Legon, 2002–2003.

Addy, Marian Ewurama. Biochemist, Biochemistry Department, University of Ghana, Legon, August 21, 2002.

Adika, Togbi. Herbalist, Ghana Federation of Traditional Medicine, Accra, August 2002.

Adusei, Kofi. Former director, Traditional and Alternative Medicine, Ministry of Health, Accra, August 13, 2002.

Agyewaa, Akosua. Food seller, market, Kwabenya, July 29, 2009.

Akajah, Sylvester. Herbalist, Ghana Federation of Traditional Medicine, August 2002.

Akotoye, Hugh Komra. Chemist, Chemistry Department, University of Cape Coast, February 5, 2003.

Ameryaw, Yaw. Chemist, Chemistry Department, University of Cape Coast, 2003.

Ampofo, Oku. Physician, cofounder Center for Scientific and Industrial Research, Mampong, Akuapem, August 21, 1997, and September 4, 1997.

Armah, Chief and Family. Royal Compound, near Asankrangwa-Enchi Road, August 2009.

Asomanig, William A. Former Dean, Faculty of Science and Chemistry Professor, University of Ghana, Legon, 2003.

Assiamah, R. K. Herbalist, Dr. Assiamah's Herbalists Association, Kwashieman, October 9, 2002.

Atiako, Djane. Herbalist, Alafia Bitters, Madina, July 28–29, 2004.

Ayim, J. S. K. Chemist, Pharmaceutical Chemistry Department, Kwame Nkrumah University of Science and Technology, Kumasi, March–April 2003.

Ayisi, Nana Kofi. Biochemist, Noguchi Memorial Institute for Medical Research, University of Ghana, Legon, April 14, 2003.

Binka. Pharmacist, Phyto-Riker Pharmaceuticals, Accra, April 23, 2003.

Dadson, Banyan Acquaye. Chemist, Chemistry Department, University of Cape Coast, March 21, 2003.

Danquah, Samuel. Psychologist, University of Ghana, Legon, February 2003.

Dzebu, Scipio. Psychiatric nurse, consultant, Alafia Bitters, July 28–29, 2004.

Eduful, Kojo Obum. Herbalist, secretary, Ghana Federation of Traditional Medicine, Accra, August 2002.

Edusei, Yusuf. Herbalist, founder, Madam Catherine Manufacturing, Kumasi, April 3, 2003.

Eiwuley, Joseph. Twi language instructor, Harvard, November 7, 2003.

Fleischer, T. C. Senior lecturer, Pharmacy Faculty, Kwame Nkrumah University of Science and Technology, Kumasi, March 31, 2003.

Gyansah, Kennedy Kwame. Tianshi distributor, Cape Coast, March 20, 2003.

Gyapong, O. K. Administrative secretary, Center for Scientific Research into Plant Medicine, Mampong-Akuapem, 2002–2003.

Laing, Ebenezer. Biologist, University of Ghana, Legon, September 24, 2002.

Marthay, Jonathan, and Ben Botwe. Policy makers, Food and Drugs Board, Accra, April 8, 11, 2003.

Mensah, Merlin. Pharmaceutical chemist, Kwame Nkrumah University of Science and Technology, 2003.

Normeshi, Anthony. Herbalist, president, Ghana Federation of Traditional Medicine, Accra, 2002–2003.

Nyarko, Alexander K. Biologist, Noguchi Memorial Institute for Medical Research, University of Ghana, Legon, February 18, 2003.

Nylander, Ladi. Director, Johnson Wax-Ghana, Tema, April 2003.

Obeng. Former employee of Ghana Export and Promotion Council, Accra, July–August 2004.

Ofori-Adjei, David. Physician, former director, Noguchi Memorial Institute for Medical Research, Legon, 2003.

Oppong-Boachie, F. K. Pharmaceutical chemist, former director, Center for Scientific Research into Plant Medicine, Mamong-Akuapem, 1997–2003.

Osafo-Mensah, Stevens. Biochemist, Center for Scientific Research into Plant Medicine, March 27, 2003.

Oteng-Yeboah, A. A. Botanist, director, Council on Scientific and Industrial Research, Accra, September 25, 2002.

Owusu-Ansah, Ernest. Chemist, University of Cape Coast, 2002–2003.

Padan. Herbalist, Padan Herbal Clinic, January 24, 2003.

Quarm, Osofo Dankama. Founder, Ghana Ethnomedical Foundation; former director Traditional Medicine, Ministry of Health, Madina, Pebase, Accra, 1997–2009.

Sarpong, Kwame. Pharmaceutical chemist, Kwame Nkrumah University of Science and Technology, April 1, 2003.

Schiff, Paul. Pharmaceutical chemist, School of Pharmacy, Pittsburgh, March 26, 2009.

Sittie, Archibald Ayitey. Chemist, acting director, Center for Scientific Research into Plant Medicine, Mampong-Akuapem, April 2003.

Tackie, Albert Nii. Pharmaceutical chemist, former director, Center for Scientific Research into Plant Medicine, at his residence in Lartebiokorshie, October 29, 2002, and April 11, 2003.

Tackie, Reggie Nii. Pharmacist, Crown Pharmacy, Accra, October 31, 2002.

Tayman, Francis. Chemist, University of Cape Coast, February 5, 2003.

Winn, Diane Robertson. Biochemist, founder, Phyto-Riker Pharmaceuticals, Accra, 1997–2004.

Yeboah, Michael Kofi. Customer, market, Kwabenya, July 29, 2009.

Regarding Plant Medicine in Madagascar

Andriamanalintsoa, Jean Joseph. Agricultural director, Société d'Exploration Agricole, Ranopiso, December 2, 2008.

Benja, Rakotonirina. Botanist, Herbarium, Institut Malgache de Recherches Appliquées, Antananarivo, November 8, 2008.

Danthu, Pascal. Center for International Cooperation in Agricultural Research for Development, Antananarivo, 2008.

Monja, Edmond. Plant Collector, Société d'Exploration Agricole, Ranopiso, December 2, 2008.

Ralimanana, Hélène. Botanist, Royal Botanic Garden Kew Madagascar Conservation Center, Antananarivo, November 25, 2008.

Ramarosandratana, Vonjy. Scientist, Institut Malgache de Recherches Appliquées and Université d'Antananarivo, November 18, 2008, and November 27, 2008.

Randrianarvo, Emmanuël. Researcher, Institut Malgache de Recherches Appliquées, Antananarivo, November 20, 2008.

Randriaovisoa, Adolphe. Biochemist, Institut Malgache de Recherches Appliquées, Antananarivo, November 28, 2008.

Randriatafika, Faly. Biologist, Rio Tinto QMM, Fort Dauphin, December 3, 2008.

Rasoanaivo, Mireille. Director of Research, Société d'Exploration Agricole de Ranopiso, December 2, 2008.

Ratsimiala Ramonta, Isabelle. Biochemist, Université d'Antananarivo, November 26, 2008.

Razafindratovo, Ndriana. Ministry of Forestry and the Environment, Antananarivo, 2008.

Razaiarimanga, Hantalalao Marie Eugenie. Plant seller, Analakely Market, Antananarivo, 2008.

Regarding Plant Medicine in South Africa

Mabaso, Mandla. Herbalist, Faraday Market, Johannesburg, July 9, 2007.

Maharaj, Vinesh. Biochemist, Council for Scientific and Industrial Research, Pretoria, July 3, 2007.

Mchunu, Nikele. Herbalist, Faraday Market, Johannesburg, July 10, 2007.

Moroka, Tshidi. Nutritionist, Council for Scientific and Industrial Research, Pretoria, July 3, 2007.

Mvubu, Goodman. Herbalist, Faraday Market, Johannesburg, July 9, 2007.

Mvubu, Samson. Herbalist, Faraday Market, Johannesburg, July 9, 2007.

Mvubu, Zinhle Nokukunga. Herbalist, Faraday Market, Johannesburg, July 9, 2007.

Nxumalo, Zanempilo. Herbalist, Faraday Market, Johannesburg, July 10, 2007.

Vergeer, Patsy. Herbalist, Patsy's Potions, Edleen, 2007.

Notes to Introduction

1. Grains of paradise, *Cryptolepis*, *Strophanthus*, and hoodia commonly appear in surveys of ethnobotanical research in Africa. Addae-Kyereme, *Cryptolepis sanguinolenta*; Carney and Rosomoff, *In the Shadow of Slavery*; Centre for Scientific and Industrial Research, Policy Research and Strategic Planning Institute, *Ghana Herbal Pharmacopoiea*; McGown, "Out of Africa."

2. "Scientists Turn to Traditional Medicinal Plants"; Beuscher et al., "Antiviral Activity"; Bickii et al., "Antimalarial Activity"; Homsy et al., "Defining Minimum Standards"; "Noguchi to Establish Unit"; "AIDS Cure Found"; Laing, "Documentation and Protection"; Jacobs, "Indigenous Plants."

3. For instance, Ghana, GMH, Reports and Studies of the Department of Traditional and Alternative Medicine, "Traditional and Alternative Medicines"; GMH, Senah et al., "Baseline Study"; Rasoanaivo, "Traditional Medicine Programmes." See also West, "Working the Borders," 21–23; Langwick, *Bodies, Politics, and African Healing*, 232–239.

4. "Workshop on Manufacturing Herbal Products."

5. Asiedu, "Drugs Board Warns Herbalists"; "Ghana National Association of Traditional Healers"; Mensah and Asamoah, "Quack Herbalists"; Saminu, "GMA Unhappy"; Ghana, GFDB, "Lists of Approved Natural Products." See also Snyman et al., "Adulteration."

6. Food and Drug Administration, "Refusal Actions."

7. Author interview with Obeng; author interview with Djane Atiako. Actual figures may be higher than those reported in "Destination of Ghana's Non-Traditional Exports."

8. For more on the history of drug regulation and patenting in Europe and North America, see Rasmussen, "Drug Industry"; Hugill and Bachmann, "Route to the Techno-Industrial World Economy."

9. This book contributes much-needed case studies from Africa to the debate on patenting phytochemicals or genetic data. On the history of controversies over the patenting of biological information, see Wirtén, *Terms of Use*, 47–76; Bugos and Kevles, "Plants as Intellectual Property"; Kevles, "Ownership and Identity"; Kevles, "Protections, Privileges, and Patents."

10. On folk medicine in early modern Europe, see Leong, "Making Medicines"; Rankin, "Becoming an Expert Practioner"; Rankin, "Duchess, Heal Thyself." On the rise of pharmaceutical culture in the twentieth century, see, for instance, Lesch, *First Miracle Drugs*; Greene, *Prescribing by Numbers*; Tone, *Age of Anxiety*; Tone and Watkins, *Medicating Modern America*; Chandler, *Shaping the Industrial Century*; Marks, *Progress of Experiment*; Swann, *Academic Scientists*.

11. On popular resistance to oral contraceptives, see Watkins, *On the Pill*; Watkins, *Estrogen Elixir*. Recent studies of oral contraceptives in global history include Kaler, *Running after Pills*; Soto Laveaga, *Jungle Laboratories*. Colonial medicine in Africa led to testing of new pharmaceuticals for sleeping sickness, malaria, syphilis, and yaws: Lyons, *Colonial Disease*; Tilley, "Ecologies of Complexity"; White, "Tsetse Visions"; White, *Speaking with Vampires*; Summers, "Intimate Colonialism"; Frenkel and Western, "Pretext or Prophylaxis?"; Vaughan, *Curing Their Ills*.

12. On the development of laboratory science, see Geison, *Private Science of Louis Pasteur*; Hannaway, "Laboratory Design"; Shapin et al., *Leviathan and the Air-Pump*; Shapin, "House of Experiment." On social intrusions on scientific life, see Latour and Woolgar, *Laboratory Life*.

13. Artifice of the genius: Mialet, *Hawking Incorporated*. Women hidden in science: Des Jardins, *Madame Curie Complex*; Watson, *Double Helix*; Maddox, *Rosalind Franklin*. Class in the history of scientists: Secord, "Science in the Pub"; Shapin, "Invisible Technician." Colonial subjects as research assistants: Jacobs, "Intimate Politics"; Noltie, *Robert Wight*. Invisible technicians during segregation in the United States: Sargent et al., *Something the Lord Made*; Timmermans, "A Black Technician."

14. Archer, "American Language," 194; Shapin, *Scientific Life*, 322; Illife, *East African Doctors*.

15. Boyle, *Shamans, Software, and Spleens*.

16. On the transfer of African agricultural and medical knowledge systems to the Americas: Janzen, *Lemba, 1650–1930*; Carney, *Black Rice*.

17. "African Traditional Medicine," 2–3, quoted in World Health Organization, "Promotion and Development of Traditional Medicine," 8.

18. "Herbal Remedies 'Do Work'"; Baker, "Does 'Indigenous Science' Really Exist." For an overview on recent debates, see Bodeker and Kronenberg, "Public Health Agenda." Historical overviews of traditional medicine in Africa: Feierman and Janzen, *Social Basis of Health and Healing*; Last, "Importance of Knowing"; Last and Chavunduka, *Professionalisation of African Medicine*; Twumasi and Warren, "Professionalisation of Indigenous Medicine."

19. Janzen, *Lemba, 1650–1930*; Allman and Parker, *Tongnaab*; Waite, *History of Traditional Medicine*.

20. On valuing local knowledge over scientific expertise, see Scott, *Seeing like a State*, 309–341. Advocates for indigenous legal rights through the Convention on Biological Diversity (CBD) gloss over questions of mobility: McClellan, "Role of International Law," 249, 56.

21. Gijswijt-Hofstra et al., *Biographies of Remedies*; Latour, *Reassembling the Social*; Appadurai, *Social Life*; Cook, *Matters of Exchange*.

22. Illife, *East African Doctors*; Flint, "Negotiating a Hybrid Medical Culture"; Flint, *Healing Traditions*; Janzen, *Quest for Therapy in Lower Zaire*; Whyte et al., *Social Lives of Medicines*; Tilley, *Africa as a Living Laboratory*.

23. Arun Agrawal shows how the differences between science and indigenous knowledge are neither incommensurable nor distinct, but that "specific strategies for protecting, systematizing and disseminating knowledge will differentially benefit different social groups and individuals." Agrawal, "Dismantling the Divide," 433.

24. Jasanoff, "Beyond Epistemology," 397.

25. Fundamental concepts in the social studies of science are directly related to ethnographic practices originally honed through colonial research in Africa. French anthropologist of science Bruno Latour based his project for an ethnographic, observatory approach to laboratory science on the image of "an intrepid explorer of the Ivory Coast, who, having studied the belief system or material production of 'savage minds' by living with tribesmen," creates a report on his experiences of cultural intimacy. Latour and his collaborator Steve Woolgar followed the activities of scientists to introduce key concepts like the scientific gift economy, in which researchers shared data and objects within a nonmonetized system. Yet, the discipline has not always fed these concepts back into global contexts. Latour and Woolgar, *Laboratory Life*, 27, 28. On the gift in academic credit systems: Packer and Webster, "Patenting Culture in Science"; Bollier, "Enclosure of the Academic Commons"; Latour, *Reassembling the Social*. The concept of scientific authorship is detailed in Biagioli and Galison, *Scientific Authorship*. On the importance of standards to international networks, see Grewal, *Network Power*.

26. Farquhar, *Knowing Practice*, 11–15; Chakrabarti, *Western Science in Modern India*; Patwardhan and Mashelkar, "Traditional Medicine-Inspired Approaches," 804–806; Gupta, "Evolution of Ayurveda"; Dev, "Ancient-Modern Concordance"; Polur et al., "Back to the Roots"; Harris et al., "Traditional Medicine Collection."

27. Shiva et al., *Enclosure and Recovery*; Shiva et al., *Campaign against Biopiracy*. The patenting of biological life forms has been especially controversial. See Bugos and Kevles, "Plants as Intellectual Property."

28. Lawyers like Roger Chennels, who were involved in the hoodia case described in Chapter 5, are now invested in providing a platform for indigenous groups to take drug companies to task. See Wynberg et al., *Indigenous Peoples, Consent and Benefit Sharing*; van den Daele, "Does the Category of Justice"; See also Sampath, *Regulating Bioprospecting*; Lesser, *Sustainable Use*; Nketiah et al., *Equity in Forest Benefit Sharing*; "Access to Forest Genetic Resources."

29. Braun and Hammonds, "Race, Populations, and Genomics." See also Fullwiley, *Enculturated Gene*.

30. "Convention on Biological Diversity"; Lesser, *Sustainable Use*; Mgbeoji, "Beyond Rhetoric"; Oguamanam, *Indigenous Knowledge*; Mathur, "Who Owns Traditional Knowledge?"

31. See, for example, Coombe, "Fear, Hope, and Longing"; Raustiala and Victor, "Regime Complex"; Corliss Karasov, "Who Reaps the Benefits?"; Dutfield, "Public and Private Domains"; Hamilton, "Long Strange T.R.I.P.S."; Dajani, "Genetic Resources," 75.

32. Rausser and Small, "Valuing Research Leads."

33. Greene, "Indigenous People Incorporated?," 214. This article was drawn from Greene, "Paths to a Visionary Politics."

34. Hayden, *When Nature Goes Public*; Hayden, "From Market to Market."

35. Dove, "Center, Periphery, and Biodiversity."

36. For a mid-1990s perspective on the policies of Shaman Pharmaceuticals, see the overview of the company in King et al., "Biological Diversity, Indigenous Knowledge"; Greene, "Indigenous People Incorporated?," 224; Stevens, "Shamans and Scientists Seek Cures"; Christensen, "Scientist at Work." For example, Shaman patent no. 5681958 issued October 28, 1997, expired October 31, 1997, after failure to pay fees; see "Notice of Expiration of Patents."

37. "We Can't Fan the Embers," newspaper editorial in response to claims that Ghana would aid the Yoruba Parapo Party in Nigeria.

38. Niezen, *Origins of Indigenism*, 17, 18.

39. Former European settlement in places like Kenya and South Africa may foster calls for indigenous rights; this is in contrast to West African nations. See Comaroff and Comaroff, *Ethnicity, Inc.*; Connah, "Transformations in Africa"; Law, "Ethnicity and the Slave Trade"; Vail, "Ethnicity in Southern African History."

40. For an overview of bioprospecting in Africa and the legal implications, see Oguamanam, *Indigenous Knowledge*. On the dilemmas of copyright in Ghana, see Boateng, "Walking the Tradition-Modernity Tightrope"; Boateng, *Copyright Thing Doesn't Work Here*.

41. Classic guides to techniques that inform my overall approach include Curtin, "Field Techniques"; Fage, "Some Notes on a Scheme"; Tonkin, "Investigating Oral Tradition." See also Kanogo, *Squatters*; Kanogo, *African Womanhood*; Miescher, *Making Men in Ghana*; Laumann, *Remembering the Germans*.

42. Chakrabarty, "Postcoloniality," 1.

43. Mafeje, *Anthropology in Post-Independence Africa*, 24, 30.

Notes to Chapter 1

1. See, for instance, Sedjo, "Property Rights, Genetic Resources," 199; Munasinghe, "Biodiversity Protection Policy," 228. An unattributed reference to US$400 million in "annual economic benefits to Americans" from periwinkle that might have included other preparations besides Eli Lilly's appeared in Myers, "Biodiversity and the Precautionary Principle," 76–77.

2. Stone, "Biodiversity Treaty"; Svoboda, "Method of Preparing Leurosine"; Svoboda et al., "Leurosidine and Leurocristine."

3. Johnson, "Drugs from Third World Plants." This statement seems to have informed past overviews of the case that highlight the wide distribution of periwinkle. See also Brown, *Who Owns Native Culture?*, 136–138.

4. Boyle, *Shamans, Software, and Spleens*, 128.

5. Parry, "Fate of Collections," 381. See also Juma, *Gene Hunters*.

6. Bloom and Walton, "Chemical Prospecting"; Eisner, "Prospecting for Nature's Chemical Riches"; Reid et al., *Biodiversity Prospecting*; Shiva et al., *Campaign against Biopiracy*; Shiva et al., *Enclosure and Recovery*; Teresko and Welter, "Save Plants"; Eisner and Beiring, "Biotic Exploration Fund"; Pearce, "Science and Technology."

7. Richardson, "Rx Makers Study."

8. Wright, "Ecological Disaster in Madagascar," 12; emphasis added.

9. A recent scholarly critique: Harper, "Not-So Rosy Periwinkle." In the early 1990s, people did note periwinkle's sandy habitat—for instance, Strauss, "Mind and Matter."

10. Author interview with Jean Joseph Andriamanalintsoa.

11. African scientists are implicated in biopiracy, in the same way that Mexican scientists sought new steroids, or Indian physicists developed a bomb; Abraham, *Making of the Indian Atomic Bomb*; Soto Laveaga, *Jungle Laboratories*.

12. I extend Duffin's research on periwinkle and priority in Canada to Madagascar and compare it with the parallel case of pennywort; Duffin, "Poisoning the Spindle."

13. Hegel and Sibree, *Lectures on the Philosophy of History*, 95; Boahen, *African Perspectives on Colonialism*.

14. Lilly, "Traitors in Our Tissues," 25.

15. Rasoanaivo, "Traditional Medicine Programmes," 1.

16. Stearn, *"Catharanthus roseus"*; Stearn, "Synopsis of the Genus Catharanthus"; Markgraf, *Flore de Madagascar*; Rasoanaivo, "Étude Chimique d'Alcaloïdes."

17. Author interview with Hantalalao Marie Eugenie Razaiarimanga.

18. Madagascar, MEF, "Medicinal Plant Exports." See also Rasoanaivo, "Rain Forests of Madagascar."

19. My translation from the French, in Markgraf, *Flore de Madagascar*, 154.

20. Flacourt, *Histoire de la Grande Isle Madagascar*, 130. On the names "Tongue" and "Tonga," see Stearn, "Synopsis of the Genus Catharanthus," 36. An early description in French of vincetoxicon is in chapter 45 of Fuchs, *Commentaires tres excellens*. Also, Dechambre et al., *Dictionnaire encyclopédique*, 609.

21. Flacourt, *Histoire de la Grande Isle Madagascar*, 35.

22. Stearn, *"Catharanthus roseus"*; Markgraf, *Flore de Madagascar*, 145; United Kingdom, RBGKEB 106.00, APOCYNACEAE *Catharanthus roseus*.

23. Miller, *Figures of the most beautiful*, 52, 124.

24. Stearn, *"Catharanthus roseus"*; Stirton, "Apocynaceae."

25. Thomas, "Plants that Soothe the Pain."

26. United Kingdom, RBGKH 40, APOCYNACEAE *Catharanthus roseus*.

27. Notes from Zambia, roseus, King, A.E., July 1, 1955; Flora of Botswana, Collector Lady Naomi Mitchison, March 15, 1967; File Mocambique, Inhaca Island, Portuguese E. Africa . . . "basic sandy soil," July 8, 1956; Plantas de mocambique, Collector P. C. M. Jansen et A. Nuvunga & G. Petrini, PJ No. 7641 Data 5-XIII 1980, ibid.

28. Seligson, "On Collecting Herbs," 27.

29. Rouillard-Guellec et al., "Étude Comparative," 489.

30. Susruta and Bhishagratna, *Sushruta Samhita*, 517; Singh et al., "Kushtha (Skin Disorders)." Orientalists translated *mandukaparni* as *Hydrocotyle asiatica*, the earlier Linnean term for pennywort; see Dutt and King, *Materia Medica*, 54, 309. On possible dates and revisions for Susruta, see Wujastyk, *Roots of Ayurveda*, 63–64.

31. Bhavna and Khatri, "*Centella asiatica.*"

32. Roxburgh, *Hortus Bengalensis*, vii, 21.

33. Biondi and Pedrosa-Macedo, "Plantas Invasoras," 136; Melo et al., "Qualidade de Produtos"; Moreira et al., "Elemental Composition"; Polonini et al., "In Vitro Photoprotective Activity."

34. Zhang et al., "Chemical Fingerprinting"; Hou et al., "Study on Invasive plants"; Zhang et al., "Genetic Diversity."

35. Earlier, a Romanian researcher also identified insulin and even filed a patent for its production in Bucharest, although it did not lead to a profitable product. See Bliss, *Discovery of Insulin;* Lilly, "Traitors in Our Tissues"; Medrzejewski, "Fiftieth Anniversary"; Wright, "Almost Famous." Some diabetics believed periwinkle was superior to insulin; see Epstein, "Action of Vinca," 35.

36. Nye and Fitzgerald, "Vinca Treatments of Diabetes," 626.

37. White, "Vinca Rosea," 143.

38. Epstein, "Action of Vinca," 38.

39. Cramp, *Nostrums and Quackery*, 9.

40. Garcia, "Treatment of Diabetes Mellitus."

41. See also Magdalena Cantoria, "Aromatic and Medicinal Herbs"; Garcia, "Distribution and Deterioration"; Garcia, "Distribution of Insulin-like Principle."

42. Noble et al., "Role of Chance Observations," 882; Noble, "Discovery of the Vinca Alkaloids," 1344; Duffin, "Poisoning the Spindle," 160; Wright, "Almost Famous," 1391.

43. Herbert A. Berens, Biddle, Sawyer & Co. Ltd., London to J. Feuell, Director, the Colonial Products Advisory Bureau (Plant and Animal), South Kensington, September 23, 1953, in United Kingdom, PRO AY4/202, "World Sources of Plant *V. rosea.*"

44. Adjanohoun, *Contribution . . . République Populaire du Congo*, 95.

45. Cramp, *Nostrums and Quackery*, 48; Kreig, *Green Medicine*, 219.

46. Interview with Johnson by Sally Smith Hughes in Johnson, "Eli Lilly."

47. Garcia, "Treatment of Diabetes Mellitus."

48. Duffin, "Poisoning the Spindle," 175.

49. See United States, BBLH, "Drug Makers," 16.

50. Interview with Johnson by Sally Smith Hughes.

51. Ibid.

52. Noble et al., "Role of Chance Observations"; Beer and Noble, "Vincaleukoblastine"; Noble, "Discovery of the Vinca Alkaloids."

53. Noble et al., "Role of Chance Observations"; Noble et al., "Further Biological Activities of Vincaleukoblastine"; Noble, "Discovery of the Vinca Alkaloids"; Noble, "*Catharanthus roseus*"; Noble, "Anti-cancer Alkaloids"; Cutts et al., "Biological Properties of Vincaleukoblastine"; Johnson et al., "Antitumor Principles"; Johnson et al., "Experimental Basis"; Svoboda et al., "Alkaloids of *Vinca rosea.*"

54. Svoboda, "Method of Preparing Leurosine"; Svoboda et al., "Leurosidine and Leurocristine"; Tyler, "Gordon Svoboda, 1922–1994"; Reid et al., *Biodiversity Prospecting.*

55. My translation from the French. Madagascar, BPBZT, "Sova Momba Ny Nahazoan."

56. Millot and Paulian, *Le Parc Botanique.*

57. Madagascar, BPBZT, Boiteau, "Travaux sur l'Asiaticoside."

58. Madagascar, BPBZT, "Les Mpsikidy."

59. Madagascar, BPBZT, Grimes and Boiteau, "Rapport sur la Thérapeutique."

60. Madagascar, BPBZT, Boiteau, "Travaux sur l'Asiaticoside," 75.

61. J. H. Sandground, parasitologist, Eli Lilly and Co., to Pierre Boiteau, June 5, 1945, in Madagascar, ANM H 291, "Lepre: Traitement de le Lepre."

62. "Advance in the Treatment of Leprosy"; Capt. V. K. Mehendale, Leprosy Specialist, Sholapur, India to Chief of Medical Services in Madagascar and the Dependencies, March 26, 1945; Crozat, Director of Medical and Sanitary Services of Madagascar to Mehendale, July 18, 1945, in Madagascar, ANM H 291, "Lepre: Traitement de le Lepre."

63. Rabemananjara, *Prince Albert Rakoto Ratsimamanga*; Allen and Covell, *Historical Dictionary of Madagascar.*

64. For more on the indigenous medical service of which Ratsimamanga was a part, see Madagascar, ANM H 261, "Organisation de l'Assistance Médicale Indigène"; ANM H 262, "Étude au Sujet d'une Réorganisation des Services Sanitaires et Médicaux de Madagascar."

65. My translation from the French, Rabemananjara, *Prince Albert Rakoto Ratsimamanga*, 214.

66. Madagascar, MIMRA, Dr. Thiroux, "Bulletin de notes." Museum provides the name Thiroux, may be Theroux.

67. Madagascar, ANM H 400, "Exposition Coloniale Internationale de Paris"; ANM H 401, "Exposition Colonial Internationale." See the list of professions of artists sent from Madagascar to France, Madascar, MIMRA, "État nominatif des artistes. "

68. Madagascar, MIMRA, "Certificat Provisoire," Faculté de Medecine, Université de Paris, December 21, 1933; MIMRA, Certificate from Professeur de Bacteriologie, Faculté de Medicine, Université de Paris, March 21, 1931.

69. Madagascar, MIMRA, "Curriculum Vitae." His research on vitamin C was read widely. MIMRA, "Telegrams to and from Ratsimamanga," includes telegrams from Herbert M. Evans, Institute of Experimental Biology, Berkeley, CA, to Ratsimamanga August 9, 1939, and from R. Tislowitz, Johns Hopkins Hospital, Baltimore, MD, to Ratsimamanga August 14, 1939. See also, for example, Buu and Ratsimamanga, "Production of Lesions of Tuberculosis"; Ratsimamanga, "Connections between Ascorbic Acid"; Ratsimamanga, "Comparative Action"; Ratsimamanga et al., "Action Hormonale."

70. Madagascar, MIMRA, "Contract between M. Marcel Laroche and M. A. Ratsimamanga, March 26, 1943."

71. Rasoanaivo, "Traditional Medicine Programmes," 73; Willis, *Malagasy Institute.*

72. Ratsimamanga and Boiteau, "Hemisuccinates and Salts." See also Boiteau et al., Boiteau et al., "Contribution a l'Étude de l'Acide"; Boiteau and Ratsimamanga, "Éffets d'un Triterpene."

73. See patent, Ratsimamanga et al., "Mixtures"; Willis, *Malagasy Institute*, 12–14.

74. Madagascar, MIMRA, Albert Rakoto Ratsimamanga Papers.

75. See, for instance, Carney, *Black Rice*; Lyons, *Colonial Disease*; Soto Laveaga, *Jungle Laboratories*.

76. Noble, "Discovery of the Vinca Alkaloids," 1346.

77. R. L. Noble, Collip Medical Research Laboratory, to Director of Research at Colonial Products Research Council, January 16, 1958, in United Kingdom, PRO AY4/202, "World Sources of Plant *Vinca rosea.*"

78. August 5, 1958, A. J. Birch, Department of Chemistry, University of Manchester replies to G. T. Bray, Esq., Colonial Products Council, Tropical Products Institute, 56/62, Gray's Inn Road, London, W.C. 1, in ibid.

79. P. C. Spensley to Noble, August 5, 1958, in ibid.

80. Noble, "Discovery of the Vinca Alkaloids," 1346–1347.

81. Lilly, "New Medicines," 11.

82. Andriamanalintsoa, "Contribution à l'Étude de la Production"; Rasoanaivo, "Prospects and Problems," 98.

83. Rasoanaivo, "Prospects and Problems," 98. On more recent cultivation, see Neimark, "Green Grabbing at the 'Pharm' Gate."

84. My translation from the French; SEAR, Arboretum Files, "Procès Verbal."

85. Author interview with Adolphe Randriaovisoa; author interview with Isabelle Ratsimiala Ramonta.

86. Rasoanaivo et al., "Alcaloïdes du *Catharanthus longifolius*," 2616; Rasoanaivo, "Étude chimique d'alcaloïdes"; Rasoanaivo, "Rain Forests of Madagascar"; Rasoanaivo, "Prospects and Problems"; Rasoanaivo, "Traditional Medicine Programmes."

87. Author interview with Faly Randriatafika.

88. de Fatima et al., "Wound Healing Agents."

89. Rouillard-Guellec et al., " Étude Comparative"; Liu et al., "Madecassoside Isolated from *Centella asiatica*"; Singh et al., "Neuronutrient Impact"; Babykutty et al., "Apoptosis Induction"; Zhang et al., "Chemical Fingerprinting."

90. Author interview with Rakotonirina Benja; author interview with Vonjy Ramarosandratana; author interview with Pascal Danthu.

Notes to Chapter 2

1. Observed at Pebase, Ghana, in August 2009 near the Quarm family residence. Author interview with Osofo Dankama Quarm.

2. Sources for the Medieval period include Setton et al., *Crusades*, 440; Ridley, *Spices*, 322; Margry, *Navigations françaises*, 26.

Sources for recipes since 1450 circulated primarily in Western Europe and North America include the following: Blake, *Europeans in West Africa*, 151; Pereira, *Elements of Materia Medica*, 243–245; Flückiger and Hanbury, *Pharmacographia*, 653; Brown, *Story of Africa*, 43; "Medical News: Treatment of Dysentery," 26; Marees, *Beschrijving en Historisch Verhaal*, 160; Tournefort, "Materia medica"; Parsons, *Microscopical Theatre*; Fenning and Collyer, *New system*; Steel, *Portable instructions*; *The compleat herbal; or, family physician*, 271; Meyrick, *New family herbal*, 203–204; Payne, *Universal geography*, 431; Anquetil, *Summary of universal history*, 405; Roscoe, *Monandrian Plants*, 330–357; Dechambre et al., *Dictionnaire encyclopédique*, 506; Bauhin et al., *Historia plantarum*,

204; Gerard and Johnson, *Herball*, 541–542; Reede tot Drakestein et al., *Hortus Indicus*; Rätsch and Müller-Ebeling, *Lexikon der Liebesmittel*, 303, 31, 428; Schweinfurth, *Heart of Africa*, 468–469; Gilman et al., *New International Encyclopaedia*, 594; Brooks, *Eurafricans*, 14, 19, 35; *Chambers's encyclopaedia*, 45; Kingsley, *West African Studies*, 57–58; Jank, *Spices*, 66; Sieberg et al., *Jahresbericht über die Fortschritte*, 145, 148; Wadström, *Essay on colonization*, 38, 199, 278; Gray and Redwood, *Grey's Supplement*, 490; Burton, *Wanderings*, 37, 154; Thicknesse, *Foreign vegetables*, 193; Leo Africanus, *History and Description of Africa*, 115; Macdonald, *Gold Coast*, 3; Martin, *History of the British colonies*, 544; Maisch, *Manual*, 334–335.

Sources for recent recipes from Algeria, Ethiopia, Cameroon, Congo, Gabon, Ghana, Mali, Morocco, Nigeria, Republic of Congo, Togo, Tunisia, and Uganda include the following: Adjanohoun, *Contribution . . . Floristiques au Gabon*, 139, 158; Sofowora, *Medicinal Plants*, 41; Sofowora, *State of Medicinal Plants Research*, 127, 162, 169; Olapade, *Herbs for Good Health*; Abbiw, *Useful Plants of Ghana*, 76, 147, 156, 191, 203; Duke et al., *CRC Handbook*, 33; Okoli et al., "Anti-inflammatory Activity"; Eyob et al., "Traditional Medicinal Uses"; Giday et al., "Medicinal Plant Knowledge"; Bellakhdar, *Pharmacopée Marocaine Traditionnelle*; Adjanohoun, *Contribution . . . République Populaire du Congo*, 377–378, 395, 403–406, 411–412, 420, 431, 436, 439, 441; Adjanohoun et al., *Contribution . . . Togo médecine traditionnelle et Pharmacopée*; Mshana et al., *Traditional Medicine and Pharmacopoeia*; Iwu, *Handbook*, 108–109; Bellakhdar et al., *Herb Drugs and Herbalists*, 41, 55, 63, 78, 87–88, 94, 109, 155, 178, 194; Adjanohoun, *Contribution . . . Floristic Studies in Western Nigeria*, 310, 311, 313, 318, 322, 324, 332, 335, 338, 347, 350, 351, 355, 356, 357; Odugbemi, *Textbook of Medicinal Plants*, 52–53, 75, 544; Adjanohoun, *Traditional Medicine and Pharmacopoeia*, 422, 424, 426–428, 434–436, 446–447, 452–455, 461, 463, 470–473, 477, 479–480, 490–491, 566–568, 572, 588, 598, 600, 604–608, 614, 616, 621, 627, 630–631, 633, 635, 638, 641–642, 644, 650, 652–655, 659–660, 666–668, 670–673, 676–677, 679, 683, 687–688, 694, 697, 703–706, 709, 718, 723, 725, 730, 733–734, 736–737, 739, 744–748; Adjanohoun, *Médecine . . . Floristiques au Mali*, 114, 120, 122, 124; Forbes, "Native Methods of Treatment," 371, 373–374, 376–377; Persinos et al., "Preliminary Pharmacognostical Study," 332.

Sources for recipes used in Guyana, Trinidad and Tobago, and Brazil include the following: Lans, *Creole Remedies*, 102, 106, 149, 179; Smith and Smith, *Memoir and Correspondence*, 373–375; Lachman-White et al., *Guide to the Medicinal Plants*, 6.

3. McGown, "Out of Africa," 5, 6.

4. Here, I use geospatial analysis of distributed herbal therapies made from *A. melegueta* to expand recent scholarship on the wide-reaching dimensions of African cultures in the Atlantic world, including botanical legacies: Carney and Rosomoff, *In the Shadow of Slavery*. See also Harris, *Global Dimensions;* Harris, "Expanding the Scope"; Rahier et al., *Global Circuits*. I take "diaspora" to include the diffusion of botanical knowledge between African communities on the continent and the appropriation of ethnobotanical information by collectors for herbaria worldwide. See Carrier, *Kenyan Khat*; Gebissa, *Leaf of Allah*; Juma, *Gene Hunters*.

5. Grenier, "Working with Indigenous Knowledge," 1. See also Brush and Stabinisky, *Valuing Local Knowledge*; Shiva et al., *Enclosure and Recovery*; King et al., "Biological Diversity, Indigenous Knowledge"; Merson, "Bio-Prospecting or Bio-Piracy";

Addae-Mensah, "Plant Biodiversity"; Bodeker, "Traditional Medical Knowledge"; Karin Timmermans, "Intellectual Property Rights."

6. Massey, "Questions of Locality."

7. Marafioti, "Meaning of Generic Names," 189, 190; Roscoe, *Monandrian Plants*; Pereira, "On the Fruit of Amomum Melegueta." Nigerian species of *Aframomum* termed "alligator pepper" also include *A. danielli, A. citratum*, and *A. exscapum*. Anthonio and Isoun, *Nigerian Cookbook*, 6; Adjanohoun, *Contribution . . . Floristic Studies in Western Nigeria*.

8. For instance, Evans-Pritchard, *Witchcraft, Oracles and Magic*; Laidler, "Magic Medicine"; Cardinall, *Natives of the Northern Territories*; Rattray, *Tribes of the Ashanti-hinterland*; Kenyatta, *Facing Mount Kenya*.

9. Horton, "African Traditional Thought," 155.

10. Connah, "Transformations in Africa"; Connah, *African Civilizations*, 118; Iliffe, *Africans: History of a Continent*, 2; Ehret, "Historical/Linguistic Evidence."

11. Lock et al., "Cultivation of Melegueta Pepper," 328.

12. Dokosi, *Herbs of Ghana*, 475.

13. On the coevolution of malaria, mosquitoes, yam cultivation, and sickle cell disease, see Webb, "Malaria." Reference to *A. melegueta* in "poorly drained soil": Clayton, "Secondary Vegetation," 229. Surveys of farmers in Western Nigeria in 1985 found that they intercropped *A. melegueta* with fruit trees on cocoa farms. Research in Ghana in the 1970s found that farmers believed cocoa trees could inhibit *A. melegueta* by providing too much shade. Lock et al., "Cultivation of Melegueta Pepper," 324.

14. Lock et al., "Cultivation of Melegueta Pepper," 322.

15. van Harten, "Melegueta Pepper," 214; de Marees, *Beschrijving en Historisch Verhaal*, 160.

16. United Kingdom, RBGKEB 170.01, "ZINGIBERACEAE *Aframomum melegueta*."

17. Abaka, *Kola Is God's Gift*; Lovejoy, *Caravans of Kola*.

18. Bühnen, "In Quest of Susu," 45; McDougall, "View from Awdaghust"; McDougall, "Sahara Reconsidered."

19. This reference in Pertz has been widely cited, although the original Latin refers to meleguet and cardamom, which could easily be spices from India, rather than *Aframomum melegueta*. See Pertz, *Monumenta Germaniae Historica*, 45–46. The first claim that meleguet in Treviso was *A. melegueta* seems to be Flückiger and Hanbury, *Pharmacographia*, 590.

20. van Harten, "Melegueta Pepper"; Warton, *History of English Poetry*, 284.

21. de Lorris, "The Garden, the Fountain," 48.

22. Beichner, "Grain of Paradise," 302; Warton, *History of English Poetry*, 284.

23. Ridley, *Spices*, 322; Taillevent and Scully, *Viandier of Taillevent*; Freedman, "Spices," 7.

24. Mauny, "Notes Historiques."

25. Davis, *Trickster Travels*, 35.

26. Hu et al., *Soup for the Qan*, 149.

27. Chang, "Africa and the Indian Ocean," 21–26.

28. Mauny, "Notes Historiques," 708.

29. Crawfurd, "On the History and Migration," 188–189; Achaya, *A Historical Dictonary*; Mauny, "Notes Historiques," 708.

30. de Marees, *Beschrijving en Historisch Verhaal,* 160. Sketches also appear in Carney and Rosomoff, *In the Shadow of Slavery.*

31. de Marees, *Beschrijving en Historisch Verhaal,* 23.

32. Comments in brackets inserted by translator: Rømer, *Reliable Account,* 199.

33. Carney and Rosomoff, *In the Shadow of Slavery.*

34. Law, "Ethnicity and the Slave Trade," 208; Klein, "Trade in African Slaves," 540.

35. Voeks, *Sacred Leaves of Candomblé,* 112, 126; Sales, *Receitas de feitiços,* 35.

36. J. S. Stutchbury, of Demerara, to the Pharmaceutical Society of London, c. 1826–1828, cited in Pereira, "On the Fruit of Amomum Melegueta," 415–416. A description of a specimen is also in a letter from W. Roscoe to Sir James Edward Smith, December 4, 1826, in Smith and Smith, *Memoir and Correspondence,* 373–374.

37. "Minutes taken (in session 1818)," 37; "Bill for improving," 170.

38. These estimates of frequency are based on a database of words appearing in a 10 percent sample of books digitized by Google, Inc. in 2010. Researchers may access graphs displaying the relatively frequency of individual words at http://books.google.com/ngrams/. Michel et al., "Quantitative Analysis of Culture."

39. Lock et al., "Cultivation of Melegueta Pepper."

40. Label on box of Sam Adams Pale Ale, purchased in Boston by the author in 2006.

41. Sofowora, *Medicinal Plants;* Adjanohoun, *Végétation des Savanes;* Adjanohoun, *Médecine . . . Floristiques au Mali;* Adjanohoun, *Contribution . . . Floristiques au Niger;* Adjanohoun, *Contribution . . . Floristiques aux Comores;* Adjanohoun, *Contribution . . . Floristiques a' Maurice;* Adjanohoun, *Contribution . . . Floristiques au Gabon;* Adjanohoun, *Contribution . . . République Populaire du Bénin;* Adjanohoun, *Traditional Medicine and Pharmacopoeia;* Adjanohoun, *Contribution . . . Floristic Studies in Western Nigeria;* Adjanohoun, *Contribution . . . Floristic Studies in Uganda;* Adjanohoun, *Contribution . . . République Populaire du Congo.* Also, Adjanohoun and Laurent, *Contribution au Recensement;* Adjanohoun et al., *Contribution . . . Togo Médecine Traditionnelle et Pharmacopée;* Adjanohoun, "Les Problèmes Soulevés."

42. These recipes incorporated both "true" grains of paradise, *Amomum granum paradise* and melegueta pepper, which may be *Aframomum melegueta.* Bellakhdar et al., *Herb Drugs and Herbalists,* 109, 120, 155, 191–192, 194, 204, 214.

43. Adjanohoun et al., *Contribution . . . Togo Médecine Traditionnelle et Pharmacopée,* 134, 135; Adjanohoun, *Traditional Medicine and Pharmacopoeia,* 604, 605; Iwu, *Handbook,* 108.

44. Adjanohoun, *Traditional Medicine and Pharmacopoeia,* 604, 605; Iwu, *Handbook,* 108.

45. Achebe, *Things Fall Apart,* 5, 14, 49.

46. Olapade, *Herbs for Good Health,* 182.

47. Kamtchouing et al., "Effects of *Aframomum melegueta.*" Author's email correspondence with Pierre Kamtchouing.

48. Much of the interest in Cameroon's biodiversity has focused on conservation of the tree *Prunus africana,* a potential treatment for prostate tumors. Mbom, "Cameroon: Scientists Fear Extinction"; Debat, "Prunus Africana Extract"; Cable et al., *Plants of Mt Cameroon;* Molisa, "Cameroon: A Third Elephant"; Page, "Political Ecology of *Prunus africana.*"

49. Betti, "Medicinal Plants Sold in Yaoundé Markets."

50. Adjanohoun, *Traditional Medicine and Pharmacopoeia*, iii, iv, 417, 652–654, 606.

51. Ibid., 452.

52. Ibid., iii, iv.

53. Ibid., i.

54. Ibid.

55. Ibid.

56. Diafouka and Bitsindou, "Medicinal Plants Sold."

57. Apparently he also mentions the use of *A. melegueta* in healing therapies in his thesis, cited in Betti, "Medicinal Plants Sold in Yaoundé Markets."

58. Allas et al., "Aframomum Seeds." Michel Ibea died in 1999 at the age of fifty, according to his daughter Gilespie. Author's email correspondence with Catherine Rivier and Jean Rivier.

59. Apparently mentioned in Cousteix, "L'Art et la Pharmacopée," although I have not seen the original, which was cited in Betti, "Medicinal Plants Sold in Yaoundé Markets."

60. There are calls for more robust analysis of such weak patents; see Gupta, "Correspondence: Surge in U.S. Patents."

61. Enti, *Rejuvenating Plants*, 21.

62. Author interview with Akosua Agyewaa.

63. Author interview with Michael Kofi Yeboah.

64. Based on my observations of cultivated *A. melegueta* near Asangregwa, Ghana. For further details on planting cycles, see Lock et al., "Cultivation of Melegueta Pepper," 324.

65. Author discussion with Chief Armah and family.

66. Addae-Mensah requested that his mother's name not be disclosed. Author interview with Ivan Addae-Mensah; author interview with Emmanuël Randrianarvo.

67. Arthur, *Cloth as Metaphor*; Konadu, "Concepts of Medicine."

68. Schiebinger, *Plants and Empire*. See also Schiebinger, "Agnotology and Exotic Abortifacients."

Notes to Chapter 3

1. United Kingdom, RBGKEB 106.00, "APOCYNACEAE *Strophanthus hispidus, S. gratus, S. divergens*."

2. Great Britain et al., *General Act of the Conference at Berlin*; Daly and Crummell, "Plan of the King."

3. The Committee for Economic Agriculture was established by the Gold Coast Legislative Council on May 16, 1887. Memo from Percival Hughes, Ag. Colonial Secretary, *Government Gazette*, Accra, Gold Coast, Western Africa, September 30, 1887, reprinted in 1889 Report on Economic Agriculture on the Gold Coast, 46, in United Kingdom, RBGKA 10.32 Ghana, "Gold Coast Colonial Products."

4. Faria e Sousa, *Africa Portuguesa*. This is referenced in the papers of a colonial medical officer. See Leitch, "Native Remedies and Poisons."

5. Hakluyt, *Voyagers' Tales*. Cited by Leitch in "Native Remedies and Poisons." See also Kutalek and Prinz, "African Medicinal Plants."

6. Vogt, *Poisons de Flèches*; Leitch, "Native Remedies and Poisons."

7. United Kingdom, RBGKA 10.32, Ghana, "Miscellaneous Gold Coast"; RBGKA 10.32, Ghana, "Gold Coast Colonial Products"; Ghana, PRAAD ADM 56/1/225, "Strophanthus Seeds."

8. Belloc and Blackwood, *Modern Traveller*; Adas, *Machines as the Measure*; Pacey, *Technology in World Civilizations*.

9. "Judy," "AMUSING: Finish 'em now."

10. Punctuation and emphasis added. Garland, "Report on the Arrow Poison Used by the Frafra," in United Kingdom, RBGKA 10.32, Ghana, "Gold Coast Colonial Products."

11. Northcott, *Report on the Northern Territories*; Irvine, *Plants of the Gold Coast*; Echenberg, "Late Nineteenth-Century."

12. United Kingdom, RBGKA 10.32, Ghana, "Miscellaneous Gold Coast"; Northcott, *Report on the Northern Territories*; Smith, "Yoruba Armament"; Mshana et al., *Traditional Medicine and Pharmacopoeia*.

13. Garland, August 23, 1899, "Medical Report on Sergeant Igala Grunshi's arrow wound at Frafra," in United Kingdom, RBGKA 10.32, Ghana, "Gold Coast Colonial Products."

14. The surname "Grunshi" suggested Gurenne or Gurunshi ethnicity and was used interchangeably with "Frafra" and "Kanjaga" to describe military recruits from the region. The term "Frafra" came to refer to the pastoral communities residing between the Rivers Sissilla and White Volta after the sound of the everyday greeting of thanks to be heard among them. Including the Nankani, Namnam, and Tallensi, "Frafra" was a loose designator for a subset of the Gur-speaking people (Gurunshi or Gurenne) based today in northeastern Ghana and southern Burkina Faso. On soldier recruitment and inoculation, see Cardinall, *Natives of the Northern Territories*; Killingray, "Military and Labour Recruitment." On the toxicity of *Strophanthus*, see Beentje, *Monograph on Strophanthus*.

15. Garland, August 23, 1899, "Medical Report."

16. Mallam, *Zabarma Conquest*, 80, 87.

17. Northcott, *Report on the Northern Territories*.

18. Garland, "Report on the Arrow Poison used by the Frafra," in United Kingdom, RBGKA 10.32, Ghana, "Gold Coast Colonial Products."

19. Ibid.

20. Leitch, "Native Remedies and Poisons."

21. Chalmers, "Further Report of Experiments," September 1, 1899, in United Kingdom, RBGKA 10.32, Ghana, "Gold Coast Cultural Products."

22. Secretary Chamberlain to Director of the Royal Gardens, Kew, February 26, 1900, in ibid.

23. Botanical report on Gold Coast arrow poison, June 13, 1899, in ibid.

24. United Kingdom, PRO FO 881/7110, "Africa: Report."

25. Labeled "not for publication": Chalmers, "Further Report of Experiments."

26. Chalmers, "Further Report of Experiments," 280.

27. Killingray and Matthews, "Beasts of Burden."

28. Chief Commissioner, Northern Territories, Gambaga to Lieut. P. J. Partridge, October 16, 1905, in Ghana, PRAAD ADM 56/1/38, "Chief Commissioner Northern Territories."

29. Ibid.

30. Ghana, PRAAD ADM 56/1/38, "Chief Commissioner Northern Territories."

31. Ibid.; Ghana, PRAAD ADM 4/1/16, "Criminal Code."

32. Ghana, PRAAD ADM 56/1/38, "Chief Commissioner Northern Territories."

33. Brukum, *Guinea Fowl*.

34. Ghana, PRAAD ADM 56/1/38, "Chief Commissioner Northern Territories."

35. Provincial Commissioner, North-Eastern Province to Chief Commissioner, Tamale, December 21, 1918, in Ghana, PRAAD ADM 56/1/236, "Zouaragu District Native Affairs."

36. Armitage to Acting Provincial Commissioner, Gambaga, June 10, 1918, in ibid.

37. Acting Col. Secretary to Armitage, July 4, 1918; A. W. Cardinall to C.N.E.P., Gambaga, May 7, 1918; Bila Moshi, Witness Statement, May 1918, all in ibid.

38. Cardinall, *Natives of the Northern Territories*, 47; emphasis added.

39. Doyle, *Lost World*, 88.

40. Beentje, *Monograph on Strophanthus*.

41. Examples of terminology for species of *Strophanthus* in Ghanaian languages include *Ajokuma* (Nzema); *Kwaman kwani* (Hausa); *Omaatwa, Omletwa* (Ga-Dangme); *Yoagbe* (Mole); *Ahoti, Matwa, Amatsiga* (Ewe); *Amamfoha, Omaatwa* (Twi); and *Oman eduapanyin* (Fante). See Mshana et al., *Traditional Medicine and Pharmacopoeia*, 83, 169, 637; Center for Scientific and Industrial Research, Policy Research and Strategic Planning Institute, *Ghana Herbal Pharmacopoiea*, 132–134.

42. Acting Chief Commissioner, Northern Territories to Tudhope, February 10, 1917, in Ghana, PRAAD ADM 56/1/225, "Strophanthus Seeds"; Fairchild, "Two Expeditions"; Mshana et al., *Traditional Medicine and Pharmacopoeia*.

43. Reported use in heart failure may be an indigenous therapy or a reference to clinical derivatives such as strophanthin.

44. Mshana et al., *Traditional Medicine and Pharmacopoeia*.

45. Leitch, "Native Remedies and Poisons."

46. Vogt, *Poisons de Flèches*; Leitch, "Native Remedies and Poisons."

47. Leitch, "Native Remedies and Poisons." In placing *Strophanthus* ahead of *Cryptolepis*, I also consider the extent to which colonial scientists expanded the diaspora of plant samples and information, the related species of *S. kombe* in Africa, and the phylogeny of the genus, which suggests historically related species in Asia.

48. Report on Economic Agriculture on the Gold Coast in United Kingdom, RBGKA 10.32, Ghana, "Gold Coast Colonial Products," 28.

49. Imperial Institute of the United Kingdom, the Colonies and India, "Report on *Strophanthus* Seed from the Gold Coast, 6 June 1917," in Ghana, PRAAD ADM 56/1/225, "Strophanthus Seeds."

50. Fraser, "*Strophanthus hispidus*," 976. On the Zambezi expedition, see Dritsas, "Civilising Missions, Natural History."

51. Fraser, "*Strophanthus hispidus*."

52. Ibid.

53. Ibid. Fraser improved on *inée*, developed by French researchers Polaillon and Carville in 1872.

54. Ibid.

55. Ibid.

56. Anon., "Note on Strophanthus."

57. Fraser, "*Strophanthus hispidus.*"

58. On William Withering's "discovery" of digitalis from folk medicine in England, see Koppanyi, "Rise of Pharmacology"; Withering, *Miscellaneous Tracts.*

59. Fraser, "*Strophanthus hispidus.*"

60. Fraenkel and Schwartz, "Abhandlungen zur Digitalistherapie."

61. See Pharmaceutical Society of Great Britain, *British Pharmaceutical Codex.* See also Desfontaines, "Extrait d'un mémoir"; Planchon, "Drogues récemment"; Stillé, *National Dispensatory*; Wood et al., *Dispensatory of the United States*; *Deutsches Arznei-buch*; Dalziel, *Useful Plants*; Irvine, *Woody Plants of Ghana.*

62. Wellcome, *From Ergot to "Ernutin,"* 165.

63. Squibb, *Squibb's Materia Medica*, 245, 340–342, 389–390.

64. Ibid.

65. During World War II, Fraenkel was asked to step down in a climate of anti-Semitism. McKenzie, "Rise and Fall of Strophanthin," 98.

66. Beentje, *Monograph on Strophanthus.*

67. Moor and Priest, "Notes," 33–34.

68. Irvine, *Plants of the Gold Coast*, xviii.

69. The Imperial Institute, "Report on *Strophanthus* Seed from Togoland," February 10, 1916, in Ghana, PRAAD ADM 56/1/225, "Strophanthus Seeds."

70. Ibid.

71. Ibid.

72. Saunders, Curator at Agricultural Department, Aburi to the Chief Commissioner, Northern Territories, November 1, 1917, in ibid.

73. Tudope, Agricultural Department, Aburi to the Colonial Secretary, Accra, October 30, 1917, in ibid.

74. District Commissioner, Yendi to Chief Commissioner, Tamale, March 30, 1918, in ibid.

75. Provincial Commissioner Northwestern Province, Wa, to Chief Commisioner, Northern Territory, Tamale, May 6, 1918, in ibid.

76. Chief Commissioner, Northern Territories to the Curator, Botanical Gardens, Tamale, April 26, 1918, in ibid.

77. Ibid.

78. Johnston, "Importance of Africa," 184.

79. Chief Commissioner, Northern Territories, Tamale, to the Commissioner of the Northern Province, Navarro, May 2, 1922, in Ghana, PRAAD ADM 56/1/225, "Strophanthus Seeds."

80. Ibid.

81. Imperial Institute, 1917, in ibid.

82. Planchon, "Drogues récemment"; Perrédès, *Contribution to the Pharmacognosy*; Jacobs, "Strophantin"; Beentje, *Monograph on Strophanthus*; McKenzie, "Rise and Fall of Strophanthin."

83. Perrédès, *Contribution to the Pharmacognosy.*

84. Tudhope, "Development of the Cocoa Industry"; Green and Hymer, "Cocoa in the Gold Coast"; Southall, "Farmers, Traders and Brokers."

85. *S. sarmentosus* was used for cortisone in the 1950s. Cantor, "Cortisone and the Politics of Empire"; McKenzie, "Rise and Fall of Strophanthin."

86. Dalziel, *Useful Plants*, 379.

87. Isaiah 2:4.

88. Duffin, *History of Medicine*, 111; Lesch, "Conceptual Change," 310–311; Bynum, "Chemical Structure," 531; Parascandola, "Alkaloids to Arsenicals," 79–81. Much of the work on drugs in African history focuses on their application rather than on discovery: Lyons, *Colonial Disease*; Tilley, "Ecologies of Complexity"; Echenberg, *Black Death, White Medicine*.

89. Said, *Culture and Imperialism*; Ashcroft et al., *Post-colonial Studies Reader*.

90. This contrasted with not only the lucrative cocoa industry, where Gold Coast plantations were developed and sustained through African leadership, but also shea-nut butter and wild rubber collection schemes in the Northern Territories. Dumett, "Rubber Trade"; Arhin, "Ashanti Rubber Trade"; Chalfin, *Shea Butter Republic*; Abaka, *Kola Is God's Gift*. See also Acting District Commissioner, Navarro to the Chief Commissioner Northern Territories, Tamale, December 31, 1907, in Ghana, PRAAD ADM 56/1/429, "District Report—Navarro."

91. Hayden, "From Market to Market."

92. Agrawal, "Dismantling the Divide."

93. Aaba, *African Herbalism*; Ghana, PRAAD ADM 29/6/5, J. Neil Leitch, "Confidential Memorandum."

Notes to Chapter 4

1. Ghana, CSRPM, Ampofo, "Autobiography"; author interview with Diane Robertson Winn.

2. Previous discussions of this case in Ghana include Addae-Kyereme, "*Cryptolepis sanguinolenta*"; Addae-Mensah, *Towards a Rational Scientific Basis*; Boye and Ampofo, "Clinical Uses of *Cryptolepis sanguinolenta* (Asclepiadaceae)"; Addy, "Cryptolepis"; Ansah and Gooderham, "Popular Herbal Antimalarial"; Winn, Gilbert Boye 1934–2003."

3. Director of Colonial Scholars, London to the Colonial Secretary, Gold Coast, July 18, 1936, in Ghana, PRAAD CSO 11/7/3, "Medical Training of Africans"; Ghana, CSRPM, Ampofo, "Autobiography."

4. Ampofo was a founding member of the arts collective, "The Akwapim Six," and became an internationally renowned sculptor. He began his studies in art while a student in Europe and was ironically inspired by exhibitions of the work of Picasso. He and his wife also starred in the 1952 British Colonial film *Boy Kumasenu*, where he played the role of a physician. Ghana, CSRPM, Ampofo, "Autobiography"; Ampofo, *My Kind of Sculpture*; July, *An African Voice*, 52–58.

5. Addae, *Evolution of Modern Medicine*, 278–279, 408–409.

6. Ghana, CSRPM, Ampofo, "Autobiography," 27.

7. Kuta-Dankwa, "Overseas Service," 1326. In East Africa, the British established medical training at Makerere by 1923. See Iliffe, *East African Doctors*, 61.

8. Ghana, CSRPM, Ampofo, "Autobiography."

9. Kuta-Dankwa, "Overseas Service," 1326.

10. Ghana, CSRPM, Ampofo, "Autobiography."

11. Chernin, "Sir Ronald Ross"; Jolly, "Suśruta on Mosquitoes."

12. Frenkel and Western, "Pretext or Prophylaxis?"

13. Edington, "Cerebral Malaria," 300.

14. Messent, "Chloroquine"; Messent, "Use and Effectiveness"; Addae, *Evolution of Modern Medicine*, 319–321. On the history of malarial treatments, see Schlitzer, "Malaria Chemotherapeutics"; Packard, *Making of a Tropical Disease*; Akyeampong, "Disease in West African History," 189–200.

15. United Kingdom, RBGKH 6, "ASCLEPIADACEAE *Cryptolepis sanguinolenta*"; RBGKEB 107.00, "ASCLEPIADACEAE *Cryptolepis sanginolenta*"; Dokosi, *Herbs of Ghana*.

16. Boye and Ampofo, "Clinical Uses," 37.

17. Adu Darko was to become a key informant for Ampofo when he opened CSRPM. He was succeeded in his position at the Center by his son Papa Adjei Darko, with whom I spoke in 1997. See also Ghana, CSRPM, Ampofo, "Autobiography."

18. Author interview with Oku Ampofo.

19. Aaba, *African Herbalism*, 6.

20. Ghana, JMPP, Moxson, "Autobiography of James Moxson," chapter 11: "Oku and Rosina."

21. Author interview with Oku Ampofo.

22. Eric Allman, Head Pharmacy Department, KNUST to the Vice Chancellor, KNUST, "The Development of the Local Herbs Programme and Its Place in the Normal Evolution of the Department of Pharmacy at the Kwame Nkrumah University of Science and Technology," in Ghana, KNUST FP/R1/199, "Research—General."

23. Nkrumah, "Speech Delivered"; emphasis added.

24. Ghana, CSRPM, Oppong-Boachie, "Research into Traditional Medicine."

25. Eric Allman, Head Pharmacy Department of Kumasi College to Dr. Duncanson, Principal of Kumasi College, February 5, 1959, in Ghana, KNUST PD/28 (vol. 1), "Departmental Letters."

26. The pharmacy school was transferred to Kumasi in 1953; United Kingdom, BL MSS. Afr. s. 2010 (4), "University of Science and Technology Kumasi."

27. Allman to Duncanson in Ghana, KNUST PD/28 (vol. 1), "Departmental Letters."

28. Allman, "Problem of the Licensed Seller of Poisons in Ghana," draft article for submission to *West African Pharmacist* (1959), in ibid.

29. F. E. V. Smith, Executive Secretary of National Research Council to Eric Allman, Head of Pharmacy Department, Kumasi College, May 4, 1959, in ibid.

30. Baako, "Opening Address."

31. Author interviews with Albert Nii Tackie.

32. "List of Herbs Processed for the Ghana Academy of Sciences Local Herbs Programme by the Pharmacy Department at the Kwame Nkrumah University of Science and Technology, Kumasi. January, 1964," in a handwritten note at the top, apparently by Allman: "Programme commenced 1961. Plants with alkaloidal content were isolated and investigated further," in United Kingdom, BL MSS Afr.s. 2010 (1), "Eric Allman Papers 'Notes and Cuttings.'"

33. Silviculturist, Forestry Department-Kumasi, to Eric Allman, Kumasi College, May 21, 1962; Eric Allman, Kumasi College, to F. Nartey, Denmark, February 15, 1961,

in Ghana, KNUST FP/R1/199, "Research—General"; Enti, "Collection of Botanical Specimens"; Enti, *Rejuvenating Plants*.

34. Eric Allman, Head, Pharmacy Department, Kumasi to E. A. Gyang, June 30, 1961; Eric Allman to A. H. Beckett, Chelsea College of Science and Technology, London, October 31, 1960, in Ghana, KNUST FP/R1/199 "Research—General."

35. "World Scientists to Study Secrets."

36. R. B. Baffour, Vice-Chancellor KNUST to Eric Allman, Head, Pharmacy Department, October 8, 1963, in Ghana, KNUST FP/R1/199, "Research—General."

37. Allman to Baffour, October 15, 1963, in ibid.

38. Tackie was listed as the second author on articles emanating from his July 1963 PhD from the University of London. For instance, Beckett and Tackie, "Structure of Rotundifoline"; Beckett and Tackie, "Structures of Alkaloids"; Beckett et al., "*Mitragyna* Species of Ghana . . . Pellegr"; Beckett et al., "*Mitragyna* Species of Ghana . . . Kuntze."

39. Arnold Beckett, Chelsea, to Editor, *Ghanaian Times*, September 17, 1963, in Ghana, KNUST FP/R1/199, "Research—General."

40. "Ghana Makes Big Medical Discovery."

41. Ghana, KNUST PD/39, "National Research Council Minutes."

42. Fosu, Head Teacher, Catholic Middle School, Dwinyama, via Kumasi, note to scientists of local herb committee, June 23, 1961, in ibid.; emphasis added.

43. S. T. Quansah, Executive Secretary, N.R.C., to J. L. Fosu, head teacher, Catholic Middle School, Dwinyama, June 19, 1961, in ibid.

44. Ghana, GPTHA, "Correspondence, Ghana Psychic and Traditional Healers Association, Western Region," 1970–1977.

45. "Healers Find New Drug."

46. Prof. A. N. Tackie, Dean Faculty of Pharmacy, to Mr. H. K. Armah, Vice District Chief Farmer, Aiyinasi, December 10, 1964. Armah sent his original offer to supply information on "medical discovery in herbs" on June 18, 1964, in Ghana, KNUST FP/R1/199, "Research—General."

47. Selormey, "Address at the Inauguration of the CSRPM," in Ghana, KNUST FP/A7c/vol, "Advisory Committee."

48. F. G. Torto, Head Chemistry Department, University of Ghana-Legon, to Dr. Yanney-Wilson, secretary Ghana Academy of Science, March 4, 1964, in Ghana, KNUST FP/R1/199, "Research—General."

49. Tackie and Ampofo, "Memorandum," in Ghana, KNUST FP/A7c/vol, "Advisory Committee."

50. Ibid.

51. Selormey, "Inauguration of the CSRPM," in ibid.

52. J. K. Apeagyei, Reduction Plant, Konogo Goldfields–State Gold Mining Corporation, to Director, CSRPM, November 14, 1973, in ibid.; emphasis added.

53. Boye and Ampofo, "Clinical Uses"; Tackie et al., "Cryptospirolepine."

54. Boye and Ampofo, "Clinical Uses," 38.

55. Author interview with Banyan Acquaye Dadson.

56. Ewusi, *Political Economy of Ghana*, 18.

57. E. Evans-Anfom, Chairman, National Council for Higher Education, "Minutes of a Meeting held at the Conference Room of the National Council for Higher Education at the State House, at 10 A.M. on Tuesday 27th May, 1975 to Discuss Co-operation

between the Centre for Scientific Research into Plant Medicine, Mampong-Akuapem, and the Universities including the Medical School and the Council for Scientific and Industrial Research," in Ghana, KNUST FP/A7c/vol, "Advisory Committee."

58. Author interview with Banyan Acquaye Dadson.

59. E. A. Gyang, Chairman of CSRPM Council, Dean Faculty of Pharmacy, University of Science and Technology (U.S.T.), in "Minutes of Meeting held between representatives of [CSRPM] and heads of Departments of the Faculty of Pharmacy, U.S.T. in the Faculty Common Room on Mon. 24th March, 1975," in Ghana, KNUST FP/A7c/vol, "Advisory Committee."

60. "Memorandum for ¢250,000.00 for Mass Spectrometer, Physiograph and Automatic Polarimeter for Medicinal Plant Research in Ghana," to Secretary, National Redemption Council, 9 April 1975, in ibid.; emphasis added.

61. "CSRPM Minutes . . . Mon. 17th Nov., 1975," in ibid.

62. "Request for Funds to Purchase Mass Spectrometers for Scientific Research into Plant Medicine," May 12, 1976, in ibid.

63. Author interview with Albert Nii Tackie.

64. K. Tete-Asiedu, secretary, CSRPM, "National Service," July 7, 1975, in Ghana, KNUST FP/A7c/vol, "Advisory Committee."

65. Boakye-Yiadom and Bamgbose, *Proceedings*.

66. Ghana, CSRPM, CPM/A/CR/59, "Council meetings—Director's papers."

67. Author interview with FrancisTayman; author interview with Archibald Sittie.

68. Ampofo, *First Aid in Plant Medicine*.

69. Addae-Mensah, *Towards a Rational Scientific Basis*, 33–36.

70. Author interviews with Albert Tackie.

71. Tackie and Schiff, "Compound and Method."

72. Author interview with Paul Schiff.

73. These advertisements were observed during conversations with Tackie's son; author interview with Reggie Nii Tackie.

74. "Plant Medicine," in *Phyto-Riker Pharmaceuticals Ltd.*, 4.

75. "Bi-Annual Report."

76. Ghana, GPTHA, "Correspondence . . . 1963–1977."

77. Kilham, *Medicine Hunting*; author interview with Sylvester Akajah.

78. Biagioli and Galison, *Scientific Authorship*.

79. Soto Laveaga, *Jungle Laboratories*. Soto Laveaga notes that the national vision in Mexico depended on research in the United States as well.

Notes to Chapter 5

1. United Kingdom, RBGKEB 107.00 ASCLEPIADACEAE *Hoodia gordoni*. Apparently, "it was collected in the Karoo by Mr. Lycett of Worcester in 1873 and named for Sir Henry Barkley who sent an account of it to . . . Kew in the following year." By 1932, "*H. Barkleyi* has not been rediscovered," making the bottled specimen even more of a novelty in its day. Watt and Breyer-Brandwijk, *Medicinal and Poisonous Plants of Southern Africa*, 1086.

2. "Hoodia Slimming Tea Box," Bija Healing Teas, purchased 2007, author's collection.

3. See Coombe, *Cultural Life*, 174–177; Vermeylen, "From Life Force to Slimming Aid."

4. On the extent to which the Hoodia case conforms to the World Trade Organization Agreement on Trade Related Aspects of Intellectual Property, see Bodeker, "Traditional Medical Knowledge," 785–814. See also Dolder, "Traditional Knowledge and Patenting"; Moyer-Henry, "Patenting Neem and Hoodia"; "Protecting Traditional Knowledge."

5. "South Africa: Feature."

6. For an overview of competing land law in countries where the San live, see Munzer and Simon, "Territory, Plants, and Land-Use Rights."

7. Wiredu, "Our Problem of Knowledge," 182; Agrawal, "Dismantling the Divide," 424.

8. Jim Chen argues that nature, although sacred, nonetheless provides raw data for industrial applications and should be made widely accessible. Chen, "Webs of Life," 589.

9. Wynberg et al., *Indigenous Peoples, Consent and Benefit Sharing*. In South Africa, the primacy of Bushmen to the region has become critical to national narratives of continuous migration. See Morgan, *Representing Bushmen*, 6; Bank, *Bushmen in a Victorian World*.

10. It is unclear whether both Khoi and San groups used these leaves. See Thunberg, *Travels in Europe, Africa and Asia*, 312; van Wyk, "Review of Khoi-San," 333.

11. On numbers of San, see Hitchcock et al., *Updating the San*. On *Pteronia*, see Hulley et al., "Ethnobotany"; Hulley et al., "Review of Pteronia Species."

12. Odendaal et al., *Richtersveld*, 52; Watt and Breyer-Brandwijk, *Medicinal and Poisonous Plants of Southern and Eastern Africa*, 138.

13. Thunberg, *Travels in Europe, Africa and Asia*, 140.

14. Ibid., 171.

15. *Use and Misuse*, 68.

16. Watt and Breyer-Brandwijk, *Medicinal and Poisonous Plants of Southern Africa*, 1082.

17. Ibid., 1076–1078, 1081, 1085.

18. Pappe, *Silva capensis*, 54–55. This is the only nineteenth-century reference cited in Hoodia patents and scientific articles, suggesting that CSIR researchers were not aware of further references I have found. See van Heerden et al., "Appetite Suppressant," 2546.

19. United Kingdom, RBGKH 147, "ASCLEPIADACEAE *Hoodia gordonii*."

20. Vylder, *Journal of Gustaf de Vylder*, 59. Reference found through a Google search for "ghaap," though I have not seen it referred to elsewhere.

21. Watt and Breyer-Brandwijk, *Medicinal and Poisonous Plants of Southern and Eastern Africa*, 138.

22. "Evening Meeting in London," 389.

23. Ibid., 389–390.

24. "Story of Our Fire."

25. Smith, *Theories about the Origin*, 8; Coombs, *South African Plants*.

26. On Khoi-Dutch relations, see, for instance, Marks, "Khoisan Resistance"; Mitchell, *Belongings*.

27. Marloth, *Flora of South Africa*, 91; Mathews, "South African Succulents," 14.

28. van Wyk, "Review of Khoi-San."

29. Osborn and Noriskin, "Data Regarding Native Diets," 605–606.

30. Marloth, *Flora of South Africa*, 94. Marloth, quoted and referenced as a key insight to Hoodia's dietary potential in van Wyk, "Review of Khoi-San," 333. Also cited in later CSIR patents and publications: van Heerden et al., "Pharmaceutical Compositions"; van Heerden et al., "Appetite Suppressant," 2546.

31. Marloth, *Dictionary of the Common Names*, 61, 125.

32. By the 1930s, thousands of herbalists and plant peddlers had sent requests for licenses, to no avail. One argument is that white doctors sought to snuff out the competition. See Flint, *Healing Traditions*.

33. Dubow, *Scientific Racism*, 20–65; Gordon, *Bushman Myth*.

34. Maingard, "Some Notes on Health and Disease," 227.

35. Ibid., 235.

36. Ibid., 232.

37. Tobias, "15 Years of Study," 74.

38. South Africa, SANA, NTS 7756/26/335n "Grahamstown."

39. Bousman, "Coping with Risk."

40. For an excellent overview of Bushmen history in Namibia and surrounding countries, including atrocities, see Gordon, *Bushman Myth*, 77–85.

41. "Bossieman" may have been shortened to "boseman" and "bushman" in English: van Wyk, "Review of Khoi-San," 333. See also Gordon, *Bushman Myth*, 4–7; Hitchcock et al., *Updating the San*.

42. Maingard, "Introduction," 234.

43. Gordon, *Bushman Myth*; Scott, *Art of Not Being Governed*, 394. For an overview of genetic debates, see Morris, "Myth of the East African 'Bushmen.'" On the debate about the historical isolation of the San, see, for example Hitchcock et al., *Updating the San*; Solway et al., "Foragers, Genuine or Spurious?"

44. Thomas, *Old Way*, 51; Thomas, "Lion/Bushman Relationship."

45. Brief overviews of phytochemical research in South Africa are found in Hutchings, "Survey and Analysis," 111; Fox and Young, *Food from the Veld*, 13–17.

46. "South Africa Crops Ruined"; "Storm Ends African Drought"; Osborn and Noriskin, "Data Regarding Native Diets," 605. See also Levy et al., "Food Value"; Fox and Back, "Preliminary Survey."

47. Levy et al., "Food Value," 699.

48. Osborn and Noriskin, "Data Regarding Native Diets," 605–606.

49. Ibid., 606–608.

50. Fox and Back, "Preliminary Survey," 13.

51. Ibid., 321; Levy et al., "Food Value." For further analysis of the "Dressed Native" concept in mining research and research on greens, see Packard, *White Plague, Black Labor*.

52. Kepe, "Social Dynamics."

53. Levy et al., "Food Value," 699.

54. For more on Fox and nutrition research in South Africa, see Wylie, *Starving on a Full Stomach*, 141–159.

55. United States, PMAH, "Diaries and Logs of Lorna and Laurence Marshall."

56. Shopping receipt No. 2973 for Mr. L. K. Marshall, August 8, 1956, in United States, PMAH, "Letters, Ephemera."

57. Robert Story, Division of Botany, Department of Agriculture, Pretoria, to L. K. Marshall, March 22, 1957, in United States, PMAH, "Southwest Africa Expedition."

58. Paul C. Mangelsdorf, Botanical Museum of Harvard University, to J. O. Brew, Director, Peabody Museum, April 1957, in United States, PMAH, "Southwest Africa Expedition."

60. Richard B. Lee papers in United States, PMAH, "Records of the Harvard-Smithsonian Expedition."

61. Lee, "What Hunters Do," 37.

62. Ibid., 53–55. On Lee's influential concept of an "affluent society," see Sahlins, *Stone Age Economics.*

63. Tanaka, "Subsistence Ecology," 107; Fox and Young, *Food from the Veld.*

64. United States, PMAH, Richard Borshay Lee Field Notes; Lee, "Eating Christmas."

65. Hansen et al., "Hunter-Gatherer to Pastoral Way," 559.

66. Jenkins et al., "Transition " 410.

67. Kingwill, *CSIR.*

68. Pretorius and Wehmeyer, "Assessment of Nutritive Value"; Pretorius et al., "Magnesium Balance Studies"; Theron et al., "State of Pyridoxine Nutrition"; Theron et al., "Histochemical Investigation"; Abramson et al., "Diet and Health."

69. CSIR, "Bremer Bread." For more on nutrition outreach programs in the 1950s, see also Packard, *White Plague, Black Labor,* 266–272; Wylie, *Starving on a Full Stomach.*

70. Wehmeyer and Rose, "Important Indigenous Plants," 615.

71. Wehmeyer, "Edible Wild Plants," 2.

72. Ibid., 3.

73. Strydom and Wehmeyer, "Preparation of Edible Wild Fruit"; Wehmeyer, "Edible Wild Plants."

74. "CSIR Twenty-Fourth Annual Report," 47.

75. Wehmeyer, "Edible Wild Plants."

76. Arnold et al., "Khoisan Food Plants."

77. Wehmeyer et al., "Nutrient Composition," 1530.

78. Wehmeyer, "Edible Wild Plants," 4–6; Kingwill, *CSIR,* 210.

79. Kingwill, *CSIR,* 198.

80. "CSIR Twenty-Second Annual Report," 586.

81. "CSIR Twenty-Third Annual Report," 51.

82. "CSIRTwenty-Eighth Annual Report 1972," 5.

83. "CSIR Annual Report 1975," 12.

84. "CSIR Annual Report 1983."

85. Retrospective accounts insist that CSIR methodology depended solely on "journal articles and books" that scientists "used to collect information on *Hoodia gordonii* and related species": van Heerden, "*Hoodia gordonii,*" 434.

86. Author interview with Vinesh Maharaj; Watt and Breyer-Brandwijk, *Medicinal and Poisonous Plants of Southern and Eastern Africa.*

87. van Heerden et al., "Appetite Suppressant," 2551.

88. van Heerden et al., "Pharmaceutical Compositions"; van Heerden et al., "Appetite Suppressant."

89. van Heerden et al., "Pharmaceutical Compositions."

90. "CSIR Annual Report (2005–06)," 99.

91. Author interview with Vinesh Maharaj.

92. Kahn, "South Africa: Unilever Dumps Plans."

93. Ibid.

94. Stahl, "African Plant"; "Can a Generation of Potent but Safe Diet Pills."

95. Author interview with Vinesh Maharaj.

96. Lee and Balick, "Indigenous Use of Hoodia gordonii."

97. Pereira et al., "Hoodia Gordonii," 2311.

98. "Phytopharm Falls 43%."

99. Ibid.

100. Makoni, "San People's Cactus Drug."

101. Rubin et al., "Extracts, Compounds and Pharmaceutical Compositions"; Ismaili et al., "Processes for Production."

102. "Hoodia Gordonii Weight Loss Pills."

103. Dentlinger, "Namibian Government"; "South Africa: Feature"; Stahl, "African Plant."

104. Desai, *Bushman's Secret.*

105. "South Africa: Feature."

106. Kahn, "South Africa: Unilever Dumps Plans."

107. van Heerden, "Hoodia Gordonii," 437; emphasis added.

108. Milton and Dean, *Karoo Veld*, 16–17. See for instance: Stahl, "African Plant"; Dentlinger, "Namibian Government"; Foley and Mesure, "Unilever Enlists Kalahari Bushmen"; "Safe Diet Pills"; "Stoneage Plant."

109. Munzer and Simon, "Territory, Plants, and Land-Use Rights," 875.

110. Comaroff and Comaroff, *Ethnicity, Inc.*, 86–116.

111. Hardin, "Concessionary Politics."

112. Chennells, "Toward Global Justice." See also Wynberg and Chennells; "Green Diamonds"; Chennells, "Vulnerability and Indigenous Communities."

113. Crouch et al., "South Africa's Bioprospecting," 357.

114. "National Environmental Management," 68.

115. Ferreira, "Red Tape."

116. Sherry, "Permission to Sip."

117. Author interview with Tshidi Moroka.

118. Crouch et al., "South Africa's Bioprospecting," 357.

119. Early optimism included: van den Daele, "Does the Category of Justice"; Odendaal et al., *Richtersveld*, 52. However, Pfizer and then Unilever abandoned licensing agreements based on Hoodia patents. See Kahn, "South Africa: Unilever Dumps Plans."

120. Kahn, "Firm Hopes to Take 'San Prozac.' "

121. Marshall and Marshall, *Bitter Melons*; United States, PMAH, "Journal of Elizabeth Marshall Thomas."

Notes to Conclusion

1. Mputhia, "State Needs to Enforce Laws"; "Environmental Management and Co-ordination Act"; "National Environmental Management"; "Constitution of Kenya."

2. For an insightful overview of the global shift in management of plant genetic resources from an open to closed framework, see Raustiala and Victor, "Regime Complex," 278.

3. Basalla, "Spread of Western Science"; Grewal, *Network Power*.

4. Jasanoff, "Heaven and Earth," 41. A God's-eye view has been furthered through satellite mapping of genetic resources; see Mooney, "Africa."

5. "Nagoya Protocol on Access," 9.

6. On early regional alliances to monitor biodiversity prospecting, see, for instance, Grajal, "Biodiversity and the Nation State," 6; Coombe, "Intellectual Property," 104.

7. For example, see Cerminara, "Class Action Suit"; Barnett, "Equitable Trusts"; Hillebrand and Torrence, "Claims Procedures." Alternatives to direct benefits include social responsibility programs to limit potential conflicts; see Heal, "Corporate Social Responsibility."

8. Moran, "Rural Space as Intellectual Property," 266.

9. Bérand and Marchenay, "Tradition, Regulation and Intellectual Property." In his depiction of neo-Europes, Alfred E. Crosby wrongly asserts that those of African descent live only on three continents; Crosby, *Ecological Imperialism*, 1, 2.

10. Boateng, *Copyright Thing Doesn't Work Here*; Sautier et al., "Geographical Indications"; Teuber, "Geographical Indications of Origin."

11. Raustiala and Munzer, "Global Struggle."

12. Grasseni, "Packaging Skills."

13. Addae-Mensah, "Plant Biodiversity," 171.

14. Mooney, "Africa."

15. Sociologist Ruha Benjamin recognizes a similar paradox in the quest for "genomic sovereignty" in the race to map population genomes in India and Mexico: Benjamin, *"A Lab of Their Own."*

Archives

Ghana

CSRPM   Centre for Scientific Research into Plant Medicine, Mampong-Akwapim
Ampofo, Oku. "Autobiography of Oku Ampofo." Mampong, c. 1990.
CPM   Council Meeting Minutes
      CPM/A/CR/59, "Council meetings—Director's papers,"
      1981–1982.
Oppong-Boachie, F. K. "Research into Traditional Medicine: The Experi-
ence of the Centre for Scientific Research into Plant Medicine."
Mampong-Akuapem: Center for Scientific Research into Plant
Medicine, c. 1997.

GFDB   Ghana Food and Drugs Board
"Lists of Approved Natural Products."

GMH   Ghana Ministry of Health
Reports and Studies of the Department of Traditional and Alternative
Medicine
"Traditional and Alternative Medicines Census Report on the Three
Northern Regions of Ghana." Accra: Ministry of Health, 2002.
Senah, Kodjo, Kofi Adusei, Sam Akor, E. N. Mensah, and Nana Offei
Agyentutu III. "A Baseline Study into Traditional Medicine Prac-
tice in Ghana." Accra: Ministry of Health, 2001.

GPTHA   Ghana Psychic and Traditional Healers Association, Michael Acquah
Personal Files, Kojokrum

"Correspondence, Ghana Psychic and Traditional Healers Association, Western Region," 1963–1970.

"Correspondence, Ghana Psychic and Traditional Healers Association, Western Region," 1970–1977.

JMPP James Moxson Personal Papers, Accra

Moxson, James. "Autobiography of James Moxson" [unpublished]. Accra, c. 2002.

KNUST Kwame Nkrumah University of Science and Technology

FP Faculty of Pharmacy Administrative Files

FP/A7c/vol, "Advisory Committee of the Centre for Scientific Research into Plant Medicine (Agenda & Minutes of Meetings)," 1973–1977.

FP/R1/199, "Research—General," 1960–1971.

PD Pharmacy Department Files

PD/28 (vol. 1), "Departmental Letters," 1959.

PD/39, "National Research Council Minutes," 1961.

PRAAD Public Records and Archives Administration Department, Accra

ADM 4 Gold Coast Laws

ADM 4/1/16, "The Criminal Code: Laws of the Gold Coast," 1892.

ADM 29 Gold Coast Administration

ADM 29/6/5, J. Neil Leitch. "Confidential Memorandum: West African Materia Medica and Toxicology, Being an Enquiry into Native Remedies and Poisons by Dr. J. Neil Leitch, Government Pathologist," 1930.

ADM 56 Northern Territories

ADM 56/1/38, "Chief Commissioner Northern Territories Gambaga 21 August 1905: Fra Fra—Proposed Establishment of a Post," 1905–1911.

ADM 56/1/225, "Strophanthus Seeds, Case No. 22/1918," 1914–1922.

ADM 56/1/236, "Zouaragu District Native Affairs."

ADM 56/1/429, "District Report—Navarro," 1907.

CSO 11 Medical

CSO 11/7/3, "Medical Training of Africans—Scholarships For," 1929–1940.

Madagascar

ANM Archives Nationale de Madagascar, Antananarivo

H Sante (Health)

H 261, "Organisation de l'Assistance Médicale Indigène."

H 262, "Étude au Sujet d'une Réorganisation des Services Sanitaires et Médicaux de Madagascar."

H 291, "Lepre: Traitement de le Lepre au Moyen d'un Extract de l'Hydrocotyle Asiatica," 1945.

H 400, "Exposition Coloniale Internationale de Paris."

H 401, "Exposition Colonial Internationale de Paris (Documents Photographiques)."

H 403, "Congrès de la Sante."

BPBZT   Bibliothèque du Parc Botanique et Zoologique de Tsimbazaza

Boiteau, Pierre. "Travaux sur l'Asiaticoside: Étude Botanique." *Huitième Rapport Annuel de la Société du Parc Botanique et Zoologique de Tsimbazaza* (1944): 71–75.

Grimes, Charles, and Pierre Boiteau. "Rapport sur la Thérapeutique de la Lèpre." *Huitième Rapport Annuel de la Société du Parc Botanique et Zoologique de Tsimbazaza* (1944): 51–59.

"Les Mpsikidy: Les Mpisikidy Faiseurs de Remèdes Ody Employés par les Ancetres contre les Maladies." *Bulletin de L'Academie Malgache* 11 (1913): 153–214.

"Sova Momba Ny Nahazoan'Ny Vazaha Ny Tany. Chanson sur l'Arrivée des Vazaha á Madagascar" [Song of the conquest of the foreigners in Madagascar]. *Bulletin de L'Académie Malgache* 11 (1913): 128–129.

MEF   Ministre de l'Environnement et des Forêts

"Medicinal Plant Exports."

MIMRA   Muséum de l'Institut Malgache de Recherches Appliquées: Albert Rakoto-Ratsimamanga Papers and Exhibition

"Certificate from Professeur de Bacteriologie," Faculté de Médicine, Université de Paris, 21 Mars, 1931.

"Certificat Provisoire," Faculté de Médecine, Université de Paris, 21 Décembre, 1933.

"Curriculum Vitae" of Professor Albert Rakoto-Ratsimamanga.

Dr. Thiroux, "Bulletin de Notes: Concernant les Médicins Indigènes, Ratsimamanga Albert," Exposition Coloniale Internationale de Paris, Commissariat de L'Afrique Occidentale Française, 6 Novembre, 1931.

"État Nominatif des Artistes, Exposition Coloniale Internationale de Paris."

"Telegrams to and from Ratsimamanga."

SEAR   Société d'Exploration Agricole de Ranopiso

Arboretum Files

"Procès Verbal (Réunion des Sociétés Atsimo-Export et Pronatex) au Motel de Libanona un Fort-Dauphin," 1975.

South Africa

CSIR   Council for Scientific and Industrial Research, Pretoria

SANA   South African National Archives, Pretoria

NTS 7756/26/335, "Grahamstown: Fingo Hottentot Location (Malnutrition among Natives)" (1928–1960).

United Kingdom

BL   Bodleian Library of Commonwealth and African Studies, Rhodes House, Oxford
    MSS Afr. s. 2010 (1), Eric Allman Papers "Notes and Cuttings: Kumasi College."
    MSS. Afr. s. 2010 (4), "University of Science and Technology Kumasi: Silver Jubilee, 1951/52–1976/77, Service to the Community."

PRO   Public Record Office, Kew
    AY4/202, "World Sources of Plant *Vinca Rosea*: University of Western Ontario, Canada," 1958.
    FO 881/7110, "Africa: Report. Poisoned Native Arrows, Etc., from Uganda (Extracts.) (Dr. Woodhead)," 1898.

RBGKA Archives of the Royal Botanic Garden, Kew
    10.32 Ghana
     "Gold Coast Colonial Products 1888–1906, B–J," 1888–1906.
     "Miscellaneous Gold Coast Cultural Products (I–W)," 1888–1906.

RBGKEB Economic Botany Library of the Royal Botanic Gardens, Kew
    106.00, APOCYNACEAE *Catharanthus roseus*
    106.00, APOCYNACEAE *Strophanthus hispidus, S. gratus, S. divergens*
    107.00, ASCLEPIADACEAE *Cryptolepis sanginolenta*
    107.00, ASCLEPIADACEAE *Hoodia gordonii*
    170.01, ZINGIBERACEAE *Aframomum melegueta*

RBGKH Herbarium of the Royal Botanic Gardens, Kew
    6, ASCLEPIADACEAE *Cryptolepis sanguinolenta*
    40, APOCYNACEAE *Catharanthus roseus*
    147, ASCLEPIADACEAE *Hoodia gordonii*

United States

BBLH   Baker Business Library Historical Collection, Harvard University
    "The Drug Makers and the Drug Distributors: Background Papers for the United States Task Force on Prescription Drugs." Office of the Secretary, U.S. Department of Health, Education, and Welfare, 1968.

PMAH   Peabody Museum Archives, Harvard University
    "Diaries and Logs of Lorna and Laurence Marshall."
    "Journal of Elizabeth Marshall Thomas."
    "Letters, Ephemera from the Marshall Family Exhibit Case Displayed at the Peabody Museum (c. 1991–2003)."
    "Records of the Harvard-Smithsonian Expedition to Botswana"
    Richard Borshay Lee Field Notes: "!Kung Bushman Food Plants and Botanical Categories," 1963–1965.
    "Southwest Africa Expedition Records, 1950–1959, Includes Correspondence by Lorna and Lawrence Marshall and Colleagues."

Selected Patents

Allas, Soray, Victor Ngoka, Neil G. Hartman, Michel Ibea, and Simon Owassa. "Aframomum Seeds for Improving Penile Activity." U.S. Patent and Trademark Office, March 9, 1999. U.S. Pat. 5879682.

Beer, Charles T., and Robert L. Noble. Canadian Patents and Development, Ltd. "Vincaleukoblastine." U.S. Patent Office, July 9, 1963. U.S. Pat. 3097137.

Debat, Jacques. Laboratoires du Dr. Debat, Paris, France. "Prunus Africana Extract." U.S. Patent Office, December 24, 1974. U.S. Pat. 3,856,946.

Ismaili, Smail Alaoui, Sybille Buchwald-Werner, Frederik Michiel Meeuse, and Kevin John Povey. Phytopharm PLC, Godmanchester, Cambridge (GB). "Processes for Production of *Hoodia* Plant Extracts Containing Steroidal Glycosides." U.S. Patent and Trademark Office, October 5, 2010. U.S. Pat. 7,807,204 B2.

Ratsimamanga, Albert Rakoto, and Pierre Boiteau. "Hemisuccinates and Salts of the Hemisuccinates of Asiatic Acid." U.S. Patent Office, January 30, 1968. U.S. Pat. 3,366,669.

Ratsimamanga, Albert Rakoto, Suzanne Rakoto Ratsimamanga, Philippe Rasoanaivo, Jean Leboul, Jean Provost, and Daniel Reisdorf. Rhone-Poulenc Rorer S. A., Antony, France; Institut Malgache de Recherches Appliquées, Antananarivo, Madagascar. "Mixtures Derived from Grains of *Eugenia jambolana* Lamarck, Preparation and Use of Said Mixtures and Some of Their Constituents as Medicaments." U.S. Patent and Trademark Office, October 26, 1999. U.S. Pat. 5,972,342.

Rubin, Ian Duncan, Jasjit Singh Bindra, and Michael Anthony Cawthorne. Phytopharm PLC, Godmanchester (GB). "Extracts, Compounds and Pharmaceutical Compositions Having Anti-diabetic Activity and Their Use." United States Patent Office, April 25, 2006. U.S. Pat. 7,033,616 B2.

Svoboda, Gordon H. "Method of Preparing Leurosine and Vincaleukoblastine." U.S. Patent Office, December 21, 1965. U.S. Pat. 3,225,030.

Svoboda, Gordon H., Albert J. Barnes, Robert J. Armstrong. "Leurosidine and Leurocristine and Their Production." United States Patent Office, September 7, 1965. U.S. Pat. 3,205,220.

Tackie, A.N., and P. Schiff. "Compound and Method of Treatment for Falciparum Malaria." U.S. Patent and Trademark Office, November 8, 1994. U.S. Pat. 5,362,726.

van Heerden, Fanie Retief Van, Robert Vleggaar, Roelof Marthinus Horak, Robin Alec Learmonth, Vinesh Maharaj, and Rory Desmond Whittal. CSIR, Pretoria (ZA). "Pharmaceutical Compositions Having Appetite Suppressant Activity." U.S. Patent and Trademark Office, April 23, 2002. U.S. Pat. 6,376,657 B1.

———. CSIR, Pretoria (ZA). "Pharmaceutical Compositions Having Appetite Suppressant Activity." U.S. Patent and Trademark Office, 2007. U.S. Pat. 7,166,611 B2.

Selected Laws and Agreements

"Bill for improving the quality of beer in Ireland, by preventing the use of unmalted corn, or of any deleterious or unwholesome ingredients therein; and for the better securing the collection of the malt duties in Ireland." House of Commons Sessional Papers, 1809.

"Constitution of Kenya." National Council for Law Reporting, 2010.

"Convention on Biological Diversity." United Nations, 1992.

"The Environmental Management and Co-ordination Act, No. 8 of 1999." Government of Kenya, 1999.

"Nagoya Protocol on Access to Genetic Resources and the Fair and Equitable Sharing of Benefits Arising from Their Utilization to the Convention on Biological Diversity." United Nations, Secretariat of the Convention on Biological Diversity, 2011.

"National Environmental Management: Biodiversity Act, 2004." Act No. 10 of 2004. *Government Gazette* 467, no. 26436, Republic of South Africa, June 7, 2004, 1–84.

Journal Articles

Abramson, J. H., C. Slome, and N. T. Ward. "Diet and Health of a Group of African Agricultural Workers in South Africa." *American Journal of Clinical Nutrition* 8, no. 6 (1960): 875–884.

Addy, Marian Ewurama. "Cryptolepis: An African Traditional Medicine that Provides Hope for Malaria Victims." *HerbalGram* no. 60 (2003): 54–59, 67.

Adjanohoun, E. "Les Problèmes Soulevés par la Conservation de la Flore en Côte d'Ivoire." *Bulletin du Jardin Botanique National de Belgique / Bulletin van de Nationale Plantentuin van Belgie* 41, no. 1 (1971): 107–133.

"An Advance in the Treatment of Leprosy (News and Notes)." *Lancet* 245, no. 6342 (1945): 357.

Agrawal, Arun. "Dismantling the Divide between Indigenous and Scientific Knowledge." *Development and Change* 26, no. 3 (1995): 413–439.

"AIDS Cure Found—Scientist Murdered." *Global Africa Pocket News (Gap News) Online Magazine* 1, no. 2 (1994): 8.

Anon. "Note on Strophanthus." *American Druggist* 16, no. 8 (1887): 141–142.

Ansah, Charles, and Nigel J. Gooderham. "The Popular Herbal Antimalarial, Extract of *Cryptolepis sanguinolenta*, Is Potently Cytotoxic." *Toxicological Sciences* 70, no. 2 (2002): 245–251.

Archer, William. "The American Language." *Pall Mall Magazine* 19 (September–December 1899): 191–196.

Arhin, Kwame. "The Ashanti Rubber Trade with the Gold Coast in the Eighteen-Nineties." *Africa: Journal of the International African Institute* 42, no. 1 (1972): 32.

Baako, Kofi. "Opening Address of the First Conference of the Ghana Science Association Held at Legon on 31st March, 1961 by Hon. Kofi Baako, M.P., Minister Responsible for Research." *Ghana Journal of Science* 1, nos. 1 and 2 (1961): 5–8.

Babykutty, S., J. Padikkala, P. P. Sathiadevan, V. Vijayakurup, T. K. A. Azis, P. Srinivas, and S. Gopala. "Apoptosis Induction of *Centella asiatica* on Human Breast Cancer Cells." *African Journal of Traditional Complementary and Alternative Medicines* 6, no. 1 (2009): 9–16.

Baker, David. "Does 'Indigenous Science' Really Exist." *Australian Science Teachers Journal* 42, no. 1 (1996): 18–21.

Barnett, Kerry. "Equitable Trusts: An Effective Remedy in Consumer Class Actions." *Yale Law Journal* 96, no. 7 (1986–1987): 1591–1614.

Basalla, George. "The Spread of Western Science." *Science* 156, no. 3775 (1967): 611–622.

Beckett, A. H., E. J. Shellard, and A. N. Tackie. "The *Mitragyna* Species of Ghana: The Alkaloids of Leaves of *Mitragyna ciliata* Aubr. et Pellegr." *Journal of Pharmacy and Pharmacology* 15, no. S1 (1963): 166T–169T.

Beckett, A. H., and A. N. Tackie. "Structure of Rotundifoline and Mitragynol (Isorotundifoline)." *Chemistry & Industry* no. 27 (1963): 1122–1123.

———. "Structures of Alkaloids from Mitragyna Species of Ghana." *Journal of Pharmacy and Pharmacology* 15, no. S1 (1963): 267T–269T.

Beckett, A. H., A. N. Tackie, and E. J. Shellard. "The *Mitragyna* Species of Ghana: The Alkaloids of the Leaves of *Mitragyna stipulosa* (D.C.) O. Kuntze." *Journal of Pharmacy and Pharmacology* 15, no. S1 (1963): 158T–165T.

Beichner, Paul E. "The Grain of Paradise." *Speculum* 36, no. 2 (1961): 302–307.

Benjamin, Ruha. "*A Lab of Their Own*: Genomic Sovereignty as Postcolonial Science Policy." *Policy and Society* 28, no. 4 (2009): 341–355.

Betti, Jean L. "Medicinal Plants Sold in Yaoundé Markets, Cameroon." *African Study Monographs* 23, no. 2 (2002): 47–64.

Beuscher, N., C. Bodinet, D. Neumannhaefelin, A. Marston, and K. Hostettmann. "Antiviral Activity of African Medicinal-Plants." *Journal of Ethnopharmacology* 42, no. 2 (1994): 101–109.

Bhavna, Dora, and Jyoti Khatri. "*Centella asiatica*: The Elixir of Life." *International Journal of Research in Ayurveda & Pharmacy* 2, no. 2 (2011): 431–438.

Bickii, J., G. R. F. Tchouya, J. C. Tchouankeu, and E. Tsamo. "Antimalarial Activity in Crude Extracts of Some Cameroonian Medicinal Plants." *African Journal of Traditional Complementary and Alternative Medicines* 4, no. 1 (2007): 107–111.

Birhanu, Fikremarkos Merso. "Challenges and Prospects of Implementing the Access and Benefit Sharing Regime of the Convention on Biological Diversity in Africa: The Case of Ethiopia." *International Environmental Agreements-Politics Law and Economics* 10, no. 3 (2010): 249–266.

Biondi, D., and J. H. Pedrosa-Macedo. "Plantas Invasoras Encontradas na Área Urbana de Curitiba (PR)" [Invasive plants observed in the urban area of Curitiba, Brazil]. *Floresta* 38, no. 1 (2008): 129–144.

Bloom, H., and H. F. Walton. "Chemical Prospecting." *Scientific American* 197, no. 1 (1957): 41–47.

Boateng, Boatema. "Walking the Tradition-Modernity Tightrope: Gender Contradictions in Textile Production and Intellectual Property Law in Ghana." *American University Journal of Gender Social Policy and Law* 15, no. 2 (2007): 341–358.

Bodeker, Gerard. "Traditional Medical Knowledge, Intellectual Property Rights and Benefit Sharing." *Cardozo Journal of International and Comparative Law* 11, no. 2 (2003): 785–814.

Bodeker, Gerard, and Fredi Kronenberg. "A Public Health Agenda for Traditional, Complementary, and Alternative Medicine." *American Journal of Public Health* 92, no. 10 (2002): 1582–1591.

Boiteau, P., M. Nigeondureuil, and A. R. Ratsimamanga. "Contribution a l'Étude de l'Acide Asiatique vis-a-vis de la Tuberculose Expérimentale de la Souris." *Comptes Rendus Hebdomadaires des Séances de l'Académie des Sciences* 232, no. 5 (1951): 450–451.

Boiteau, P., and A. R. Ratsimamanga. "Éffets d'un Titerpene (Asiaticoside) de la Série des Amyrines sur la Germination et la Croissance des Végétaux." *Comptes Rendus des Séances de la Société de Biologie et de Ses Filiales* 152, no. 7 (1958): 1106–1110.

Bollier, David. "The Enclosure of the Academic Commons." *Academe* 88, no. 5 (2002): 18–22.

Bousman, C. Britt. "Coping with Risk: Later Stone Age Technological Strategies at Blydefontein Rock Shelter, South Africa." *Journal of Anthropological Archaeology* 24, no. 3 (2005): 193–226.

Braun, L., and E. Hammonds. "Race, Populations, and Genomics: Africa as Laboratory." *Social Science & Medicine* 67, no. 10 (2008): 1580–1588.

Bugos, G. E., and D. J. Kevles. "Plants as Intellectual Property: American Practice, Law, and Policy in World Context." *Osiris*, 2nd ser., 7 (1992): 75–104.

Bühnen, Stephan. "In Quest of Susu." *History in Africa* 21 (1994): 1–47.

Buu, Hoi, and A. R. Ratsimamanga. "Production of Lesions of Tuberculosis Type Accompanied by Toxic Phenomena by Di-substituted Ethylenic Acids Alpha Alpha." *Comptes Rendus des Séances de la Société de Biologie et de Ses Filiales* 137 (1943): 369–370.

Bynum, William. "Chemical Structure and Pharmacological Action: A Chapter in the History of 19th Century Molecular Pharmacology." *Bulletin of the History of Medicine* 44, no. 6 (1970): 518–538.

Cantor, D. "Cortisone and the Politics of Empire: Imperialism and British Medicine, 1918–1955." *Bulletin of the History of Medicine* 67, no. 3 (1993): 463–493.

Cantoria, Magdalena. "Aromatic and Medicinal Herbs of the Philippines." *Quarterly Journal of Crude Drug Research* 14 (1976): 97–128.

"Celebrating 60 Years of Excellence in Research and Innovation." *Science Scope: Newsletter of the CSIR* 1, no. 1 (2005): 4.

Cerminara, Kathy. "The Class Action Suit as a Method of Patient Empowerment in the Managed Care Setting." *American Journal of Law and Medicine* 24, no. 1 (1988): 7–58.

Chakrabarty, Dipesh. "Postcoloniality and the Artifice of History: Who Speaks for 'Indian' Pasts?" in "Imperial Fantasies and Postcolonial Histories," special issue, *Representations* no. 37 (Winter 1992): 1–26.

Chalmers, Albert J. "A Further Report of Experiments upon the Fra Fra Arrow Poison." *Medical and Sanitary Reports of the Gold Coast 1898–1908* (1899): 274–280.

Chang, Kuei-Sheng. "Africa and the Indian Ocean in Chinese Maps of the Four-teenth and Fifteenth Centuries." *Imago Mundi* 24 (1970): 21–30.

Chen, Jim. "Webs of Life: Biodiversity Conservation as a Species of Information Policy." *Iowa Law Review* 89, no. 2 (2004): 495.

Chennells, Roger. "Toward Global Justice through Benefit-Sharing." *Hastings Center Report* 40, no. 1 (2010): 3

———. "Vulnerability and Indigenous Communities: Are the San of South Africa a Vulnerable People?" *Cambridge Quarterly of Healthcare Ethics* 18, no. 2 (2009): 147–154.

Chernin, Eli. "Sir Ronald Ross vs. Sir Patrick Manson: A Matter of Libel." *Journal of the History of Medicine and Allied Sciences* 43, no. 3 (1988): 262–274.

Clayton, W. D. "Secondary Vegetation and the Transition to Savanna Near Ibadan, Nigeria." *Journal of Ecology* 46, no. 2 (1958): 217–238.

Coombe, Rosemary J. "Fear, Hope, and Longing for the Future of Authorship and a Revitalized Public Domain in Global Regimes of Intellectual Property." *De Paul Law Review* 52 (2003): 1171–1191.

———. "Intellectual Property, Human Rights, and Sovereignty: New Dilemmas in International Law Posed by the Recognition of Indigenous Knowledge and the Conservation of Biodiversity." *Indiana Journal of Global Legal Studies* 6 (1998): 59–115.

Crawfurd, John. "On the History and Migration of Cultivated Plants Used as Condiments." *Transactions of the Ethnological Society of London* 6 (1868): 188–206.

Crouch, N. R., E. Douwes, M. M. Wolfson, G. F. Smith, and T. J. Edwards. "South Africa's Bioprospecting, Access and Benefit-Sharing Legislation: Current Reali-ties, Future Complications, and a Proposed Alternative." *South African Journal of Science* 104, no. 9–10 (2008): 355–366.

CSIR. "Bremer Bread: The Following Article Appeared in an East London News-paper on 11 December 1952." *Science Scope: Newsletter of the CSIR* 1, no. 1 (2005): 6.

Curtin, Phillip. "Field Techniques for Collecting and Processing Oral Data." *Journal of African History* 9, no. 3 (1968): 367–385.

Cutts, J. H., C. T. Beer, and R. L. Noble. "Biological Properties of Vincaleukoblas-tine, an Alkaloid in *Vinca rosea* Linn, with Reference to Its Antitumor Action." *Cancer Research* 20, no. 7 (1960): 1023–1031.

Daly, Charles P., and Alexander Crummell. "The Plan of the King of Belgium for the Civilization of Central Africa, and the Suppression of the Slave Trade." *Journal of the American Geographical Society of New York* 9 (1877): 88–103.

de Fatima, A., L. V. Modolo, A. C. C. Sanches, and R. R. Porto. "Wound Healing Agents: The Role of Natural and Non-natural Products in Drug Development." *Mini-Reviews in Medicinal Chemistry* 8, no. 9 (2008): 879–888.

Desfontaines, René Louiche. "Extrait d'un mémoire du citoyen de Candolle, sur le genre *Strophanthus*." *Annales Muséum National D'Histoire Naturelle, Paris* 1 (1802): 408–412.

Dev, Sukh. "Ancient-Modern Concordance in Ayurvedic Plants: Some Examples." *Environmental Health Perspectives* 107, no. 10 (1999): 783–789.

Diafouka, A., and M. Bitsindou. "Medicinal Plants Sold in Brazzaville Markets (Congo)." *Acta Horticulturae* 332 (1993): 95–104.

Dolder, F. "Traditional Knowledge and Patenting: The Experience of the Neem-fungicide and the Hoodia Cases." *Biotechnology Law Report* 26, no. 6 (2007): 583–590.

Dritsas, L. "Civilising Missions, Natural History and British Industry: Livingstone in the Zambezi." *Endeavour* 30, no. 2 (2006): 50–54.

Duffin, Jacalyn. "Poisoning the Spindle: Serendipity and Discovery of the Anti-Tumour Properties of the Vinca Alkaloids." *Canadian Bulletin of Medical History* 17, no. 1 (2000): 155–192.

———. "Poisoning the Spindle: Serendipity and Discovery of the Anti-Tumour Properties of the Vinca Alkaloids (illustrated reprinting) Part I." *Pharmacy in History* 44, no. 2 (2002): 64–76.

———. "Poisoning the Spindle: Serendipity and Discovery of the Anti-Tumour Properties of the Vinca Alkaloids (illustrated reprinting) Part II." *Pharmacy in History* 44, no. 3 (2002): 105–118.

Dumett, Raymond. "The Rubber Trade of the Gold Coast and Asante in the Nineteenth Century: African Innovation and Market Responsiveness." *Journal of African History* 12, no. 1 (1971): 79.

Dutfield, Graham. "The Public and Private Domains: Intellectual Property Rights in Traditional Knowledge." *Science Communication* 21, no. 3 (2000): 274–295.

Echenberg, Myron J. "Late Nineteenth-Century Military Technology in Upper Volta." *Journal of African History* 12, no. 2 (1971): 241–254.

Edington, G. M. "Cerebral Malaria in the Gold Coast African: Four Autopsy Reports." *Annals of Tropical Medicine and Parasitology* 48, no. 3 (1954): 300–306.

Eisner, Thomas. "Prospecting for Nature's Chemical Riches." *Issues in Science and Technology* 6, no. 2 (1990): 31–34.

Eisner, Thomas, and Elizabeth A. Beiring. "Biotic Exploration Fund: Protecting Biodiversity through Chemical Prospecting." *BioScience* 44, no. 2 (1994): 95.

Enti, A. A. "The Collection of Botanical Specimens and Herbarium Work in Ghana with Special Emphasis on the Forest Herbarium at Kumasi." *Ghana Journal of Science* 3, no. 2 (1963): 139–141.

Epstein, David. "The Action of Vinca on the Blood Sugar of Rabbits." *South African Medical Record* 24, no. 2 (1926): 35–38.

"Evening Meeting in London: Wednesday, November 13, 1889." *Pharmaceutical Journal and Transactions* 20, no. 1012 (1889): 389–391.

Eyob, S., M. Appelgren, J. Rohloff, A. Tsegaye, and G. Messele. "Traditional Medicinal Uses and Essential Oil Composition of Leaves and Rhizomes of Korarima (Aframomum corrorima [Braun] P. C. M. Jansen) from Southern Ethiopia." *South African Journal of Botany* 74, no. 2 (2008): 181–185.

Fage, J. D. "Some Notes on a Scheme for the Investigation of Oral Tradition in the Northern Territories of the Gold Coast." *Journal of the Historical Society of Nigeria* 1, no. 1 (1956): 15–19.

Fairchild, David. "Two Expeditions after Living Plants." *Scientific Monthly* 26, no. 2 (1928): 97–127.

Forbes, J. Graham. "Native Methods of Treatment in West Africa: With Notes on the Tropical Diseases Most Prevalent among the Inhabitants of the Gold Coast Colony." *Journal of the Royal African Society* 3, no. 12 (1904): 361–380.

Fraenkel, A., and G. Schwartz. "Abhandlungen zur Digitalistherapie Über intravenöse Strophanthininjektionen bei herzkranke" [Essays on Digitalis Therapy. About intravenous Strophanthine injections in cardiac patients]. *Archiv für Experimentelle Pathologie und Pharmakologie* 57, no. 1/2 (1907): 79–122.

Fraser, Thomas R. "*Strophanthus hispidus*: Its Natural History, Chemistry, and Pharmacology." *Transactions of the Royal Society of Edinburgh* 35, no. 4 (1890–1891): 955–1027.

Freedman, Paul. "Spices and Late-Medieval European Ideas of Scarcity and Value." *Speculum* 80, no. 4 (2005): 1209–1227.

Frenkel, Stephen, and John Western. "Pretext or Prophylaxis? Racial Segregation and Malarial Mosquitos in a British Tropical Colony: Sierra Leone." *Annals of the Association of American Geographers* 78, no. 2 (1988): 211–228.

Garcia, Faustino L. "Distribution and Deterioration of Insulin-like Principle in *Lagerstroemia speciosa (banabá)*." *Acta Medica Philippina* 3 (1941): 99–104.

———. "Distribution of Insulin-like Principle in Different Plants and Its Therapeutic Application to a Few Cases of Diabetes Mellitus." *Philippine Journal of Science* 76 (1944): 3–19.

Giday, Mirutse, Zemede Asfaw, Zerihun Woldu, and Tilahun Teklehaymanot. "Medicinal Plant Knowledge of the Bench Ethnic Group of Ethiopia: An Ethnobotanical Investigation." *Journal of Ethnobiology and Ethnomedicine* 5, November 13 (2009): 34–43.

Grajal, Alejandro. "Biodiversity and the Nation State: Regulating Access to Genetic Resources Limits Biodiversity Research in Developing Countries." *Conservation Biology* 13, no. 1 (1999): 6–10.

Green, R. H., and S. H. Hymer. "Cocoa in the Gold Coast: A Study in the Relations between African Farmers and Agricultural Experts." *Journal of Economic History* 26, no. 3 (1966): 299.

Greene, Shane. "Indigenous People Incorporated? Culture as Politics, Culture as Property in Pharmaceutical Bioprospecting." *Current Anthropology* 45, no. 2 (2004): 211–224.

Gupta, R. K. "Correspondence: Surge in U.S. Patents on Botanicals." *Nature Biotechnology* 22, no. 6 (2004): 653–654.

Hannaway, Owen. "Laboratory Design and the Aim of Science: Andreas Libavius versus Tycho Brahe." *Isis* 77, no. 4 (1986): 585–610.

Hansen, J. D. L., D. S. Dunn, R. B. Lee, P. J. Becker, and T. Jenkins. "Hunter-Gatherer to Pastoral Way of Life: Effects of the Transition on Health, Growth and Nutrititional Status." *South African Journal of Science* 89, no. 11–12 (1993): 559–564.

Hardin, Rebecca. "Concessionary Politics: Property, Patronage, and Political Rivalry in Central African Forest Management: with CA comment by Serge Bahuchet." *Current Anthropology* 52, no. S3 (2011): S113–S125.

Harper, Janice. "The Not-So Rosy Periwinkle: Political Dimensions of Medicinal Plant Research." *Ethnobotany Research and Applications* 3 (2005): 295–308.

Harris, E. S. J., S. D. Erickson, A. N. Tolopko, S. G. Cao, J. A. Craycroft, R. Scholten, Y. L. Fu, W. Q. Wang, Y. Liu, Z. Z. Zhao, J. Clardy, C. E. Shamu, and D. M. Eisenberg. "Traditional Medicine Collection Tracking System (TM-CTS): A Database for Ethnobotanically Driven Drug-Discovery Programs." *Journal of Ethnopharmacology* 135, no. 2 (2011): 590–593.

Harris, Joseph E. "Expanding the Scope of African Diaspora Studies: The Middle East and India, a Research Agenda." *Radical History Review* 87, no. 1 (2003): 157–168.

Hayden, Cori. "From Market to Market: Bioprospecting's Idioms of Inclusion." *American Ethnologist* 30, no. 3 (2003): 359–371.

Heal, Geoffrey. "Corporate Social Responsibility: An Economic and Financial Framework." *Geneva Papers on Risk and Insurance* 30, no. 3 (2005): 387–409.

Hillebrand, Gail, and Daniel Torrence. "Claims Procedures in Large Consumer Class Actions and Equitable Distribution of Benefits." *Santa Clara Law Review* 28, no. 4 (1988): 747–774.

Homsy, Jaco, Rachel King, Joseph Tenywa, Primrose Kyeyune, Alex Opio, and Dorothy Balaba. "Defining Minimum Standards of Practice for Incorporating African Traditional Medicine into HIV/AIDS Prevention, Care, and Support: A Regional Initiative in Eastern and Southern Africa." *Journal of Alternative and Complementary Medicine* 10, no. 5 (2004): 905–910.

Horton, Robin. "African Traditional Thought and Western Science." *Africa* 37, no. 2 (1967): 155–187.

Hou, ZhiYong, YongHong Xie, XinSheng Chen, Xu Li, Feng Li, Ying Pan, and ZhengMiao Deng. "Study on Invasive Plants in Dongting Lake Wetlands." *Research of Agricultural Modernization* 32, no. 6 (2011): 744–747.

Hugill, Peter J., and Veit Bachmann. "The Route to the Techno-Industrial World Economy and the Transfer of German Organic Chemistry to America before, during, and Immediately after World War I." *Comparative Technology Transfer and Society* 3, no. 2 (2005): 158–186.

Hulley, I. M., B. E. Van Wyk, P. M. Tilney, G. P. Kamatou, A. M. Viljoen, and S. F. Van Vuuren. "A Review of Pteronia Species Used in Traditional Medicine in South Africa." *South African Journal of Botany* 76, no. 2 (2010): 395–396.

Hulley, I. M., A. M. Viljoen, P. M. Tilney, S. F. Van Vuuren, G. P. P. Kamatou, and B. E. Van Wyk. "Ethnobotany, Leaf Anatomy, Essential Oil Composition and Antibacterial Activity of *Pteronia onobromoides* (Asteraceae)." *South African Journal of Botany* 76, no. 1 (2010): 43–48.

Hutchings, Anne. "A Survey and Analysis of Traditional Medicinal Plants as Used by the Zulu, Xhosa and Sotho," *Bothalia* 19, no. 1 (1989): 111–123.

Jacobs, Nancy J. "The Intimate Politics of Ornithology in Colonial Africa." *Comparative Studies in Society and History* 48, no. 3 (2006): 564–603.

Jacobs, W. A. "Strophantin. III. Crystalline Kombe Strophanthin—Preliminary Note." *Journal of Biological Chemistry* 57, no. 2 (1923): 569–572.

Jasanoff, Sheila. "Beyond Epistemology: Relativism and Engagement in the Politics of Science." *Social Studies of Science* 26, no. 2 (1996): 393–418.

Jenkins, T., B. I. Joffe, V. R. Panz, M. Ramsay, and H. C. Seftel. "Transition from a Hunter-Gatherer to a Settled Life-Style among the Kung San (Bushmen): Effect on Glucose Tolerance and Insulin Secretion." *South African Journal of Science* 83, no. 7 (1987): 410–412.

Johnson, I. S., H. F. Wright, and G. H. Svoboda. "Antitumor Principles Derived from *Vinca rosea* Linn." *Proceedings of the American Association for Cancer Research* 3, no. 2 (1960): 122.

Johnson, Irving S. "Drugs from Third World Plants." *Science* 257, no. 5072 (1992): 860.

Johnson, Irving S., Gordon H. Svoboda, and Howard F. Wright. "Experimental Basis for Clinical Evaluation of Two New Alkaloids from *Vinca rosea* Linn." *Proceedings of the American Association for Cancer Research* 3, no. 4 (1962): 331.

Johnston, Harry H. "The Importance of Africa." *Journal of the Royal African Society* 17, no. 67 (1918): 177.

Jolly, J. "Suśruta on Mosquitoes." *Journal of the Royal Asiatic Society of Great Britain and Ireland* (January 1906): 222–224.

Kamtchouing, P., G. Y. F. Mbongue, T. Dimo, P. Watcho, H. B. Jatsa, and S. D. Sokeng. "Effects of *Aframomum melegueta* and *Piper guineense* on Sexual Behaviour of Male Rats." *Behavioural Pharmacology* 13, no. 3 (2002): 243–247.

Karasov, Corliss. "Who Reaps the Benefits of Biodiversity?" *Environmental Health Perspectives* 109, no. 12 (2001): A582–A587.

Kepe, Thembela. "Social Dynamics of the Value of Wild Edible Leaves (Imifino) in a South African Rural Area." *Ecology of Food and Nutrition* 47, no. 6 (2008): 531–558.

Kevles, D. J. "Ownership and Identity." *Scientist* 17, no. 1 (2003): 22–23.

———. "Protections, Privileges, and Patents: Intellectual Property in American Horticulture, The Late Nineteenth Century to 1930." *Proceedings of the American Philosophical Society* 152, no. 2 (2008): 207–213.

Killingray, David. "Military and Labour Recruitment in the Gold Coast during the Second World War." *Journal of African History* 23, no. 1 (1982): 83–95.

Killingray, David, and James Matthews. "Beasts of Burden: British West African Carriers in the First World War." *Canadian Journal of African Studies* 13, no. 1/2 (1979): 5–23.

Klein, Herbert S. "The Trade in African Slaves to Rio de Janeiro, 1795–1811: Estimates of Mortality and Patterns of Voyages." *Journal of African History* 10, no. 4 (1969): 533–549.

Koppanyi, Theodore. "The Rise of Pharmacology." *Scientific Monthly* 41, no. 4 (1935): 316–324.

Kuta-Dankwa, A. K. "Overseas Service." *British Medical Journal* 2, no. 4849 (1953): 1326.

Laidler, P. W. "The Magic Medicine of the Hottentots." *South African Journal of Science* 25 (1928): 433–447.

Last, Murray. "The Importance of Knowing about Not Knowing." *Social Science & Medicine. Part B: MedicalAnthropology* 15, no. 3 (1981): 387–392.

Law, Robin. "Ethnicity and the Slave Trade: 'Lucumi' and 'Nago' as Ethnonyms in West Africa." *History in Africa* 24 (1997): 205–219.

Lee, R. A., and M. J. Balick. "Indigenous Use of *Hoodia gordonii* and Appetite Suppression." *Explore—The Journal of Science and Healing* 3, no. 4 (2007): 404–406.

Lee, Richard B. "Eating Christmas in the Kalahari." *Natural History* (December 1969): 14–22, 60–63.

Leong, Elaine. "Making Medicines in the Early Modern Household." *Bulletin of the History of Medicine* 82, no. 1 (2008): 145–168.

Lesch, John E. "Conceptual Change in an Empirical Science: The Discovery of the First Alkaloids." *Historical Studies in the Physical Sciences* 11, no. 2 (1981): 305–328.

Levy, L.F., D. Weintroub, and Francis W. Fox. "The Food Value of Some Common Edible Leaves." *Journal of the South African Medical Association* 10, no. 10 (October 1936): 699–707.

Lilly, Eli. "New Medicines from Cancer Research." *Report to the Shareholders* (1961): 8–11.

———. "Traitors in Our Tissues." *Report to the Shareholders* (1965): 25–26.

Liu, M., Y. Dai, Y. Li, Y. B. Luo, F. Huang, Z. N. Gong, and Q. Y. Meng. "Madecassoside Isolated from *Centella asiatica* Herbs Facilitates Burn Wound Healing in Mice." *Planta Medica* 74, no. 8 (2008): 809–815.

Lock, J. M., J. B. Hall, and Daniel K. Abbiw. "The Cultivation of Melegueta Pepper." *Economic Botany* 31, no. 3 (1977): 321–330.

Marafioti, Richard Lynn. "The Meaning of Generic Names of Important Economic Plants." *Economic Botany* 24, no. 2 (1970): 189–207.

Marks, Shula. "Khoisan Resistance to the Dutch in the Seventeenth and Eighteenth Centuries." *Journal of African History* 13, no. 1 (1972): 55–80.

Massey, Doreen. "Questions of Locality." *Geography* 78, no. 2 (1993): 142–149.

Mathews, J. W. "South African Succulents." *Journal of the Botanical Society of South Africa* 13 (1927): 13–17.

Mathur, Ajeet. "Who Owns Traditional Knowledge?" *Economic and Political Weekly* 38, no. 42 (2003): 4471–4481.

Mauny, Raymond. "Notes Historiques Autour des Principales Plantes Cultivées d'Afrique Occidentale." *Bulletin de l'Institut Français d'Afrique Noire* 15, no. 2 (1953): 684–730.

McClellan, Traci L. "The Role of International Law in Protecting the Traditional Knowledge and Plant Life of Indigenous Peoples." *Wisconsin International Law Journal* 19, no. 2 (2000): 249–266.

McDougall, E. Ann. "The Sahara Reconsidered: Pastoralism, Politics and Salt from the Ninth through the Twelfth Centuries." *African Economic History* no. 12 (1983): 263–286.

———. "The View from Awdaghust: War, Trade and Social Change in the Southwestern Sahara, from the Eighth to the Fifteenth Century." *Journal of African History* 26, no. 1 (1985): 1–31.

"Medical News: Treatment of Dysentery on the Gold Coast." *British Medical Journal* 1, no. 1 (1861): 26.

Medrzejewski, W. "The Fiftieth Anniversary of Insulin Discovery: The Pioneer Work of the Rumanian Physiologist Nicolas Paulesco." *Materia Medica Polona. Polish Journal of Medicine and Pharmacy* 4, no. 3 (1972): 152–153.

Melo, J. G. de, J. D. G. da R. Martins, E. L. C. de Amorim, and U. P. de Albuquerque. "Qualidade de Produtos a Base de Plantas Medicinais Comercializados no Brasil: Castanha-da-índia (*Aesculus hippocastanum* L.), Capim-limão (*Cymbopogon citratus* [DC.] Stapf), Centela (*Centella asiatica* [L.] Urban), Quality of Products Made from Medicinal Plants Commercialized in Brazil: Horsechestnut (*Aesculus hippocastanum* L.), Lemongrass (*Cymbopogon citratus* [DC.] Stapf), and Gotu Kola (*Centella asiatica* [L.] Urban)]." *Acta Botanica Brasilica* 21, no. 1 (2007): 27–36.

Merson, John. "Bio-Prospecting or Bio-Piracy: Intellectual Property Rights and Biodiversity in a Colonial and Postcolonial Context." *Osiris* 15 (2000): 282–296.

Messent, J. J. "Chloroquine." *British Medical Journal* 1, no. 4710 (1951): 818–819.

———. "Use and Effectiveness of Antimalarial Drugs." *British Medical Journal* 2, no. 4836 (1953): 629.

Mgbeoji, Ikechi. "Beyond Rhetoric: State Sovereignty, Common Concern, and the Inapplicability of the Common Heritage Concept to Plant Genetic Resources." *Leiden Journal of International Law* 16, no. 4 (2003): 821–837.

Michel, Jean-Baptiste, Yuan Kui Shen, Aviva Presser Aiden, Adrian Veres, Matthew K. Gray, Google Books Team, Joseph P. Pickett, Dale Hoiberg, Dan Clancy, Peter Norvig, Jon Orwant, Steven Pinker, Martin A. Nowak, and Erez Lieberman Aiden. "Quantitative Analysis of Culture Using Millions of Digitized Books." *Science* 331, January 14, 2011: 176–182.

Moor, C. G., and Martin Priest. "Notes on Certain British Pharmacopoeia Tests." *Analyst, a Monthly Journal Devoted to the Advancement of Analytical Chemistry* 26 (1901): 29b–35.

Moran, Warren. "Rural Space as Intellectual Property." *Political Geography* 12, no. 3 (1993): 263–277.

Moreira, H. S., M. B. A. Vasconcellos, E. R. Alves, F. M. Santos, and M. Saiki. "Elemental Composition of Herbal Medicines Sold Over-the-Counter in São Paulo City, Brazil." *Journal of Radioanalytical and Nuclear Chemistry* 290, no. 3 (2011): 615–621.

Morris, Alan G. "The Myth of the East African 'Bushmen.'" *South African Archaeological Bulletin* 58, no. 178 (2003): 85–90.

Moyer-Henry, K. "Patenting Neem and Hoodia: Conflicting Decisions Issued by the Opposition Board of the European Patent Office." *Biotechnology Law Report* 27, no. 1 (2008): 1–10.

Munasinghe, Mohan. "Biodiversity Protection Policy: Environmental Valuation and Distribution Issues." *Ambio* 21, no. 3 (1992): 227–236.

Munzer, Stephen R., and Phyllis Chen Simon. "Territory, Plants, and Land-Use Rights among the San of Southern Africa: A Case Study in Regional Biodiversity, Traditional Knowledge, and Intellectual Property." *William and Mary Bill of Rights Journal* 17 no. 3 (2009): 831–891.

Myers, Norman. "Biodiversity and the Precautionary Principle." *Ambio* 22, no. 2/3 (1993): 74–79.

Neimark, Benjamin. "Green Grabbing at the 'Pharm' Gate": Rosy Periwinkle Production in Southern Madagascar." *Journal of Peasant Studies* 39, no. 2 (2012): 423–245.

Nkrumah, Kwame. "Speech Delivered by Osagyefo the President at the Laying of the Foundation Stone of Ghana's Atomic Reactor at Kwabenya on 25th November, 1964." *Ghana Journal of Science* 5, no. 1 (1965): 1–5.

Noble, R. L. "Anti-cancer Alkaloids of *Vinca rosea.*" *Biochemical Pharmacology,* Suppl., 12 (1963).

———. "*Catharanthus roseus (Vinca rosea)*—Importance and Value of a Chance Observation." *Lloydia* 27, no. 4 (1964): 280+.

———. "The Discovery of the Vinca Alkaloids—Chemotherapeutic Agents against Cancer." *Biochemistry and Cell Biology-Biochimie et Biologie Cellulaire* 68, no. 12 (1990): 1344–1351.

Noble, R. L., C. T. Beer, and J. H. Cutts. "Further Biological Activities of Vincaleukoblastine—An Alkaloid Isolated from *Vinca rosea* (L.)" *Biochemical Pharmacology* 1, no. 4 (1958): 347–348.

———. "Role of Chance Observations in Chemotherapy—*Vinca rosea.*" *Annals of the New York Academy of Sciences* 76, no. 3 (1958): 882–894.

Nye, L. J. Jarvis, and Margaret E. Fitzgerald. "Vinca Treatments of Diabetes." *Medical Journal of Australia* 2 (1928): 626–627.

Okoli, C. O., P. A. Akah, S. V. Nwafor, U. U. Ihemelandu, and C. Amadife. "Anti-inflammatory Activity of Seed Extracts of *Aframomum melegueta.*" *Journal of Herbs, Spices & Medicinal Plants* 13, no. 1 (2007): 11–21.

Osborn, T. W. B., and J. N. Noriskin. "Data Regarding Native Diets in Southern Africa." *South African Journal of Science* 33 (1937): 605–610.

Osseo-Asare, Abena Dove. "Bioprospecting and Resistance: Transforming Poisoned Arrows into Strophanthin Pills in Colonial Gold Coast, 1885–1922." *Social History of Medicine* 21, no. 2 (2008): 269–290.

Packer, Kathryn, and Andrew Webster. "Patenting Culture in Science: Reinventing the Scientific Wheel of Credibility." *Science, Technology & Human Values* 21, no. 4 (1996): 427–453.

Page, B. "The Political Ecology of *Prunus africana* in Cameroon." *Area* 35, no. 4 (2003): 357–370.

Patwardhan, Bhushan, and Raghunath Anant Mashelkar. "Traditional Medicine-Inspired Approaches to Drug Discovery: Can Ayurveda Show the Way Forward?" *Drug Discovery Today* 14, no. 15–16 (2009): 804–811.

Pereira, C. A., L. L. S. Pereira, and A. D. Correa. "*Hoodia gordonii* in the Treatment of Obesity: A Review." *Journal of Medicinal Plants Research* 4, no. 22 (2010): 2305–2312.

Pereira, Jonathan. "On the Fruit of Amomum Melegueta—(Roscoe.)." *Pharmaceutical Journal* 6, no. 9 (1847): 412–419.

Persinos, Georgia J., Maynard W. Quimby, and John W. Schermerhorn. "A Preliminary Pharmacognostical Study of Ten Nigerian Plants." *Economic Botany* 18, no. 4 (1964): 329–341.

Planchon, Louis. "Les Drogues récemment inscrites au codex: 1 Les Strophanthus." *Bulletin de Pharmacie du Sud-Est* (1896).

Polonini, H. C., C. A. Caneschi, M. A. F. Brandao, and N. R. B. Raposo. "In Vitro Photoprotective Activity of Plants Extracts from Zona da Mata Mineira (Brazil)." *Latin American Journal of Pharmacy* 30, no. 3 (2011): 604–607.

Polur, H., T. Joshi, C. T. Workman, G. Lavekar, and I. Kouskoumvekaki. "Back to the Roots: Prediction of Biologically Active Natural Products from Ayurveda Traditional Medicine." *Molecular Informatics* 30, no. 2–3 (2011): 181–187.

Pretorius, P. J., J. J. Theron, and A. S. Wehmeyer. "Magnesium Balance Studies in South African Bantu Children with Kwashiorkor." *American Journal of Clinical Nutrition* 13, no. 5 (1963): 331.

Pretorius, P. J., and A. S. Wehmeyer. "Assessment of Nutritive Value of Fish Flour in the Treatment of Convalescent Kwashiokor Patients." *American Journal of Clinical Nutrition* 14, no. 3 (1964): 147–155.

"Protecting Traditional Knowledge: The San and Hoodia." *Bulletin of the World Health Organization* 84, no. 5 (2006): 345.

Rankin, Alisha. "Becoming an Expert Practitioner: Court Experimentalism and the Medical Skills of Anna of Saxony (1532–85)." *Isis: Journal of the History of Science Society* 98, no. 1 (2007): 23–53.

———. "Duchess, Heal Thyself: Elisabeth of Rochlitz and the Patient's Perspective in Early Modern Germany." *Bulletin of the History of Medicine* 82, no. 1 (2008): 109–144.

Rasmussen, N. "The Drug Industry and Clinical Research in Interwar America: Three Types of Physician Collaborator." *Bulletin of the History of Medicine* 79, no. 1 (2005): 50–80.

Rasoanaivo, Philippe. "Rain Forests of Madagascar: Sources of Industrial and Medicinal Plants." *Ambio* 19, no. 8 (1990): 421–424.

Rasoanaivo, Philippe, N. Langlois, and P. Potier. "Partie 8 Plantes Malgaches: Alcaloïdes du *Catharanthus longifolius*" [Malagasy plants part 8: Alkaloids from *Catharanthus longifolius*]. *Phytochemistry* 11, no. 8 (1972): 2616–2617.

Ratsimamanga, A. R. "The Comparative Action of Cortical and Desoxycorticosterone Extracts on the Adrenal Insufficiency during Scurvy." *Comptes Rendus des Séances de la Société de Biologie et de Ses Filiales* 138 (1944): 19–20.

———. "The Connections between Ascorbic Acid and Muscular Activity." *Comptes Rendus des Séances de la Société de Biologie et de Ses Filiales* 126 (1937): 1134–1136.

Ratsimamanga, A. R., P. Boiteau, G. Costessodigne, and M. Nigeondureuil. "Action Hormonale de Type Corticosteroidique des Deux Triterpenes l'Acide Arjunolique et l'Acide Asiatique." *Comptes Rendus des Séances de la Société de Biologie et de Ses Filiales* 153, no. 12 (1959): 1989–1991.

Rausser, Gordon C., and Arthur A. Small. "Valuing Research Leads: Bioprospecting and the Conservation of Genetic Resources." *Journal of Political Economy* 108, no. 1 (2000): 173.

Raustiala, Kal, and Stephen R. Munzer. "The Global Struggle over Geographic Indications." *European Journal of International Law* 18, no. 2 (2007): 337–365.

Raustiala, Kal, and David G. Victor. "The Regime Complex for Plant Genetic Resources." *International Organization* 58, no. 2 (2004): 277–309.

Richardson, Elizabeth. "Rx Makers Study Potential Rainforest Plant Medical Disaster." *Drug Store News* (1991).

Rouillard-Guellec, F., J. R. Robin, A. Rakoto-Ratsimamanga, S. Ratsimamanga, and P. Rasoanaivo. "Étude Comparative de *Centella asiatica* d'Origine Malgache et d'Origine Indienne" [Comparative study of *Centella asiatica* of Madagascan origin and Indian origin]. *Acta Botanica Gallica* 144, no. 4 (1997): 489–493.

Schiebinger, Londa L. "Agnotology and Exotic Abortifacients: The Cultural Production of Ignorance in the Eighteenth-Century Atlantic World." *Proceedings of the American Philosophical Society* 149, no. 3 (2005): 316–343.

Schlitzer, M. "Malaria Chemotherapeutics Part 1: History of Antimalarial Drug Development, Currently Used Therapeutics, and Drugs in Clinical Development." *ChemMedChem* 2, no. 7 (2007): 944–986.

Secord, A. "Science in the Pub: Artisan Botanists in Early Nineteenth-Century Lancashire." *History of Science, an Annual Review of Literature, Research and Teaching* 32, no. 97 (1994): 269–315.

Sedjo, Roger A. "Property Rights, Genetic Resources, and Biotechnological Change." *Journal of Law and Economics* 35, no. 1 (1992): 199–213.

Seligson, David. "On Collecting Herbs in Madagascar." *Arnoldia* 32, no. 1 (1972): 23–29.

Shapin, Steven. "The House of Experiment in Seventeenth-Century England." *Isis* 79, no. 3 (1988): 373–404.

———. "The Invisible Technician." *American Scientist* 77, no. 6 (1989): 554–563.

Singh, O. P., B. Das, M. M. Padhi, and N. S. Tewari. "Kushtha (Skin Disorders) in Vedic and Other Religious Literatures—A Review." *Bulletin of the Indian Institute of History of Medicine, Hyderabad* 32, no. 1 (2002): 51–55.

Singh, R., K. Narsimhamurthy, and G. Singh. "Neuronutrient Impact of Ayurvedic Rasayana Therapy in Brain Aging." *Biogerontology* 9, no. 6 (2008): 369–374.

Smith, Robert. "Yoruba Armament." *Journal of African History* 8, no. 1 (1967): 87–106.

Snyman, T., M. J. Stewart, A. Grove, and V. Steenkamp. "Adulteration of South African Traditional Herbal Remedies." *Therapeutic Drug Monitoring* 27, no. 1 (2005): 86–89.

Solway, Jacqueline S., Richard B. Lee, Alan Barnard, M. G. Bicchieri, Alec C. Campbell, James Denbow, Robert Gordon, Mathias Guenther, Henry Harpending, Patricia Draper, Robert K. Hitchcock, Tim Ingold, L. Jacobson, Susan Kent, Pnina Motzafi-Haller, Thomas C. Patterson, Carmel Schrire, Bruce G. Trigger, Polly Wiessner, Edwin N. Wilmsen, John E. Yellen, and Aram A. Yengoyan. "Foragers, Genuine or Spurious? Situating the Kalahari San in History (and Comments and Reply)." *Current Anthropology* 31, no. 2 (1990): 109–146.

Southall, Roger J. "Farmers, Traders and Brokers in the Gold Coast Cocoa Economy." *Canadian Journal of African Studies* 12, no. 2 (1978): 185.

Stearn, William T. "*Catharanthus roseus*, the correct name for the Madagascar periwinkle." *Lloydia* 29, no. 3 (1966): 196–200.

Stirton, C. H. "Apocynaceae: *Catharanthus* and *Vinca* in Southern Africa." *Bothalia* 14, no. 1 (1982): 69.

Stone, Richard. "The Biodiversity Treaty: Pandora's Box or Fair Deal?" *Science* 256, no. 5064 (1992): 1624.

"The Story of Our Fire." *Pharmaceutical Journal and Transactions* 20, no. 2011 (1889): vii–xiv.

Strydom, E. S., and A. S. Wehmeyer. "The Preparation of Edible Wild Fruit and Plant Samples for Analysis and Some Difficulties Encountered in Such Analyses." *South African Medical Journal* 43, no. 50 (1969): 1530–1532.

Summers, Carol. "Intimate Colonialism: The Imperial Production of Reproduction in Uganda, 1907–1925." *Signs* 16, no. 4 (1991): 787–807.

Svoboda, Gordon H., Marvin Gorman, Norbert Neuss, Albert J. Barnes Jr. "Alkaloids of *Vinca rosea* Linn. (*Catharanthus roseus* G. Don.) VIII. Preparation and Characterization of New Minor Alkaloids." *Journal of Pharmaceutical Sciences* 50, no. 5 (1961): 409–413.

Tackie, A. N., G. L. Boye, M. H. M. Sharaf, P. Schiff, R. C. Crouch, T. D. Spitzer, R. L. Johnson, J. Dunn, D. Minick, and G. Martin. "Cryptospirolepine, a Unique Spiro-Nonacyclic Alkaloid Isolated from *Cryptolepis sanguinolenta*." *Journal of Natural Products* 56, no. 5 (1993): 653–670.

Teresko, John, and Therese Welter. "Save Plants for their Chemicals." *Industry Week*, April 2, 1990, 81.

Teuber, Ramona. "Geographical Indications of Origin as a Tool of Product Differentiation: The Case of Coffee." *Journal of International Food & Agribusiness Marketing* 22, no. 3–4 (2010): 277–298.

Theron, J. J., V. Coetzee, and P. J. Pretorius. "A Histochemical Investigation of the Leucocytes in Kwashiorkor." *Journal of Clinical Pathology* 12, no. 5 (1959): 454–458.

Theron, J. J., C. P. Joubert, P. J. Pretorius, and H. Wolf. "State of Pyridoxine Nutrition in Patients with Kwashiorkor." *Journal of Pediatrics* 59, no. 3 (1961): 439.

Thomas, Elizabeth Marshall. "The Lion/Bushman Relationship in Nyae Nyae in the 1950s: A Relationship Crafted in the Old Way." *Anthropologica* 45, no. 1 (2003): 73–78.

Tilley, Helen. "Ecologies of Complexity: Tropical Environments, African Trypanosomiasis, and the Science of Disease Control in British Colonial Africa, 1900–1940." *Osiris* 19 (2004): 21–38.

Timmermans, Karin. "Intellectual Property Rights and Traditional Medicine: Policy Dilemmas at the Interface." *Social Science & Medicine* 57, no. 4 (2003): 745–756.

Timmermans, Stefan. "A Black Technician and Blue Babies." *Social Studies of Science* 33, no. 2 (2003): 197–229.

Tobias, Philipp V. "15 Years of Study on the Kalahari Bushmen or San: A Brief History of Kalahari Research Committee" *South African Journal of Science* 71, no. 3 (1975): 74–78.

Tonkin, Elizabeth. "Investigating Oral Tradition." *Journal of African History* 27 no. 2 (1986): 203–213.

Tudhope, W. T. D. "The Development of the Cocoa Industry in the Gold Coast and Ashanti." *Journal of the Royal African Society* 9, no. 33 (1909): 34.

Tyler, Varro E. "Gordon Svoboda, 1922–1994." *HerbalGram* 32 (1994): 62.

van Harten, A. M. "Melegueta Pepper." *Economic Botany* 24, no. 2 (1970): 208–216.

van Heerden, Fanie R. "*Hoodia gordonii*: A Natural Appetite Suppressant." *Journal of Ethnopharmacology* 119, no. 3 (2008): 434–437.

van Heerden, Fanie R., Roelof Marthinus Horak, Vinesh J. Maharaj, Robert Vleggaar, Jeremiah V. Senabe, and Philip J. Gunning. "An Appetite Suppressant from *Hoodia* Species." *Phytochemistry* 68 (2007): 2545–2553.

van Wyk, Ben-Erik. "A Review of Khoi-San and Cape Dutch Medical Ethnobotany." *Journal Of Ethnopharmacology* 119, no. 3 (2008): 331–341.

Vermeylen, S. "From Life Force to Slimming Aid: Exploring Views on the Commodification of Traditional Medicinal Knowledge." *Applied Geography* 28, no. 3 (2008): 224–234.

Webb, James L.A. "Malaria and the Peopling of Early Tropical Africa." *Journal of World History* 16, no. 3 (2005): 269–291.

Wehmeyer, A. S., R. B. Lee, and M. Whiting. "The Nutrient Composition and Dietary Importance of Some Vegetable Foods Eaten by the Kung Bushmen." *South African Medical Journal* 43 (1969): 1529–1530.

Wehmeyer, A. S., and E. F. Rose. "Important Indigenous Plants Used in the Transkei South Africa as Food Supplements." *Bothalia* 14, no. 3–4 (1983): 613–616.

White, C. T. "Vinca Rosea—A Reputed Cure for Diabetes." *Queensland Agricultural Journal* 23 (1925): 143–144.

White, Luise. "Tsetse Visions—Narratives of Blood and Bugs in Colonial Northern Rhodesia, 1931–9." *Journal of African History* 36, no. 2 (1995): 219–245.

Winn, Diane Robertson. "Gilbert Boye 1934–2003." *HerbalGram* 60 (2003): 76–77.

Wright, James R. "Almost Famous: E. Clark Noble, the Common Thread in the Discovery of Insulin and Vinblastine." *Canadian Medical Association Journal* 167, no. 12 (2002): 1391–1396.

Zhang, X. G., T. Han, Q. Y. Zhang, H. Zhang, B. K. Huang, L. L. Xu, and L. P. Qin. "Chemical Fingerprinting and Hierarchical Clustering Analysis of Centella asiatica from Different Locations in China." *Chromatographia* 69, no. 1–2 (2009): 51–57.

Zhang, Xiao-Gang, Ting Han, Zhi-Gao He, Qiao-Yan Zhang, Lei Zhang, Khalid Rahman, and Lu-Ping Qin. "Genetic Diversity of *Centella asiatica* in China Analyzed by Inter-simple Sequence Repeat (ISSR) Markers: Combination Analysis with Chemical Diversity." *Journal of Natural Medicines* 66, no. 1 (2012): 241–247.

Books and Book Chapters

Aaba, J. A. Kwesi. *African Herbalism: A Mine of Health, Part I*. Sekondi: Surwunku Industries, 1934.

Abaka, Edmund Kobina. *"Kola Is God's Gift": Agricultural Production, Export Initiatives and the Kola Industry in Asante and the Gold Coast, c. 1820–1950*. Oxford: James Currey, 2005.

Abbiw, Daniel K. *Useful Plants of Ghana: West African Uses of Wild and Cultivated Plants*. Kew: Royal Botanic Gardens, 1990.

Abraham, Itty. *The Making of the Indian Atomic Bomb: Science, Secrecy and the Postcolonial State*. London: Zed Books, 1998.

Achaya, K. T. *A Historical Dictionary of Indian Food*, New Delhi: Oxford University Press, 1998.

Achebe, Chinua. *Things Fall Apart*. Oxford: Heinemann Educational, 1996.

Adas, Michael. *Machines as the Measure of Men: Science, Technology, and Ideologies of Western Dominance*. Ithaca, N.Y.: Cornell University Press, 1989.

Addae, Steven. *The Evolution of Modern Medicine in a Developing Country: Ghana 1880–1960*. Durham, N.C.: Durham Academic Press, 1996.

Addae-Kyereme, Jonathan. *"Cryptolepis sanguinolenta."* In *Traditional Medicinal Plants and Malaria*, ed. Merlin Willcox, Gerard Bodeker, and Philippe Rasoanaivo, 151–161. Boca Raton, Fla.: CRC Press, 2004.

Addae-Mensah, Ivan. "Plant Biodiversity, Herbal Medicines, Intellectual Property Rights and Industrially Developing Countries: Socio-economic, Ethical and Legal Implications." In *Ghana: Changing Values/Changing Technology, Ghanaian Philosophical Studies II*, ed. Helen Lauer, 165–182. Washington, D.C.: Council for Research in Values and Philosophy, 2000.

———. *Towards a Rational Scientific Basis for Herbal Medicine: A Phytochemist's Two-Decade Contribution.* Accra: Ghana Universities Press, 1992.

Adjanohoun, Édouard. *Contribution aux Études Ethnobotaniques et Floristiques à Maurice: Îles Maurice et Rodrigues: Rapport Présenté à l'A.C.C.T.* Paris: Agence de Coopération Culturelle et Technique, 1983.

———. *Contribution aux Études Ethnobotaniques et Floristiques au Gabon.* Paris: Agence de Coopération Culturelle et Technique, 1984.

———. *Contribution aux Études Ethnobotaniques et Floristiques aux Comores: Rapport.* Paris: Agence de Coopération Culturelle et Technique, 1982.

———. *Contribution aux Études Ethnobotaniques et Floristiques en République Populaire du Bénin.* Paris: Agence de Coopération Culturelle et Technique, 1989.

———. *Contribution aux Études Ethnobotaniques et Floristiques en République Populaire du Congo.* Paris: Agence de Coopération Culturelle et Technique, 1988.

———. *Contribution to Ethnobotanical and Floristic Studies in Uganda.* Lagos: Organization of African Unity, Scientific, Technical & Research Commission, 1993.

———. *Contribution to Ethnobotanical and Floristic Studies in Western Nigeria.* Lagos: Organization of African Unity, Scientific, Technical & Research Commission, 1991.

———. *Médecine Traditionnelle et Pharmacopée: Contribution aux Études Ethnobotaniques et Floristiques au Mali.* Paris: Agence de Coopération Culturelle et Technique, 1979.

———. *Médecine Traditionnelle et Pharmacopée: Contribution aux Études Ethnobotaniques et Floristiques au Niger.* Paris: Agence de Coopération Culturelle et Technique, 1980.

———. *Traditional Medicine and Pharmacopoeia: Contribution to Ethnobotanical and Floristic Studies in Cameroon.* Porto-Novo, Benin: Organization of African Unity, Scientific, Technical & Research Commission, 1996.

———. *Végétation des Savanes et des Rochers Découverts en Côte d'Ivoire Centrale.* Paris: ORSTOM, 1964.

Adjanohoun, Édouard, and Ake Assi Laurent. *Contribution au Recensement des Plantes Médicinales de Côte d'Ivoire.* Abidjan, Ivory Coast] Centre Nationale de Floristique, 1979.

Adjanohoun, É. J., M. R. A. Ahyi, L. Ake Assi, K. Akpagana, P. Chibon, A. El-Hadj Watara, J. Eyme, M. Garba, J.-N. Gassita, M. Gbeassor, E. Goudote, S. Guinko, K.-H. Hodouto, P. Houngnon, A. Keita, Y. Keoula, W. P. Kluga-Ocloo, I. Lo, K. M. Siamevi, and K. K. Taffame . *Contribution aux Études Ethnobotaniques* et Floristiques au Togo Médecine Traditionnelle et Pharmacopée. Paris: Agence de Coopération Culturelle et Technique, 1986.

Africanus, Leo. *The History and Description of Africa: And of the Notable Things Therein Contained written by Al-Hassan Ibn-Mohammed Al-Wezaz Al-Fasi* [original English translation 1600], trans. John Pory, ed. Robert Brown. Vol. 1. London: Hakluyt Society, 1896.

Akyeampong, Emmanuel Kwaku. "Disease in West African History." In *Themes in West Africa's History*, ed. Emmanuel Kwaku Akyeampong, 186–207. Athens: Ohio University Press, 2006.

Allen, Philip M., and Maureen Covell. *Historical Dictionary of Madagascar*, ed. Jon Woronoff. Lanham, Md.: Scarecrow Press, 2005.

Allman, Jean Marie, and John Parker. *Tongnaab: The History of a West African God.* Bloomington: Indiana University Press, 2005.

Ampofo, Oku. *First Aid in Plant Medicine.* Accra: Waterville Publishing House, 1983.

———. *My Kind of Sculpture.* Mampong-Akuapem: O. Ampofo, 1991.

Anquetil, Louis-Pierre. *A summary of universal history. Exhibiting the rise, decline, and revolutions of the different nations of the world from creation to the present time.* Vol. 6. London: G. G. and J. Robinson, 1800.

Anthonio, H. O., and M. Isoun. *Nigerian Cookbook.* London: Macmillian, 1982.

Appadurai, Arjun, ed. *The Social Life of Things: Commodities in Cultural Perspective.* Cambridge: Cambridge University Press.

Arnold, T. H., M. J. Wells, and A. S. Wehmeyer. "Khoisan Food Plants: Taxa with Potential for Future Economic Exploitation." In *Plants for Arid Lands: Proceedings of the Kew International Conference on Economic Plants for Arid Lands Held in the Jodrell Laboratory, Royal Botanic Gardens, Kew, England, 23–27 July 1984*, ed. G. E. Wickens, J. R. Goodin, and D. V. Field, 69–86. London: Allen & Unwin, 1985.

Arthur, G. F. Kojo. *Cloth as Metaphor: (Re)reading the Adinkra Cloth Symbols of the Akan of Ghana.* Accra: Center for Indigenous Knowledge Systems, 1999.

Ashcroft, Bill, Gareth Griffiths, and Helen Tiffin. *The Post-colonial Studies Reader.* London: Routledge, 1995.

Bank, Andrew. *Bushmen in a Victorian World: The Remarkable Story of the Bleek-Lloyd Collection of Bushmen Folklore.* Cape Town: Double Storey, 2006.

Bauhin, Johann, Johann Heinrich Cherler, and Franz Ludwig von Graffenried. *Historia plantarum universalis, nova, et absolutissima: cum consensu et dissensu circa eas. 2.* N.p., 1651.

Bellakhdar, Jamal. *La Pharmacopée Marocaine Traditionnelle: Médecine Arabe Ancienne et Savoirs Populaires.* Paris: Ibis Press, 1997.

Bellakhdar, Jamal, Gishō Honda, and Wataru Miki. *Herb Drugs and Herbalists in the Maghrib.* Vol. 19. Tokyo: Institute for the Study of Languages and Cultures of Asia and Africa, 1982.

Belloc, Hilaire "H. B.," and Basil Temple "B. T. B." Blackwood. *The Modern Traveller.* London: Edward Arnold, 1898.

Bérand, Laurence, and Philippe Marchenay. "Tradition, Regulation, and Intellectual Property: Local Agricultural Products and Foodstuffs in France." In *Valuing Local Knowledge: Indigenous People and Intellectual Property Rights*, ed.

Stephen Brush and Doreen Stabinisky, 230–242. Washington, D.C.: Island Press, 1996.

Biagioli, Mario, and Peter Louis Galison. *Scientific Authorship: Credit and Intellectual Property in Science*. New York: Routledge, 2003.

Blake, John W. *Europeans in West Africa, 1450–1560: Documents to Illustrate the Nature and Scope of Portuguese Enterprise in West Africa, the Abortive Attempt of Castilians to Create an Empire There, and the Early English Voyages to Barbary and Guinea*. London: Hakluyt Society, 1942.

Bliss, Michael. *The Discovery of Insulin*. Chicago: University of Chicago Press, 1982.

Boahen, A. Adu. *African Perspectives on Colonialism*. Baltimore: Johns Hopkins University Press, 1987.

Boakye-Yiadom, K., and S. O. A. Bamgbose, eds. *Proceedings of the First International Symposium on Cryptolepine*. Kumasi, Ghana: University of Science and Technology, 1983.

Boateng, Boatema. *The Copyright Thing Doesn't Work Here: Adinkra and Kente Cloth and Intellectual Property in Ghana*. Minneapolis: University of Minnesota Press, 2011.

Boye, Gilbert L., and Oku Ampofo. "Clinical Uses of *Cryptolepis sanguinolenta* (Asclepiadaceae)." In *Proceedings of the First International Symposium on Cryptolepine*, ed. K. Boakye-Yiadom and S. O. A. Bamgbose. Kumasi, Ghana: University of Science and Technology, 1983.

Boyle, James. *Shamans, Software, and Spleens: Law and the Construction of the Information Society*. Cambridge, Mass.: Harvard University Press, 1996.

Brooks, George E. *Eurafricans in Western Africa: Commerce, Social Status, Gender, and Religious Observance from the Sixteenth to the Eighteenth Century*. Athens: Ohio University Press, 2003.

Brown, Michael. *Who Owns Native Culture?* Cambridge, Mass.: Harvard University Press, 2003.

Brown, Robert. *The story of Africa and its explorers*. Vol. 1. London: Cassell, 1892.

Brukum, N. J. K. *The Guinea Fowl, Mango and Pito Wars: Episodes in the History of Northern Ghana, 1980–1999*. Accra: Ghana Universities Press, 2001.

Brush, Stephen, and Doreen Stabinisky, eds. *Valuing Local Knowledge: Indigenous People and Intellectual Property Rights*. Washington, D.C.: Island Press, 1996.

Burton, Richard F. *Wanderings in West Africa from Liverpool to Fernando Po*. Vol. 2. London: Tinsley Brothers, 1863.

Cable, Stuart, Kew Royal Botanic Gardens, and Project Mount Cameroon. *The Plants of Mt Cameroon: A Conservation Checklist*. Richmond, U.K.: Royal Botanic Gardens, Kew, 1998.

Cardinall, A. W. *The Natives of the Northern Territories of the Gold Coast, Their Customs, Religion and Folklore*. London: Routledge, 1920.

Carney, Judith Ann. *Black Rice: The African Origins of Rice Cultivation in the Americas*. Cambridge, Mass.: Harvard University Press, 2001.

Carney, Judith Ann, and Richard Nicholas Rosomoff. *In the Shadow of Slavery: Africa's Botanical Legacy in the Atlantic World*. Berkeley: University of California Press, 2009.

Carrier, Neil C. M. *Kenyan Khat: The Social Life of a Stimulant*. Vol. 15. Leiden: Brill, 2007.

Centre for Scientific and Industrial Research, Policy Research and Strategic Planning Institute. *Ghana Herbal Pharmacopoiea.* Accra: Technology Transfer Centre/Advent Press, 1992.

Chakrabarti, Pratik. *Western Science in Modern India: Metropolitan Methods, Colonial Practices.* Delhi: Permanent Black, 2004.

Chalfin, Brenda. *Shea Butter Republic: State Power, Global Markets, and the Making of an Indigenous Commodity.* New York: Routledge, 2004.

Chambers's encyclopaedia: A dictionary of universal knowledge for the people. Vol. 5. Philadelphia: J. B. Lippincott & Co., 1863.

Chandler, Alfred Dupont. *Shaping the Industrial Century: The Remarkable Story of the Evolution of the Modern Chemical and Pharmaceutical Industries.* Cambridge, Mass.: Harvard University Press, 2005.

Comaroff, John L., and Jean Comaroff. *Ethnicity, Inc.* Chicago: University of Chicago Press, 2009.

The compleat herbal; or, family physician: Giving an account of all such plants as are now used in the practice of physic. With their descriptions and virtues. Vol. 1. Manchester: G. Swindells, Hanging-Bridge, 1787.

Connah, Graham. *African Civilizations: An Archaeological Perspective.* Cambridge: Cambridge University Press, 2001.

———. *Transformations in Africa: Essays on Africa's Later Past.* London: Leicester University Press, 1998.

Cook, Harold John. *Matters of Exchange: Commerce, Medicine, and Science in the Dutch Golden Age.* New Haven: Yale University Press, 2007.

Coombe, Rosemary J. *The Cultural Life of Intellectual Properties: Authorship, Appropriation, and the Law.* Durham, N.C.: Duke University Press, 1998.

Coombs, Sarah V. *South African Plants for American Gardens.* New York: Frederick A. Stokes, 1936.

Cramp, Arthur Joseph. *Nostrums and Quackery: Articles on the Nostrum Evil and Quackery Reprinted from the Journal of the American Medical Association.* Chicago: American Medical Association, 1936.

Crosby, Alfred W. *Ecological Imperialism: The Biological Expansion of Europe, 900–1900.* Cambridge: Cambridge University Press, 1986.

Dahl, Øyvind. *Meanings in Madagascar: Cases of Intercultural Communication.* Westport, Conn.: Bergin & Garvey, 1999.

Dalziel, J. M. *The Useful Plants of West Tropical Africa.* London: Published under the authority of the Secretary of State for the colonies by the Crown agents for the colonies, 1937.

Davis, Natalie Zemon. *Trickster Travels: A Sixteenth-Century Muslim between Worlds.* New York: Hill and Wang, 2006.

Dechambre, Amédée, Léon Lereboullet, and Louis Hahn. *Dictionnaire encyclopédique des sciences médicales.* Paris: A. Lahure, 1887.

de Faria e Sousa, Manuel. *Africa Portuguesa.* Lisbon: A. Craesbeeck de Mello, 1681.

de Flacourt, Étienne. *Histoire de la Grande Isle Madagascar.* Paris: Alexandre Lesselin, 1658.

———. *Histoire de la Grande Isle Madagascar, Édition annotée, augmentée et présentée par Claude Allibert.* Paris: INALCO-Karthala, 2007.

de Lorris, Guillaume. "The Garden, the Fountain, and the Rose (1–1680) in Part I: The Dream of Love." In *The Romance of the Rose [Roman de la Rose]*. Princeton: Princeton University Press, 1995 [1230].

de Marees, Pieter. *Beschrijving en Historisch Verhaal van het Gouden Koninkrijk van Guinea* [Description and historical account of the Gold Kingdom of Guinea), trans. Adam Jones, ed. British Academy, Fontes historiae Africane. Oxford: Oxford University Press, 1602.

Des Jardins, Julie. *The Madame Curie Complex: The Hidden History of Women in Science*. New York: Feminist Press at the City University of New York.

Deutsches Arzneibuch [German pharmacopoeia]. Berlin: Decker, 1910.

Dokosi, O. B. *Herbs of Ghana*, ed. Council for Scientific and Industrial Research. Accra: Ghana Universities Press, 1998.

Dove, Michael. "Center, Periphery, and Biodiversity: A Paradox of Governance and a Developmental Challenge." In *Valuing Local Knowledge: Indigenous People and Intellectual Property Rights*, ed. Stephen Brush and Doreen Stabinsky, 41–67. Washington, D.C.: Island Press, 1996.

Doyle, Sir Arthur Conan. *The Lost World*.New York: Hodder & Stoughton, 1912.

Dubow, Saul. *Scientific Racism in Modern South Africa*. Cambridge: Cambridge University Press, 1995.

Duffin, Jacalyn. *History of Medicine: A Scandalously Short Introduction*. Toronto: University of Toronto Press, 1999.

Duke, James A., Mary Jo Bogenschutz-Godwin, Judi duCellier, and Peggy-Ann K. Duke. *CRC Handbook of Medicinal Spices*. Boca Raton, Fla.: CRC Press, 2003.

Dutt, Udoy Chand, and George King. *The Materia Medica of the Hindus: Compiled from Sanskrit Medical Works*. Calcutta: Thacker, Spink, 1877.

Echenberg, Myron. *Black Death, White Medicine: Bubonic Plague and the Politics of Public Health in Colonial Senegal, 1914–1945*. Portsmouth, N.H.: Heinemann, 2001.

Ehret, Christopher. "Historical/Linguistic Evidence for Early African Food Production." In *From Hunters to Farmers: The Causes and Consequences of Food Production in Africa*, ed. J. Desmond Clark and Steven A. Brandt, 26–39. Berkeley: University of California Press, 1984.

Enti, Albert A. *The Rejuvenating Plants of Tropical Africa*, ed. Anthony Kweku Andoh. San Francisco: North Scale Institute, 1988.

Evans-Pritchard, Edward Evan. *Witchcraft, Oracles and Magic among the Azande*. Oxford: Clarendon Press, 1937.

Ewusi, Kodwo. *The Political Economy of Ghana in the Post Independence Period: Description and Analysis of the Decadence of the Political Economy of Ghana and the Survival Techniques of Her Citizens*. Vol. 14. Accra: University of Ghana–Institute of Statistical, Social and Economic Research, 1984.

Farquhar, Judith. *Knowing Practice: The Clinical Encounter of Chinese Medicine*. Boulder, Colo.: Westview Press, 1994.

Feierman, Steven, and John M. Janzen, eds. *The Social Basis of Health and Healing in Africa*. Berkeley: University of California Press, 1992.

Fenning, Daniel, and Joseph Collyer. *A new system of geography or, a general description of the world . . . Embellished with a new and accurate set of maps . . . and a great variety of copper-plates*. Vol. 1. London: Printed for J. Johnson, 1778.

Flint, Karen Elizabeth. *Healing Traditions: African Medicine, Cultural Exchange, and Competition in South Africa, 1820–1948*. Athens: Ohio University Press, 2008.

Flückiger, Friedrich A., and Daniel Hanbury. *Pharmacographia: A History of the Principal Drugs of Vegetable Origin, Met with in Great Britain and British India*. 2nd ed. London: MacMillan and Co., 1879.

Fox, Francis William, and Marion Emma Norwood Young. *Food from the Veld: Edible Wild Plants of Southern Africa*. Johannesburg: Delta Books, 1982.

Fuchs, Leonard. *Commentaires très excellents de l'hystoire des plantes, Composez premirement en latin par Leonarth Fousch, medécin tres renommé* [French translation of the original Latin]. Paris: Jacques Gazeau, 1549.

Fullwiley, Duana. *The Enculturated Gene: Sickle Cell Health Politics and Biological Difference in West Africa*. Princeton: Princeton University Press, 2012.

Garcia, Faustino L. "The Treatment of Diabetes Mellitus by the Use of Different Philippine Medicinal Plants and a Preliminary Report on the Use of Plantisul." Paper presented at the Eighth Pacific Science Congress of the Pacific Science Association, University of the Philippines Diliman, Quezon City, 1953.

Gebissa, Ezekiel. *Leaf of Allah: Khat and Agricultural Transformation in Harerge, Ethiopia 1875–1991*. Oxford: James Currey, 2004.

Geison, Gerald L. *The Private Science of Louis Pasteur*. Princeton: Princeton University Press, 1995.

General Medical Council. *The British Pharmacopoeia*. London: Spottiswoode, 1885.

Gerard, John, and Thomas Johnson. *The Herball: or generall historie of plantes*. London: Printed by A. I. J. Norton and R. Whitakers, 1636.

Gijswijt-Hofstra, Marijke, G. M. van Heteren, and E. M. Tansey. *Biographies of Remedies: Drugs, Medicines, and Contraceptives in Dutch and Anglo-American Healing Cultures*. Amsterdam: Rodopi, 2002.

Gilman, Daniel Coit, Harry Thurston Peck, and Frank Moore Colby. *The New International Encyclopaedia*. New York: Dodd, Mead, 1903.

Gordon, Robert J. *The Bushman Myth: The Making of a Namibian Underclass*. Boulder, Colo.: Westview Press, 2000.

Grasseni, Cristina. "Packaging Skills: Calibrating Cheese to the Global Market." In *The Commodification of Everything*, 259–288. Bloomington: University of Indiana Press, 2003.

Gray, Samuel Frederick, and Theophilus Redwood. *Gray's Supplement to the Pharmacopoeia: being a concise but comprehensive dispensatory and manual of facts and formulae, for the chemist and druggist and medical practitioner*. London: Longman, S. Highley, Simpkin, J. Churchill, H. Bohn, and H. Renshaw, 1848.

Great Britain, Austria-Hungary, Belgium, Denmark, France, Germany, Italy, the Netherlands, Portugal, Russia, Spain, Sweden, Norway, Turkey, and United States. *General Act of the Conference at Berlin of the plenipotentiaries of Great Britain, Austria-Hungary, Belgium, Denmark, France, Germany, Italy, the Netherlands, Portugal, Russia, Spain, Sweden and Norway, Turkey and the United States respecting: (1) Freedom of Trade in the Basin of the Congo; (2) The Slave Trade; (3) Neutrality of the Territories in the Basin of the Congo; (4) Navigation of the Congo; (5) Navigation of the Niger; and (6) Rules for Future Occupation on the Coast of the African Continent*. 1885.

Greene, Jeremy A. *Prescribing by Numbers: Drugs and the Definition of Disease.* Baltimore: Johns Hopkins University Press, 2007.

Grewal, David Singh. *Network Power: The Social Dynamics of Globalization.* New Haven: Yale University Press, 2008.

Gupta, Brahmananda. "Evolution of Ayurveda through the Ages." In *History of Medicine in India: The Medical Encounter,* ed. Chittabrata Palit and Achintya Dutta, 207–218. Kolkata: Corpus Research Institute, 2005.

Gyekye, Kwame. *Tradition and Modernity: Philosophical Reflections on the African Experience.* New York: Oxford University Press, 1997.

Hakluyt, Richard. *Voyagers' Tales from the Collections of Richard Hakluyt.* 1589.

Harris, Joseph E. *Global Dimensions of the African Diaspora.* Washington, D.C.: Howard University Press, 1982.

Hayden, Cori. *When Nature Goes Public: The Making and Unmaking of Bioprospecting in Mexico.* Princeton: Princeton University Press, 2003.

Hecht, David, and A. M. Simone. *Invisible Governance: The Art of African Micro-Politics.* Brooklyn, N.Y.: Autonomedia, 1994.

Hegel, Georg Wilhelm Friedrich, and John Sibree. *Lectures on the Philosophy of History.* London: H. G. Bohn, 1857.

Hitchcock, Robert K., Kazunoba Ikeya, Megan Biesele, and Richard B. Lee. *Updating the San: Image and Reality of an African People in the 21st Century.* Osaka: National Museum of Ethnology, 2006.

Hu, Sihui, Paul D. Buell, Eugene N. Anderson, and Charles Perry. *A Soup for the Qan: Chinese Dietary Medicine of the Mongol Era as Seen in Hu Sihui's Yinshan zhengyao: Introduction, Translation, Commentary, and Chinese Text.* London: Kegan Paul International, 2000.

Iliffe, John. *Africans: The History of a Continent.* Cambridge: Cambridge University Press, 1995.

———. *East African Doctors: A History of the Modern Profession.* Cambridge: Cambridge University Press, 1998.

Irvine, F. R. *Plants of the Gold Coast.* London: Oxford University Press, 1930.

Irvine, Frederick Robert. *Woody Plants of Ghana, with Special Reference to Their Uses.* London: Oxford University Press, 1961.

"Isaiah 2:4." In *The Holy Bible: New International Version,* 610. Grand Rapids, Mich.: Zondervan, 1989.

Iwu, Maurice M. *Handbook of African Medicinal Plants.* Boca Raton, Fla.: CRC Press, 1993.

Jacobs, T.V. "Indigenous Plants as Complementary Immunomodulators in HIV/AIDS Patients." In *Health Knowledge and Belief Systems in Africa,* ed. Toyin Falola and Matthew Heaton, 355–360. Durham, N.C.: Carolina Academic Press, 2007.

Jank, Joseph K. *Spices: Their Botanical Origin, Their Composition, Their Commercial Use.* St. Louis, Mo.: Joseph K. Jank, 1915.

Janzen, John. *Lemba, 1650–1930: A Drum of Affliction in Africa and the New World.* New York: Garland Publishers, 1982.

Janzen, John, with the collaboration of William Arkinstall. *The Quest for Therapy in Lower Zaire.* Berkeley: University of California Press, 1978.

Jasanoff, Sheila. "Heaven and Earth: The Politics of Environmental Images." In *Earthly Politics: Local and Global in Environmental Governance*, ed. Sheila Jasanoff and Marybeth Long Martello, 31–54. Cambridge, Mass.: MIT Press, 2004.

July, Robert. *An African Voice: The Role of the Humanities in African Independence*. Durham, N.C.: Duke University Press, 1987.

Juma, Calestous. *The Gene Hunters: Biotechnology and the Scramble for Seeds*. Princeton: Princeton University Press, 1989.

Kaler, Amy. *Running after Pills: Politics, Gender, and Contraception in Colonial Zimbabwe*. Portsmouth, N.H.: Heinemann, 2003.

Kanogo, Tabitha M. *African Womanhood in Colonial Kenya, 1900–50*. Oxford: James Currey, 2005.

———. *Squatters and the Roots of Mau Mau, 1905–63*. London: James Currey, 1987.

Kenyatta, Jomo. *Facing Mount Kenya: The Tribal Life of Gikuyu*. London: Secker and Warburg, 1938.

Kilham, Chris. *Kava: Medicine Hunting in Paradise*. Rochester, Vt.: Park Street Press, 1996.

King, Steven R., Thomas J. Carlson, and Katy Moran. "Biological Diversity, Indigenous Knowledge, Drug Discovery, and Intellectual Property Rights." In *Valuing Local Knowledge: Indigenous People and Intellectual Property Rights*, ed. Stephen Brush and Doreen Stabinisky, 167–185. Washington, D.C.: Island Press, 1996.

Kingsley, Mary Henrietta. *West African Studies*. London: Macmillan and Co, 1901.

Kingwill, D. G. *The CSIR: The First 40 Years*. Pretoria: CSIR, 1990.

Kreig, Margaret. *Green Medicine: The Search for Plants that Heal*. New York: Bantam Books, 1966.

Kutalek, Ruth, and Armin Prinz. "African Medicinal Plants." In *Handbook of Medicinal Plants*, ed. Zohara Yaniv and Uriel Bachrach. Binghamton, N.Y.: Haworth Press, 2005.

Lachman-White, Deborah A., Charles Dennis Adams, and Ulric O'D Trotz. *A Guide to the Medicinal Plants of Coastal Guyana*. London: Commonwealth Science Council, 1992.

Langwick, Stacey Ann. *Bodies, Politics, and African Healing: The Matter of Maladies in Tanzania*. Bloomington: Indiana University Press, 2011.

Lans, Cheryl. *Creole Remedies of Trinidad and Tobago*: Lulu.com, 2007.

Last, Murray, and G. L. Chavunduka. *The Professionalisation of African Medicine*. Manchester, U.K.: Manchester University Press 1986.

Latour, Bruno. *Reassembling the Social: An Introduction to Actor-Network-Theory*. Oxford: Oxford University Press, 2005.

Latour, Bruno, and Steve Woolgar. *Laboratory Life: The Construction of Scientific Facts*. Princeton: Princeton University Press, 1986.

Laumann, Dennis. *Remembering the Germans in Ghana*. New York: Peter Lang, 2007.

Lee, Richard B. "What Hunters Do for a Living, or, How to Make Out on Scarce Resources." In *Man the Hunter*, ed. Richard B. Lee and Irven DeVore, 30–43. Chicago: Aldine Publishing Company, 1968.

Lesch, John E. *The First Miracle Drugs: How the Sulfa Drugs Transformed Medicine*. Oxford: Oxford University Press, 2007.

Lesser, William. *Sustainable Use of Genetic Resources under the Convention on Biological Diversity: Exploring Access and Benefit Sharing Issues.* Wallingford, U.K.: CAB International, 1998.

Lovejoy, Paul E. *Caravans of Kola: The Hausa Kola Trade, 1700–1900.* Zaria, Nigeria: Ahmadu Bello University Press, 1980.

Lyons, Maryinez. *The Colonial Disease: A Social History of Sleeping Sickness in Northern Zaire, 1900–1940.* Cambridge: Cambridge University Press, 1992.

Macdonald, George. *The Gold Coast, past and present; a short description of the country and its people.* London: Longmans, Green, & Co., 1898.

Maddox, Brenda. *Rosalind Franklin: The Dark Lady of DNA.* New York: HarperCollins, 2002.

Mafeje, Archie. *Anthropology in Post-Independence Africa: End of an Era and the Problem of Self-Redefinition. Part 1.* Nairobi: Heinrich Böll Foundation, 2001.

Maingard, J. F. "Introduction." In *Bushmen of the Southern Kalahari,* ed. J. D. Rheinallt Jones and C. M. Doke. Johannesburg: University of the Witwatersrand Press, 1937.

———. "Some Notes on Health and Disease among the Bushmen of the Southern Kalahari." In *Bushmen of the Southern Kalahari,* ed. J. D. Rheinallt Jones and C. M. Doke, 227–236. Johannesburg: University of the Witwatersrand Press, 1937.

Maisch, John Michael. *A manual of organic materia medica.* Philadelphia: Henry C. Lea's Son & Co., 1882.

Mallam, Abu. *The Zabarma Conquest of North-West Ghana and Upper Volta: A Hausa Narrative "Histories of Samory and Babatu and Others,"* ed. Stanisław Piłaszewicz, Warsaw: Polish Scientific Publishers, 1992.

Margry, Pierre. *Les navigations françaises et la révolution maritime du XIVe au XVIe siècle, d'après les documents inédits tirés de France, d'Angleterre, d'Espagne et d'Italie.* Paris: Librairie Tross, 1867.

Markgraf, F. *Flore de Madagascar et des Comores.* Paris: Museum National d'Histoire Naturelle, 1976.

Marks, Harry M. *The Progress of Experiment: Science and Therapeutic Reform in the United States, 1900–1990.* Cambridge: Cambridge University Press, 1997.

Marloth, Rudolf. *Dictionary of the Common Names of Plants with List of Foreign Plants Cultivated in the Open.* Cape Town: Specialty Press of South Africa, 1917.

———. *The Flora of South Africa: With a Synopsis of the South African Genera of Phanerogamous Plants.* Vol. 3: Sympetalae. Cape Town: Darter Bros. & Co., 1932.

Martin, R. Montgomery. *History of the British colonies IV, Possessions in Africa and Austral-Asia.* London: Cochrane, 1835.

McKenzie, A. G. "The Rise and Fall of Strophanthin." In *The History of Anesthesia: Proceedings of the Fifth International Symposium on the History of Anesthesia, Santiago, Spain, 19–23 September 2001,* ed. J. C. Diz, A. Franco, D. R. Bacon, J. Rupreht and J. Alvarez. Amsterdam: Elsevier, 2002.

Meyrick, William. *The new family herbal or, domestic physician.* Birmingham: Thomas Pearson, 1790.

Mialet, Hélène. *Hawking Incorporated: Stephen Hawking and the Anthropology of the Knowing Subject.* Chicago: University of Chicago Press, 2012.

Miescher, Stephan. *Making Men in Ghana.* Bloomington: Indiana University Press, 2005.

Miller, Philip. *Figures of the most Beautiful, Useful, and Uncommon Plants Described in the Gardeners Dictionary, exhibited on Three Hundred Copper Plates, accurately engraven after Drawings taken from Nature.* 2 vols. London: Printed for the author, 1760.

Millot, J., and R. Paulian. *Le Parc Botanique et Zoologique de Tananarive-Tsimbazaza.* Antananarivo, Madagascar: Société des Amis du Parc Botanique et Zoologique, 1949.

Milton, Sue J., and W. Richard J. Dean. *Karoo Veld: Ecology and Management.* Lynn East, South Africa: ARC-Range and Forage Institute, 1996.

Mitchell, Laura Jane. *Belongings: Property, Family, and Identity in Colonial South Africa, an Exploration of Frontiers, 1725–c. 1830.* New York: Columbia University Press, 2008.

Morgan, Shane. *Representing Bushmen: South Africa and the Origin of Language.* Rochester, N.Y.: University of Rochester Press, 2009.

Mshana, N. R., D. K. Abbiw, I Addae-Mensah, E. Adjanouhoun, M. R. A. Ahyi, J. A. Ekpere, E. G. Enow-Orock, Z. O. Gbile, G. K. Noamesi, M. A. Odei, H. Odunlami, A. A. Oteng-Yeboah, K. Sarpong, A. Sofowora, and A. N. Tackie. *Traditional Medicine and Pharmacopoeia: Contribution to the Revision of Ethnobotanical and Floristic Studies in Ghana.* Accra: Organization of African Unity/Scientific, Technical and Research Commission, 2000.

Niezen, Ronald. *The Origins of Indigenism: Human Rights and the Politics of Identity.* Berkeley: University of California Press, 2003.

Nketiah, K. S., J. A. S. Ameyaw, and B. Owusu. *Equity in Forest Benefit Sharing: Stakeholders' Views.* Wageningen: Tropenbos International, 2005.

Noltie, Henry J. *Robert Wight and the Botanical Drawings of Rungiah and Govindoo.* Edinburgh: Royal Botanic Garden, 2007.

Northcott, H. P. *Report on the Northern Territories of the Gold Coast: from reports furnished by officers of the administration.* London: Intelligence Division, War Office, 1899.

Odendaal, François, Helen Suich, and Rojas Velsquez. *Richtersveld: The Land and Its People.* Cape Town: Struik, 2007.

Odugbemi, Tolu. *A Textbook of Medicinal Plants from Nigeria.* Lagos: University of Lagos Press, 2008.

Oguamanam, Chidi. *Indigenous Knowledge in International Law: Intellectual Property, Plant Biodiversity and Traditional Medicine.* Toronto: University of Toronto Press, 2006.

Olapade, Ebenezer O. *The Herbs for Good Health: The 50th Anniversary Lecture of the University of Ibadan.* Ibadan: NARL Specialist Clinic, 2002.

Pacey, Arnold. *Technology in World Civilization: A Thousand-Year History.* Cambridge, Mass.: MIT Press, 1993.

Packard, Randall. *The Making of a Tropical Disease: A Short History of Malaria*, ed. Charles Rosenberg. Baltimore: Johns Hopkins University Press, 2007.

Packard, Randall M. *White Plague, Black Labor: Tuberculosis and the Political Economy of Health and Disease in South Africa.* Berkeley: University of California Press, 1989.

Pappe, K. W. L. *Silva capensis; or, A description of South African forest-trees and arborescent shrubs used for technical and economical purposes by the colonists of the Cape of Good Hope.* Cape Town: van de Sandt de Villiers & Co., 1862.

Parascandola, John. "Alkaloids to Arsenicals: Systematic Drug Discovery before the First World War." In *The Inside Story of Medicines: A Symposium*, ed. Gregory Higby and Elaine C. Stroud, 77–92. Madison, Wis.: American Institute of the History of Pharmacy, 1997.

Parry, Bronwyn. "The Fate of Collections: Social Justice and the Annexation of Plant Genetic Resources." In *People, Plants and Justice: The Politics of Nature Conservation*, ed. Charles Zerner, 374–400. New York: Columbia University Press, 2000.

Parsons, James. *The Microscopical Theatre of Seeds: Being a Short View of the Particular Marks, Characters, Contents, and Natural Dimensions of All the Seeds of the Shops, Flower and Kitchen-Gardens &c: With Many Other Curious Observations and Discoveries, which Could Not be Known Without the Assistance of the Microscope.* London: F. Needham, 1745.

Payne, John. *Universal geography formed into a new and entire system describing Asia, Africa, Europe, and America; . . . also giving a general account of . . . the history of man . . . the state of arts, sciences, commerce . . . To which is added, a short view of astronomy.* Vol. 1. London: Printed for the author, 1791.

Pereira, Jonathan. *The Elements of Materia Medica and Therapeutics: Third American Edition, Enlarged and Improved by the Author, Including notices of most of the medical substances in use in the Civilized world and forming an Encyclopaedia of Materia Medica*, ed. Joseph Carson. Vol. 2. Philadelphia: Blanchard and Lea, 1854.

Perrédès, Pierre Élie Félix. *A Contribution to the Pharmacognosy of Official Strophanthus Seed.* London: Wellcome Chemical Research Laboratories, 1900.

Pertz, Georg Heinrich. *Monumenta Germaniae Historica/1/5: inde ab anno Christi quingentesimo usque ad annum millesimum et quingentesimum 19 Annales aevi Suevici.* Hannover: Hahn, 1866.

Pharmaceutical Society of Great Britain. *The British Pharmaceutical Codex: An Imperial Dispensatory for the Use of Medical Practitioners and Pharmacists.* London: Printed at the St. Clements' Press Ltd. for the Pharmaceutical Society, 1907.

Rabemananjara, Raymond William. *Prince Albert Rakoto Ratsimamanga: Un fils de la Lumière, Au service de l'Homme, de la Science et de la Paix.* Vol. 1. Paris: L'Harmattan, 2003.

Rahier, Jean, Percy C. Hintzen, and Felipe Smith. *Global Circuits of Blackness: Interrogating the African Diaspora.* Urbana: University of Illinois Press, 2010.

Ralibera, Rémy. *Vazaha et Malgaches en Dialogue.* Fianarantsoa, Madagascar: Imprimerie Catholique, 1966.

Rasoanaivo, Philippe. "Prospects and Problems of Trading in Medicinal Plants: A Case Study of Madagascar." In *Proceedings of the International Symposium: Biodiversity and Health, Focusing Research to Policy*, ed. J. T. Arnason, P. M. Catling, E. Small, P. T. Dang, and J. D. H. Lambert, 97–114. Ottawa: National Research Council of Canada, 2003.

———. "Traditional Medicine Programmes in Madagascar." In *Proceedings of the International Symposium: Biodiversity and Health, Focusing Research to Policy*, ed. J.T. Arnason, P. M. Catling, E. Small, P. T. Dang, and J. D. H. Lambert, 73–76. Ottawa: National Research Council of Canada, 2003.

Rätsch, Christian, and Claudia Müller-Ebeling. *Lexikon der Liebesmittel: pflanzliche, mineralische, tierische und synthetische Aphrodisiaka* [The encyclopedia of aphrodisiacs: Psychoactive substances for use in sexual practices]. Aarau: AT Verlag, 2003.

Rattray, R. S. *The Tribes of the Ashantihinterland*. Oxford: Clarendon Press, 1932.

Reede tot Drakestein, Hendrik van Johann Casearius, and Arnold Syen. *Hortus Indicus Malabaricus: Latinis, Malabaricis, Arabicis & Brachmanum Characteribus nominibusque expressis*. Vol. 11. Amstelodami: Someren [u.a.], 1692.

Reid, Walter V., Sarah A. Laird, Carrie A. Meyer, Rodrigo Gámez, Ana Sittenfeld, Daniel H. Janzen, Michael Gollin, and Calestous Juma. *Biodiversity Prospecting: Using Genetic Resources for Sustainable Development*. Washington, D.C.: World Resources Institute, USA, 1993.

Ridley, Henry Nicholas. *Spices*. London: Macmillan and Co., 1912.

Rømer, Ludvig Ferdinand. *A reliable account of the coast of Guinea (1760)*, trans. Selena Axelrod Winsnes. Oxford: Oxford University Press, 2000.

Roscoe, William. *Monandrian Plants of the Order* Scitamineae, *chiefly drawn from living specimens in the botanic garden at Liverpool. Arranged according to the system of Linnaeus. With descriptions and observations*. Liverpool: Printed by G. Smith, 1828.

Roxburgh, W. *Hortus Bengalensis or A Catalogue of the Plants Growing in the Honourable East India Company's Botanic Garden at Calcutta*. N.p.: Serampore, 1814.

Sahlins, Marshall David. *Stone Age Economics*. Chicago: Aldine-Atherton, 1972.

Said, Edward W. *Culture and Imperialism*. New York: Knopf, 1994.

Sales, Nívio Ramos. *Receitas de Feitiços e Encantos Afro-Brasileiros*. Rio de Janeiro: Pallas Editora, 2007.

Sampath, Padmashree Gehl. *Regulating Bioprospecting: Institutions for Drug Research, Access, and Benefit Sharing*. Geneva: United Nations, 2005.

Sautier, Denis, Estelle Biénabe, and Claire Cerdan. "Geographical Indications in Developing Countries." In *Labels of Origin for Food: Local Development, Global Recognition*, ed. Elizabeth Barham and Bertil Sylvander, 138–153. Oxfordshire, U.K.: CAB International, 2011.

Schiebinger, Londa L. *Plants and Empire: Colonial Bioprospecting in the Atlantic World*. Cambridge, Mass.: Harvard University Press, 2004.

Schweinfurth, Georg August. *The heart of Africa: Three years' travels and adventures in the unexplored regions of Central Africa from 1868 to 1871*. Vol. 1. New York: Harper & Brothers, 1874.

Scott, James C. *The Art of Not Being Governed: An Anarchist History of Upland Southeast Asia*. New Haven: Yale University Press, 2009.

———. *Seeing Like a State: How Certain Schemes to Improve the Human Condition Have Failed*. New Haven: Yale University Press, 1999.

Setton, Kenneth M., Norman Peter Zacour, and Harry Williams Hazard. *A History of the Crusades*. Vol. 5: *The Impact of the Crusades on the Near East*. Madison: University of Wisconsin Press, 1985.

Shapin, Steven. *The Scientific Life: A Moral History of a Late Modern Vocation*. Chicago: University of Chicago Press, 2008.

Shapin, Steven, and Simon Schaffer. *Leviathan and the Air-Pump: Hobbes, Boyle, and the Experimental Life*. Princeton: Princeton University Press, 1985.

Shiva, Vandana, Afsar H. Jafri, and Shalini Bhutani. *Campaign against Biopiracy.* New Delhi: Research Foundation for Science Technology and Ecology, 1999.

Shiva, Vandana, Asfar H. Jafiri, Gitanjali Bedi, and Radha Holla-Bhar. *The Enclosure and Recovery of the Commons: Biodiversity, Indigenous Knowledge, and Intellectual Property Rights.* New Delhi: Research Foundation for Science, Technology, and Ecology, 1997.

Sieberg, Martius, and Scherer. *Jahresbericht über die Fortschritte der Pharmacie in allen Ländern.* Vol. 2. Erlangen: Verlag von Ferdinand Enke, 1844.

Smith, James Edward, and Pleasance Reeve Smith. *Memoir and Correspondence of the Late Sir James Edward Smith.* London: Longman, Rees, Orme, Brown, Green and Longman, 1832.

Smith, Johannes Jacobus. *Theories about the Origin of Afrikaans.* Johannesburg: Witwatersrand University Press, 1952.

Sofowora, Abayomi. *Medicinal Plants and Traditional Medicine in Africa.* New York: Wiley, 1982.

———. *The State of Medicinal Plants Research in Nigeria.* Ibadan: University Press, 1966.

Soto Laveaga, Gabriela. *Jungle Laboratories: Mexican Peasants, National Projects, and the Making of the Pill.* Durham, N.C.: Duke University Press, 2009.

Squibb, E. R. *Squibb's Materia Medica: 1910 Price-List.* New York: E. R. Squibb & Sons, 1910.

Stearn, William T. "A Synopsis of the Genus Catharanthus (Apocynaceae)." In *The Catharanthus Alkaloids,* ed. William I. Taylor and Norman R. Farnsworth, 9–44. New York: Marcel Dekker, 1975.

Steel, Henry Draper. *Portable instructions for purchasing the drugs and spices of Asia and the East-Indies pointing out the distinguishing characteristics of those that are genuine, and the arts practised in their adulteration . . . To which is prefixed, a table of the custom-house duties at London. Compiled from respectable authority by H.D.S.* London: Printed for D. Steel, 1779.

Stillé, Alfred. *The National Dispensatory: Containing the Natural History, Chemistry, Pharmacy, Actions, and Uses of Medicines : including those recognized in the pharmacopœias of the United States, Great Britain, and Germany, with numerous references to the French codex.* 5th ed. Philadelphia: Lea Brothers & Co., 1896.

Susruta.. *An English Translation of the Sushruta Samhita Based on Original Sanskrit Text,* trans and ed. Kaviraj Kunja Lal Bhishagratna. Vol. 1. Calcutta, 1907.

Swann, John Patrick. *Academic Scientists and the Pharmaceutical Industry: Cooperative Research in Twentieth-Century America.* Baltimore: Johns Hopkins University Press, 1988.

Taillevent, and Terence Scully. *The Viandier of Taillevent: An Edition of All Extant Manuscripts.* Ottawa: University of Ottawa Press, 1988.

Tanaka, Jiro. "Subsistence Ecology of Central Kalahari San." In *Kalahari Hunter-Gatherers: Studies of the !Kung San and Their Neighbors,* ed. Richard B. Lee and Irven DeVore, 98–119. Cambridge, Mass.: Harvard University Press, 1998.

Thicknesse, Ralph. *A treatise on foreign vegetables, containing an account of such as are now commonly used in the practice of physick.* London: J. Clarke, 1749.

Thomas, Elizabeth Marshall. *The Old Way: A Story of the First People*. New York: Farrar Straus Giroux, 2006.

Thunberg, Charles Petter. *Travels in Europe, Africa and Asia made between the years 1770 and 1779*. Vol. 2. London: Printed for F. and C. Rivington, 1795.

Tilley, Helen. *Africa as a Living Laboratory: Empire, Development, and the Problem of Scientific Knowledge, 1870–1950*. Chicago: University of Chicago Press, 2011.

Tone, Andrea. *The Age of Anxiety: A History of America's Turbulent Affair with Tranquilizers*. New York: Basic Books, 2009.

Tone, Andrea, and Elizabeth Siegel Watkins. *Medicating Modern America: Prescription Drugs in History*. New York: New York University Press, 2007.

Tournefort, Joseph Pitton de. "Materia medica or, a description of simple medicines generally us'd in physick." London: printed by W. H. for Andrew Bell, 1716.

Twumasi, Patrick A., and Dennis Michael Warren. "The Professionalisation of Indigenous Medicine: A Comparative Study of Ghana and Zambia." In *The Professionalisation of African Medicine*, ed. Murray Last and G. L. Chavunduka, 117–135. Manchester, U.K.: Manchester University Press for International African Institute, 1986.

The Use and Misuse of Shrubs and Trees as Fodder: With Tables Showing Composition and Digestibility. Aberystwyth: Imperial Bureau of Pastures and Field Crops, 1947.

Vail, Leroy. "Ethnicity in Southern African History." In *Perspectives on Africa: A Reader in Culture, History, and Representation*, ed. Roy Richard Grinker and Christopher B. Steiner. Cambridge, Mass.: Blackwell, 1997, 52–68.

van den Daele, Wolfgang. "Does the Category of Justice Apply to Drug Research Based on Traditional Knowledge? The Case of the Hoodia Cactus and the Politics of Biopiracy." In *Who Owns Knowledge? Knowledge and the Law*, ed. Nico Stehr, Christoph Henning, and Bernd Weiler, 255–264. New Brunswick, N.J.: Transaction Publishers, 2006.

Vaughan, Megan. *Curing Their Ills: Colonial Power and African Illness*. Palo Alto, Calif.: Stanford University Press, 1991.

Voeks, Robert A. *Sacred Leaves of Candomblé: African Magic, Medicine, and Religion in Brazil*. Austin: University of Texas Press, 1997.

Vylder, Gustaf De, ed. *The Journal of Gustaf de Vylder: Naturalist in South-Western Africa, 1873–1875*, ed. Ione Rudner and Jalmar Rudner. Goodwood, Western Cape: Printed by National Book Printers for the Van Riebeeck Society, 1998.

Wadström, Carl Bernhard. *An essay on colonization, particularly applied to the western coast of Africa: with some free thoughts on cultivation and commerce*. London: For the author by Darton and Harvey, 1794.

Waite, Gloria Martha. *A History of Traditional Medicine and Health Care in Pre-Colonial East-Central Africa*. Lewiston, N.Y.: E. Mellen Press, 1992.

Warton, Thomas. *The History of English Poetry, from the Eleventh to the Seventeenth Century*. New York: G. P. Putnam & Sons, 1870.

Watkins, Elizabeth Siegel. *The Estrogen Elixir: A History of Hormone Replacement Therapy in America*. Baltimore: Johns Hopkins University Press, 2007.

———. *On the Pill: A Social History of Oral Contraceptives, 1950–1970*. Baltimore: Johns Hopkins University Press, 1998.

Watson, James D. *The Double Helix: A Personal Account of the Discovery of the Structure of DNA*. New York: Atheneum, 1968.

Watt, John Mitchell, and Maria Gerdina Breyer-Brandwijk. *The Medicinal and Poisonous Plants of Southern Africa, Being an Account of Their Medicinal Uses, Chemical Composition, Pharmacological Effects and Toxicology in Man and Animal*. Edinburgh: E. & S. Livingstone, 1932.

———. *The Medicinal and Poisonous Plants of Southern and Eastern Africa, Being an Account of Their Medicinal Uses, Chemical Composition, Pharmacological Effects and Toxicology in Man and Animal*. Edinburgh: E. & S. Livingstone, 1962.

Wellcome, Henry Solomon. *From Ergot to "Ernutin": An Historical Sketch*, Lecture Memoranda. Chicago: American Medical Association, 1908.

West, Harry G. "Working the Borders to Beneficial Effect: The Not-So-Indigenous Knowledge of Not-So-Traditional Healers in Northern Mozambique." In *Borders and Healers: Brokering Therapeutic Resources in Southeast Africa*, ed. Tracy J. Luedke and Harry G. West, 21–41. Bloomington: Indiana University Press, 2006.

Whewell, William. *The Philosophy of the Inductive Sciences, Founded upon their History*. London: J. W. Parker, 1840.

White, Luise. *Speaking with Vampires: Rumor and History in Colonial Africa*. Berkeley: University of California Press, 2000.

Whyte, Susan Reynolds, Sjaak van der Geest, and Anita Hardon. *Social Lives of Medicines*. Cambridge: Cambridge University Press, 2002.

Willis, Kate. *Malagasy Institute for Applied Research, Antananarivo, Madagascar*. Excellence in Science: Profiles of Research Institutions in Developing Countries. Trieste: Academy of Sciences for the Developing World, n.d.

Wiredu, Kwasi. "Our Problem of Knowledge: Brief Reflections on Knowledge and Development in Africa." In *African Philosophy as Cultural Inquiry*, ed. Ivan Karp and D. A. Masola, 181–186. Bloomington: Indiana University Press, 2000.

Wirtén, E. H. *Terms of Use: Negotiating the Jungle of the Intellectual Commons*. Toronto: University of Toronto Press, 2008.

Withering, William. *The Miscellaneous Tracts of the late William Withering, M.D. F.R.S.* London: Longman, Hurst, Rees, Orme, and Brown, 1822.

Wood, George B., Franklin Bache, H. C. Wood, Joseph Price Remington, and Samuel P. Sadtler. *The Dispensatory of the United States of America*. 18th ed. Philadelphia: Lippincott, 1899.

Wright, Patricia C. "Ecological Disaster in Madagascar and the Prospects for Recovery." In *Ecological Prospects: Scientific, Religious, and Aesthetic Perspectives*, ed. Christopher Chapple, 11–24. Albany: State University of New York Press, 1994.

Wujastyk, Dominik. *The Roots of Ayurveda: Selections from Sanskrit Medical Writings*. London: Penguin Books, 2003.

Wylie, Diana. *Starving on a Full Stomach: Hunger and the Triumph of Cultural Racism in Modern South Africa*. Charlottesville: University Press of Virginia, 2001.

Wynberg, Rachel, and Roger Chennells. "Green Diamonds of the South: An Overview of the San-*Hoodia* Case." In *Indigenous Peoples, Consent and Benefit Sharing Lessons from the San-Hoodia Case*, ed. Rachel Wynberg, Doris Schroeder, and Roger Chennells, 89–124. Dordrecht: Springer Verlag, 2009.

Wynberg, Rachel , Doris Schroeder, and Roger Chennells. *Indigenous Peoples, Consent and Benefit Sharing Lessons from the San-hoodia Case.* Dordrecht: Springer Verlag, 2009.

Reports and Oral Interviews

"Access to Forest Genetic Resources and Benefit Sharing: Potential Opportunities and Challenges for Governments and Forest Stakeholders." Ottawa, 2008. http://www.ec.gc.ca/Publications/default.asp?lang=En&xml=FE5C3334-8655 -462B-9190-B9AE1E7D5D92.

"African Traditional Medicine. Report of the Regional Expert Committee." AFRO Technical Report Series, 3–4, 1976.

"Bi-Annual Report." Mampong-Akuapem: Centre for Scientific Research into Plant Medicine, 2001.

"CSIR Annual Report 1975." Pretoria: Council for Scientific and Industrial Research, 1975.

"CSIR Annual Report 1983." Pretoria: Council for Scientific and Industrial Research, 1983.

"CSIR Annual Report 2005–06." Pretoria: Council for Scientific and Industrial Research, 2006.

"CSIR Twenty-Second Annual Report." Pretoria: Council for Scientific and Industrial Research, 1966.

"CSIR Twenty-Third Annual Report." Pretoria: Council for Scientific and Industrial Research, 1967.

"CSIR Twenty-Fourth Annual Report." Pretoria: Council for Scientific and Industrial Research, 1968.

"CSIR Twenty-Eighth Annual Report 1972." Pretoria: Council for Scientific and Industrial Research, 1973.

"Destination of Ghana's Non-Traditional Exports, 1984–2004." Accra: Ghana Exports and Promotion Council, 2004.

Fox, Francis William, and Douglas Back. "A Preliminary Survey of the Agricultural and Nutritional Problems of the Ciskei and Transkeian Territories: With Special Reference to Their Bearing on the Recruiting of Labourers for the Gold Mining Industry." Pietermaritzburg: s.n., 1941.

Grenier, Louise. "Working with Indigenous Knowledge: A Guide for Researchers." Ottawa: International Development Research Center, 1998.

Johnson, Irving S. "Eli Lilly and the Rise of Biotechnology Interviews Conducted by Sally Smith Hughes, in 2004," ed. Bancroft Library Regional Oral History Office, University of California. Berkeley: Regents of the University of California, 2006.

Laing, Ebenezer. "Documentation and Protection of Biodiversity, with Comments on Protecting the Intellectual Rights of the Traditional Medical Practitioner." Paper presented at the Traditional Medicine and Modern Health Care: Partnership for the Future—Report on a Two-Day National Consensus Building Symposium on the Policies on Traditional Medicine in Ghana, March 15–16, 1995, Accra, 1995.

McGown, Jay. "Out of Africa: Mysteries of Access and Benefit Sharing," ed. Beth Burrows. Edmonds, Wash. and Richmond, South Africa: Edmonds Institute and African Center for Biosafety, 2006.

"Minutes taken (in session 1818) before the committee to whom the petition of several inhabitants of London and its vicinity, complaining of the high price and inferior quality of beer, was referred, to examine the matter thereof, and report the same, with their observations thereupon, to the House." House of Commons Parliamentary Papers, Reports of Committees, 1819.

"Notice of Expiration of Patents Due to Failure to Pay Maintenance Fee [on October 31, 2001]." Alexandria, Va.: United States Patent and Trademark Office, 2002.

Phyto-Riker Pharmaceuticals, Ltd. n.p., c. 1999.

Wehmeyer, A. S. "Edible Wild Plants of Southern Africa: Data on the Nutrient Contents of over 300 Species." Pretoria: Council for Scientific and Industrial Research, 1986.

World Health Organization. "The Promotion and Development of Traditional Medicine." In World Health Organization Technical Report Series. Geneva: World Health Organization 1978.

Selected Theses

Andriamanalintsoa, Jean Joseph. "Contribution à l'Étude de la Production de le Pervenche de Madagascar ou *Catharanthus roseus*, Cas d'Ambovombe, d'Amboasary-Sud, de Beloha et de Tsihombe" [Contribution to the study of the production of Madagascar Periwinkle or *Catharanthus roseus*, case of Ambovombe, of South Amboasary, of Beloha and of Tsihombe]. PhD diss., University of Antananarivo, 1995.

Beentje, Henk J. "A Monograph on Strophanthus DC (Apocynaceae)." PhD diss., Wageningen: H. Veenman & Zonen, 1982.

Cousteix, Pierre Jean. "L'Art et la Pharmacopée des Guérisseurs du Cameroun: (Tribu Ewondo-Région de Yaounde)." PhD diss., University of Paris, 1962.

Dajani, Ola Fouad. "Genetic Resources under the C.B.D. and T.R.I.P.S.: Issues on Sovereignty and Property." LLM thesis, McGill University, 2003.

Flint, Karen Elizabeth. "Negotiating a Hybrid Medical Culture: African Healers in Southeastern Africa from the 1820s to the 1940s." PhD diss., University of California, Los Angeles, 2001.

Greene, Shane. "Paths to a Visionary Politics: Customizing History and Transforming Indigenous Authority in the Peruvian Selva." PhD diss., University of Chicago, 2004.

Hamilton, Chris. "Long Strange T.R.I.P.S.: Intellectual Property, the World Trade Organization and the Developing World." MA thesis, Dalhousie University, 2003.

Konadu, Kwasi Bodua. "Concepts of Medicine as Interpreted by Akan healers and Indigenous Knowledge Archives among the Bono-Takyiman of Ghana, West Africa: A Case Study." PhD diss., Howard University, 2004.

Leitch, J. Neil. "The Native Remedies and Poisons of West Africa." PhD thesis (rejected), London School of Hygiene and Tropical Medicine, 1938.

Osseo-Asare, Abena Dove. "Bitter Roots: African Science and the Search for Healing Plants in Ghana, 1885–2005." PhD diss., Harvard, 2005.

Rasoanaivo, Philippe. "Étude Chimique d'Alcaloïdes de *Catharanthus longifolius* Pich. (Apocynacées malgaches): Hémisynthèse d'Alcaloïdes Bis-indoliques." PhD diss., University of Paris, 1974.

Vogt, Emile F. "Les Poisons de Flèches et les Poisons d'Épreuve des Indigenènes de l'Afrique." PhD diss., École Supérieure de Pharmacie, Université de Paris. Lons-le-Saunier: impr. et lithographie L. Declume, 1912.

Audiovisual

Desai, Rehad. *Bushman's Secret One Cactus Stands between Hope and Hunger.* Johannesburg: Film Resource Unit [distributor], 2006.

Marshall, Lorna, and John Marshall. *Bitter Melons.* Watertown, Mass.: Documentary Educational Resources, 1986.

Sargent, Joseph, Julian Krainin, Alan Rickman, Def Mos, Kyra Sedgwick, Gabrielle Union, Merritt Wever, and Christopher Young. *Something the Lord Made.* New York: HBO Video, 2004.

News Articles

Asiedu, William A. "Food and Drugs Board Warns Herbalists." *Mirror,* August 12, 2000, 4.

"Can a Generation of Potent but Safe Diet Pills Rescue Us from Obesity?" *Stanford Daily,* February 24, 2005.

Christensen, Jon. "Scientist at Work: Mark J. Plotkin; A Romance with a Rain Forest and Its Elusive Miracles," *New York Times,* November 30, 1999, F3 (1).

Dentlinger, Lindsay. "Namibian Government Is to Enter into an Agreement with British Pharmaceutical Company Phytopharm on the Cultivation and Supply of the Hoodia Plant." Uhuru Policy Group, http://www.uhurugroup.com/news/121504.htm.

Ferreira, Anton. "Red Tape Blocks Production of Blockbuster Green Drugs." *Sowetan,* November 17, 2008.

Foley, Stephen, and Susie Mesure. "Unilever Enlists Kalahari Bushmen in Effort to Revive Fortunes of Slim-Fast: San Tribesmen of the Kalahari Will Spend Any Royalties on Boosting Community Facilities." *Independent,* December 16, 2004, 44.

Food and Drug Administration. "Refusal Actions by FDA as Recorded in OASIS for Nigeria." August 5, 2009. www.accessdata.fda.gov.

"Ghana Makes a Big Medical Discovery." *Ghanaian Times,* June 16, 1964.

"Ghana National Association of Traditional Healers." *Ghanaian Chronicle,* September 11, 2006.

"Healers Find New Drug." *Daily Graphic,* October 18, 1965.

"Herbal Remedies 'Do Work,'" BBC News, September 28, 2004. http://news.bbc.co.uk/go/pr/fr/-/1/hi/health/3698348.stm.

"Hoodia Gordonii Weight Loss Pills, Hoodia Gordonii Stoneage Plant—'Miracle Obesity Cure' or Fictitious Tale?" *Stanford Daily*, February 24, 2005.

"Judy." "AMUSING: Finish 'em now—On the Recent Fight with the Jebus." *Gold Coast Chronicle*, June 6, 1892, 4.

Kahn, Tamar. "Firm Hopes to Take 'San Prozac' to Market." *BusinessDay*, September 4, 2010.

———. "South Africa: Unilever Dumps Plans for Hoodia Diet Pill." *Business Day*, December 22, 2008.

Makoni, Munyaradzi. "San People's Cactus Drug Dropped by Phytopharm." *SciDev. Net*, December 20, 2010.

Mbom, Francis Tim. "Cameroon: Scientists Fear Extinction of Medicinal Plants on Mt. Cameroon." *Post News Line*, November 23, 2007.

Mensah, Gifty, and Cindy Asamoah. "Quack Herbalists to Be Flushed Out by August 31." *Public Agenda*, August 20, 2010.

Molisa, Janet. "Cameroon: A Third Elephant Collared on Mt. Cameroon." *Post News Line*, January 19, 2009.

Mooney, Pat. "Africa: Of Infraredd and Inforedd." *Pambazuka News: Weekly Forum for Social Justice in Africa*, October 7, 2010.

Mputhia, Cathy. "State Needs to Enforce Laws to End Runaway Biopiracy." *Business Daily*, October 4, 2010

"Noguchi to Establish Unit to Test Traditional Medicine for HIV/AIDS." Southern African Humanitarian Information Network, 2005, http://www.sahims.net /doclibrary/10_03/10/zambia/Noguchi%20To%20Establish%20Unit%20To %20Test%20Traditional%20Medicine%20For%20HIV.doc.

Pearce, Fred. "Science and Technology/Bargaining for the Life of the Forest: Poor Nations Want Drug and Food Companies to Pay for the Plants They Plunder. Fred Pearce on the Dangers of Selling Out a Heritage." *Independent (London)*, March 17, 1991, 37.

"Phytopharm Falls 43% as Unilever Drops Hoodia." *Pharma Marketletter*, November 18, 2008.

Saminu, Zambaga Rufai. "GMA Unhappy with Fake Drug Advertisement." *Ghanaian Chronicle*, February 2, 2010.

"Scientists Turn to Traditional Medicinal Plants to Find New Tools for Fighting Malaria" (press release). The Multilateral Initiative on Malaria, November 14, 2005, http://allafrica.com/stories/200511140247.html.

Sherry, Shannon. "Permission to Sip: Indigenous Resources—Law Slip-Up." *Financial Mail*, October 28, 2010.

"South Africa Crops Ruined by Drought: Harrowing Tales Left in Wake of Dry Period." *Washington Post*, October 23, 1933.

"South Africa: Feature—Marginalised San Win Royalties from Diet Drug." *IRIN News* (U.N. Office for the Coordination of Humanitarian Affairs), March 26, 2003.

Stahl, Lesley. "African Plant May Help Fight Fat (CBS 60 Minutes)." *CBSNews.com*, November 21, 2004.

Stevens, William K. "Shamans and Scientists Seek Cures in Plants." *New York Times*, January 28, 1992, C1.

"Storm Ends African Drought, 15 Die." *New York Times*, August 13, 1930.

Strauss, Stephen. "Mind and Matter: Uncovering the Truth behind the Rape of Madagascar's Rosy Periwinkle." *Globe and Mail* (Canada), September 12, 1992.

Thomas, Richard. "Plants that Soothe the Pain: Richard Thomas Discovers Why Drug Companies Are Beating a Path to a 300-Year-Old Garden in the Heart of London." *Guardian*, October 9, 1992, 31.

"We Can't Fan the Embers of Tribalism." *Ghanaian Times*, August 31, 1963.

"Workshop on Manufacturing of Herbal Products in Ghana" [Advertisement]. *Daily Graphic*, September 23, 2002.

"World Scientists to Study Secrets of Ghanaian Tree." *Ghanaian Times*, 1963.

Page references followed by *f* denote figures; page references followed by *t* denote tables. References to endnotes give the endnote-page number followed by n., the note number, and the text page of endnote citation in parentheses.

Afrikaans language, 172
Agence de Coopération Culturelle et
	Technique (ACCT), 88
Agrawal, Arun, 128, 227n. 23 (18)
Agyewaa, Akosua, 98
Akadzah, Sylvester, 161
Akanyele (Bolgatanga resident), 116
alcoholic beverages, grains of paradise
	in, 86–87
Alkaloid Group, in Ghana, 140–44,
	146–48
alkaloids: from *Cryptolepis*, 153, 155,
	203; from periwinkle, 32, 46, 50–
	52, 64, 68; screening of Ghanaian
	local herbs, 140–48, 145*f*; in South
	African plants, 186
Allibert, Claude, 41
alligator peppers. *See* grains of paradise
Allman, Eric, 141, 142–43, 144, 146
aloe cultivation, 194
al-Wazzan, Al-Hasan, 80
Amarula oil production, 194
Americas, grains of paradise in, 82–86
Ampofo, Chief Kwesi, 134
Ampofo, D. A., 153
Ampofo, Oku: Alkaloid Group and, 141,
	146; art and, 135, 240n. 4 (135);
	autobiography (unpublished), 135;
	bioprospecting by, 140; *Cryptolepis*
	and, 137–38, 150, 153; CSRPM and,
	148–49, 161; Diane Winn and, 131–
	32, 150, 155, 157; documentation of
	traditional herbal recipes, 133, 137–
	40, 153–54; education of, 131, 134–
	35; establishment of Mampong clinic,
	136; Ghana Rural Reconstruction
	Movement, 153; guide to medicinal
	plants, 153–54; malaria patients of,
	136, 137; *Obi Kyere* Herbal Center,
	138*f*; as Tetteh Quarshie Memorial
	Hospital founder, 134
Andriamanalintsoa, Jean Joseph, 33,
	65, 66
animal experiments: with grains of
	paradise, 91–92; with *Strophanthus*,
	91–92
Annona senegalensis, 90
Ansah, Twum Ampofo, 152
apartheid, 167, 173, 187, 188, 190, 193, 196
Apeagyei, J. K., 150

aphrodisiac, grains of paradise for, 71,
	78, 79, 87, 89–90
appellation d'origine, 208–9
appetite suppressant, hoodia as, 166,
	183, 184, 188, 195
appropriation of plant-based therapies, 17,
	18, 108–9, 127, 128. *See also* biopiracy
Armah (Ghanaian chief), 100, 236n.
	65 (100)
arrow poison plant. See *Strophanthus*
arrow poisons, 108, 109, 110–17, 111*f*,
	119, 125–29
artesunate, 8
asiatic acid, 56, 58
asiaticoside, 55–56, 58
Assiamah, R. K., 9, 13, 208
Assiamah bitters, 9
Atiako, Djane, 9, 13
Atiako bitters, 9
Atlantic trade, diffusion of West African
	forest products by, 81–87
Avumatsado, S. K., 152
Awuku, Kwame, 138

Baffour, Robert Patrick, 144
banaba, 47–48
Barkly, Sir Henry, 165, 170
Beckett, Arnold H., 143, 144
Becuanaland Protectorate. *See* Botswana
Beer, Charles, 50, 52, 61
Belloc, Hilaire, 111
benefit sharing, 19–23, 69, 204–5, 210;
	African experiences with, 20–21;
	argument for common benefits, 23;
	as catalyst for community solidarity,
	19; compulsory licensing for
	production of generics, 22; contracts
	and agreements, 19; Convention
	on Biological Diversity (CBD), 20,
	22, 166, 201, 211, 212; Council for
	Scientific and Industrial Research
	(CSIR), South African, 167, 173,
	190–92, 196; with descendants of
	First Peoples groups, 33; difficulty
	in assigning benefits, 19–20; global
	justice and, 194; grains of paradise
	and, 72, 104; herbal knowledge as
	national goods, 21; hoodia and, 19,
	166–67; indigenous rights and, 21–22;
	in South Africa, 167, 173, 190–97, 204

Benoit, Dan, 89
Betti, Jean, 92
Biampamba, Samuel, 48
Biney, Alhaji Sidi, 138
Biodiversity Act, South African, 166, 194, 201
biodiversity prospecting. *See* bioprospecting
biological diversity: in Cameroon, 235n. 48 (92); conservation of, 23; Convention on Biological Diversity (CBD), 20, 22, 166, 201, 211, 212
biopiracy, 19, 34, 38, 68, 104, 174, 191, 194, 202, 210
bioprospecting: Africanization of, 212; by African scientists, 32, 53, 100–101; agreements, 20; Cartagena Accord and, 207; by Chris Kilham, 161; discreet, 103; by Malagasy scientists, 53; by Oku Ampofo, 140; origin of term, 11, 32; of periwinkle, 41; by San people of South Africa, 167–68. *See also* drug discovery
bioprosperity, 202, 205, 210
Biovigora, 95*f*
Birch, A. J., 61
Bitter Melons (film), 197
bitter roots. See *Cryptolepis sanguinolenta*
Biya, Paul, 92
blood circulation, *Strophanthus* effect on, 120–21
blood pressure, grains of paradise and, 94
Boiteau, Pierre, 32, 54–56, 58, 68
Bossou, Amana Ossou, 89
Botswana, 174, 180–83, 185, 192, 196, 197
Boye, Gilbert, 150–51, 153
Boyle, James, 32
Brazil: ban on hoodia commercialization, 189; grains of paradise in, 84–85; pennywort in, 44–45
Bremer Bread, 246n. 69 (185)
Brennan, Kennon A., 158
British Technical Assistance scheme, 152
Burroughs Wellcome and Company, 120, 172
Bushman's Secrets (film), 191
Bushmen, 21, 166–68, 177, 180–83, 185, 191–92, 195. *See also* San
Byrsocarpus coccineus, 89

Cameroon: African plant medicine research in, 91–92; biological diversity in, 235n. 48 (92); grains of paradise in, 91–93
cancer therapy, periwinkle and, 31, 33, 36, 45–46, 50–52, 61
Cape Town University, 180
cardamom, 74, 80, 81
Cardinall, A. W., 116–17
Carney, Judith, 84
Cartagena Accord, 207
cassava, 77
Catharanthus roseus. See periwinkle
Centella asiatica. See pennywort
Center for International Cooperation in Agricultural Research for Development (CIRAD), 68
Centre for Scientific Research into Plant Medicine (CSRPM), 132, 148–57, 159–62; assistance from Diane Winn, 157; collaboration with foreign researchers, 162; *Cryptolepis* and, 150–51, 154, 160; establishment of, 148–49, 151–52; financial issues, 151–52, 153, 160–62; inauguration of, 149; openness of, 149–50, 154, 159–60; Phyto-Riker relationship with, 159; structure of, 149; university collaboration, 152–53
Chakrabarty, Dipesh, 24
Chalmers, Albert, 114, 124*f*
Chaucer, Geoffrey, 79–80
Chennels, Roger, 190–92, 194, 197
China: grains of paradise in, 80–81; pennywort in, 45
chinchona bark, 133, 137
chloroquine, 137, 150
Chu Ssu-Pen, 81
CIRAD (Center for International Cooperation in Agricultural Research for Development), 68
class-action suits against corporations, 208
climate change, 206
closed access to plant resources, 203, 206, 207, 210
cocoa, 126, 240n. 90 (128)
colonial archives, as information resource, 24

seller's mistrust of, 69; relationship/ conflict with healers, 1, 7, 25. *See also* African scientists; *specific individuals*
Scott, James C., 177
SEAR (Société d'Exploration Agricole, Ranopiso), 63, 65, 66
secondary metabolites, 188
secrecy, cultures of, 7
Selormey, A. H., 149
Shaman Pharmaceuticals, 20, 160, 228n. 36 (20)
shea butter, 149, 240n. 90 (128)
Shiva, Vandana, 19, 104
Sibisi, Sibusiso, 184, 191
Simone, Abdou Maliqalim, 199
Sittie, Archibald, 153
slave trade, trans-Atlantic, 82–86
Smith, Kline, and French (Glaxo-Smith-Kline), 42, 186
socialism in Ghana, 141
Solanum torvum, 149
Soto Laveaga, Gabriela, 60
South Africa: apartheid, 167, 173, 187, 188, 190, 193, 196; benefit sharing in, 167, 173, 190–97; Black Economic Empowerment, 193; Council for Scientific and Industrial Research (CSIR), 167–68, 173, 179, 184–97; diet of miners, 178–79; European settlers in, 168–70, 172, 177; !Kung people, 180–83; nutrition research in, 184–86; Scientific Research Council Act (1945), 184; wild greens consumed in, 178–79, 185; Working Group of Indigenous Minorities in South Africa (WIMSA), 191, 192
South African Biodiversity Act (2004), 166, 194, 201
South African Council for Scientific and Industrial Research (CSIR), 167–68, 173, 179, 184–97
South African Institute for Medical Research (SAIMR), 178–79
South African San Council, 190–92
Southwest Africa. *See* Namibia
Spensley, P. C., 62
spice. *See* grains of paradise
Stahl, Leslie, 189
State Pharmaceutical Corporation, 157
Steyn, P. S., 187

strophanthin, 108–10, 119–21, 122f, 126
Strophanthus, 107–29; arrow poisons from, 108, 109, 110–17, 111f, 119, 125–29; collection points of *Strophanthus hispidus* samples, 118t; cultivation efforts, 123–26; export efforts, 121–26; geographical distribution, 4t, 117; healers' use of, 117–18, 129; patents associated with, 4t; pod appearance, 106f, 124f; strophanthin prepared from, 108–10, 119–21, 122f, 126; terminology for species, 238n. 41 (117)
Svoboda, Gordon, 31, 45, 49, 51, 52, 64
syphilis, 18

Tackie, Albert Nii, 203–4; Alkaloid Group and, 141–44, 145f; CSRPM and, 132, 141–44, 145f, 148–50, 152, 155, 157, 161; Malaherb, 156; media attention to, 144, 145f, 148; patents by, 155–56; University of Nsukka in Nigeria, 153
Tackie, Reggie Nii, 156
Tanaka, Jiro, 183
Tayman, Francis, 153
Tetteh Quarshie Memorial Hospital, 134
Things Fall Apart (Achebe), 90
Thiroux (colonial physician in Madagascar), 57
Thomas, Elizabeth Marshall, 177, 181
threatened species, 45
tongue (tonga). *See* periwinkle
tooth decay, in South Africa, 178
Torre, Raimond della, 81
Torto, F. G. T. O'B, 141, 142, 148, 154
traditional knowledge, ownership of, 207, 214
traditional medicine: African Traditional Medicine logo (WHO), 34, 34f; coevolution with biochemistry, 18; Decade of African Traditional Medicine (declaration by African Union), 8; definition of, 15; efficacy of, 15; historical geographies of, 13–16; as local medicine, 72; renewed interest in herbal therapies, 7–8; safety of, 15; "scientific" authority privileged over traditional medical expertise, 6; WHO popularization of concept, 14, 15; wide distribution of, 214

32230708R00190

Made in the USA
Lexington, KY
28 February 2019